ENGINEERING MECHANICS

of

COMPOSITE STRUCTURES

V. Z. Parton
École Centrale Paris

B. A. Kudryavtsev
Moscow Institute of Chemical Engineering

Translated by
E. G. Strel'chenko

CRC Press
Boca Raton   Ann Arbor   London   Tokyo

Library of Congress Cataloging-in-Publication Data

Parton, V. Z. (Vladimir Zalmanovich)
  Engineering mechanics of composite structures / V. Z. Parton and B. A. Kudryavtsev;
  translated from the Russian by E. G. Strel'chenko.
    p.   cm.
  Includes bibliographical references.
  ISBN 0-8493-9302-7
  1. Composite materials—Mechanical properties. 2. Continuum mechanics. 3. Plates (En-
gineering) 4. Shells (Engineering)
I. Kudryavtsev, B. A. (Boris Aleksandrovich) II. Title.
TA418.9.C6P374   1993
624.1'8--dc20   92-5816
                                                                                    CIP

Direct all inquiries to CRC Press, Inc., 2000 Corporate Blvd., N.W., Boca Raton, Florida 33431.

©1993 by CRC Press, Inc.

International Standard Book Number 0-8493-9302-7

Library of Congress Card Number 92-5816

Printed in the United States of America   1 2 3 4 5 6 7 8 9 0

Printed on acid-free paper

Foreword

This book provides a detailed account of recent advances in the theoretical study of the mechanical behavior of highly nonhomogeneous composite materials and structural elements that have a regular structure as their distinguishing feature. Materials and elements of this type are becoming increasingly important in mechanical and civil engineering, aviation, rocketry, shipbuilding, and other fields.

Emphasis is placed on the authors' own results in the theory of elasticity as well as coupled-field mechanics (thermoelasticity and electroelasticity) of composites and in constructing general homogenized models for composite materials and reinforced plates and shells.

Among the many practical applications of the results obtained are the calculation of the effective and local properties of laminated and fiber composites and of ribbed, wafer-type, honeycomb, networks and two-way reinforced plates and shells.

Background material, such as the basic concepts of the mechanics of nonhomogeneous anisotropic media and the mathematical formalism of the asymptotic homogenization method, is also provided.

This book is intended for scientists in the fields of continuum mechanics, mechanics of composites, and plate and shell theory; engineers engaged in the design, manufacture, and practical use of structures made of composites; undergraduate and postgraduate students in the mechanics of deformable media and mathematically related fields.

Preface

Little excuse is necessary for a book on composite materials in general and on the mechanics of composite materials in particular. The dramatic increase in the use of composite materials for almost all conceivable applications increasingly calls for the development of rigorous mathematical models capable of predicting the materials' mechanical behavior under any given set of conditions, whether they are considered "per se" or in the form of structural elements.

The composites with a regular structure that are the subject matter of this book are worth considering not only because of their numerous practical uses in a wide range of fields, but also because they have an obvious advantage over many other composite material types in that they are more amenable to theoretical analysis. A very effective tool that has been developed for treating such systems is the asymptotic method of the homogenization of differential equations with rapidly oscillating coefficients. The method has been given a strict mathematical justification and yields an asymptotically correct solution when used in its domain of applicability. By applying this method, we discuss the elastic, thermoelastic, and electroelastic properties of the composite materials with a regular structure.

This book is based largely on the authors' original work, carried out over the last few years, and the results obtained are presented either in a numerical form or, when possible, as formulas that are—or so the authors hope—easy to interpret and are amenable to design evaluation and application.

Some results given in this book were due to financial support of the industry to concrete technical problems. Substantial contribution to this work as well as to the preparation of the manuscript of this book has been done by our learners who are now professors and doctors: V. Boriskovsky, A. Kalamkarov, V. Miloserdova, N. Senik, and A. Zobnin. We wish to thank them for their cooperation.

The Russian version of this book, which is planned to be published, will differ from this book by an additional chapter on the nonlinear theory of composite plates and shells written by A. Kalamkarov.

We would like to hope that this book will contribute to the development of the science of composite materials and structures, the importance of which is difficult to overevaluate.

Paris, December 1992   V. Z. Parton
                                                            B. A. Kudryavtsev

Vladimir Z. Parton graduated from Moscow State University in 1959, received his candidate degree in 1966, and was awarded his doctoral degree from the Academy of Sciences of the Union of Soviet Socialists Republic in 1970. Since 1972 he has been Head of the Department of Higher Mathematics in the Moscow Institute of Chemical Engineering and since 1992 has been professor of Ecole Centrale des Arts et Manufactures Paris. Professor Parton is one of the leading figures in the mechanics of deformable solids and material strength, both in Russia and abroad. He is a member of the International Academy of Astronautics. In 1983 he was awarded the Prize of the Union of Soviet Socialists Republic Council of Ministers. He has authored more than 200 scientific papers and 18 books, including *Mechanics of Elastic–Plastic Fracture* (Hemisphere, 1988) and *Dynamic Fracture Mechanics*, Volumes 1 and 2 (Hemisphere, 1989).

Boris A. Kudryavtsev graduated from Moscow Aviation Institute in 1960, received his candidate degree from Moscow State University in 1971, and received his doctorate in 1983 from Leningrad State University. He is a professor of the Department of Higher Mathematics at the Moscow Institute of Chemical Engineering. His main professional and research interests are in the mechanics of continuous media and fracture mechanics. He is the author or a coauthor of more than 100 scientific papers on various areas of the mechanics of solid deformable bodies. Together with Prof. V. Z. Parton he has published in Russian and English the book *Electromagneto-elasticity Pieroelectronics and Electrically Conductive Solids* (Gordon and Breach, 1988).

Introduction

If it has been true throughout history that the evolution of technology has been controlled by the materials available, it is increasingly so today—when progress in aviation, rocketry, shipbuilding, and mechanical and civil engineering crucially depends on the reasonable compromise between high strength, stiffness, corrosion resistance, and perhaps other material properties on the one hand and reduced weight and cost on the other. The use of composite materials, that is, continuous or discontinuous objects embedded in suitable metallic or nonmetallic matrices, is therefore receiving ever wider attention in these and other areas of technology due to the inherent tailoring of the properties of these materials.

Composite materials actually have been known since the dawn of recorded history, for certainly it is to this class that we must relate mixtures of clay, sand, straw, and other naturally occurring components—those early building materials that, however primitive, antic-ipated in many respects the basic composite material concepts developed on a more sophisticated level in our technologically ambitious age.

In modern times, it was the advent of composite materials such as concrete and then reinforced concrete that was a major breakthrough in construction. The development of new building materials is one of the most identifiable trends in today's technology. In a growing number of applications, composite materials have taken over the work that was previously done by metals and metal alloys. In the United States, for example, advanced nonmetallic materials account for about 15 to 20% of all structural materials. In Japan, it is anticipated that the figure will climb to fully 50% somewhere around the year 2000.

Plans made in the Soviet Union for economic and social development of the country envisaged an accelerated introduction of composite materials into various branches of the national economy. As of now (1989), a considerable degree of success in this respect has been achieved in the aerospace industry. The *Ruslan*, one of the world's greatest airplanes now at work, contains 5.5 tons of composites, with a total area of 1500 m^2, which enabled the design team to reduce by 2 tons the weight of the machine, saving 15 tons of metal and 18,000 tons of fuel (per service life) and to obtain an impressive economic efficiency of 5 million rubles. Similar results have been achieved with new-generation IL-96 and TU-204 designs. In the Kamov design office, use of polymer composites in the latest helicopter designs has reached 53%; this provides a weight saving of 25 to 30%, doubles (or even triples) the service life, and reduces by a factor of 1.5 to 3 the working hours per aircraft. In some space-oriented designs, a percentage of composites as high as 80% has been obtained. In the space shuttle *Buran*, test data and design calculations suggest a large-scale change to new manufacturing technologies for many structural elements, with a major role being given to honeycomb composites (which may provide a weight reduction of 50 to 70%. No less promising is the use of composites in more traditional industries, such as transport, consumer services, and health services. In the Central Institute of Traumatology and Orthopedics, for example, there is evidence that the use of composite materials may result in a threefold to fivefold reduction in treatment period and will permit doctors to restore their patients to health by operations that would be prohibitively complex or even totally impossible with traditional techniques.

The large-scale introduction of composite materials has created a need for further progress in such classical areas of mechanics as the theory of anisotropic and nonhomogeneous deformable solids and the theory of optimization. One can agree with Friedlander and Bratukhin (1988) that "it is the intensification of basic research in the field of composite materials which determines the future of our national economy."

It is a well-known fact that, given the same external conditions, the response of a composite material differs from that of a homogeneous (or monolithic) material and so too, of course, do the strength and service life of the structural elements fabricated of these respective materials. It is also well known—and quite evident—that the experimental determination of the properties of composite materials for all possible reinforcement types is often impracticable because of the volume—and often the cost—of the measurements needed (Obraztsov et al. 1977). Hence the need for theoretical models capable of predicting both the averaged characteristics of the composite materials and the local structure of the processes occurring in them under various types of environment.

The mathematical framework from which to predict the material behavior of a highly nonhomogeneous medium is generally provided by equations, or sets of equations, with rapidly oscillating coefficients, these latter characterizing the properties of the individual phases of the composite material. Taken in this rigorous formulation, the corresponding boundary value problems are difficult to handle even with the help of a high-speed modern computer, and it is quite natural, therefore, to look for mathematical models that would describe a highly nonhomogeneous medium by equations with some averaged coefficients. Clearly, there is a necessary condition that must be fulfilled in constructing such averaged (or "homogenized") equations, namely, that the solution of the resulting boundary value problem be sufficiently close to the solution of the original problem. To develop a rigorous mathematical model of a composite medium, it is, of course, necessary to specify (at least to some extent) the structural configuration of the problem. Because composite materials with a regular (or nearly regular) structure are in extremely wide use today, we seem to be fully justified in choosing them as the subject matter of this book. It should be noted—although this point will be taken up later in the book—that the scale of nonhomogeneity in a particular problem is always assumed to be small as compared with the dimensions of the body as a whole.

Figure 1 shows a photograph of a typical microstructure, a filament composite fabricated on the basis of a metal matrix by the method of continuous impregnation of the filament assembly.

As far as the mathematical description of a composite with a regular structure is concerned, the essential point is that the rapidly oscillating coefficients involved in the pertinent equations are periodic functions of spatial coordinates and these equations therefore can be treated using the asymptotic method of homogenization, theoretically justified and well developed by a number of authors during the decade of the 1970s (Babuška 1976,1977, Bensoussan 1978, Sanchez-Palencia 1980, Lions 1981; see also Bakhvalov and Panasenko 1984). The method makes it possible to predict

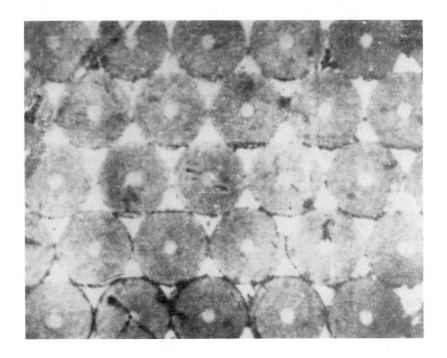

FIGURE 1.

both the overall and local properties of composites by first solving appropriate local problems set on the unit cell of the material and subsequently solving a boundary value problem for a homogeneous (or quasihomogeneous) material with effective material properties obtained at the first step. The mathematical justification of the homogenization method crucially depends on the proof that the solutions of the homogenized problem converge to those of the original problem in the limit as the unit cell of the structure tends to zero. This having been proved, an error estimate can be given (Duvaut 1976, Sanchez-Palencia 1980, Caillerie 1982, Iosifyan et al. 1982, Paşa 1983, Bakhvalov and Panasenko 1984, Kohn and Vogelius 1985).

As pointed out in Bakhvalov and Panasenko (1984), the method of homogenization yields an asymptotically correct representation of the solution, and the complexity of the method is fully compensated for its accuracy, which, in terms of the closeness of the approximation to the "true" solution, cannot be achieved by any other method of solution within the model accepted.

Considering the amount of work that has been done on purely mathematical aspects of the homogenization method, the application of the method to practical situations—particularly to the mechanics of composites and of structures of regular configurations—appears to be disproportionately limited. Most recently, however, quite a number of important results of various degrees of generality have been obtained that seem to be capable of raising the theoretical study of the mechanical behavior of composite material to a qualitatively higher level. In fact, collecting these results together, with an understandable emphasis on the authors' own work, is the raison d'être of this book.

We cover a wide range of subjects, as a look at the contents will reveal. Included are linear and geometrically nonlinear elasticity, heat conduction properties, thermoelasticity, and electroelasticity of composite materials, structurally nonhomogeneous materials, and structural elements, all possessing a regular structure with nonhomogeneity scale much less than the dimensions of the body. Much attention is given to the construction of homogenized models of composite material plates and shells. We calculate both the overall (effective) properties and local properties of various types of composites and thin-walled structural members now widely used in many fields (e.g., laminated and fibrous composites; ribbed, wafer, honeycomb, and two-way reinforced plates and shells; plates and shells with corrugated surfaces). A detailed exposition of modern state and perspectives of development of this direction of the mechanics of composite materials can be found in the review of results by Kalamkarov, Kudryavtsev, and Parton (1987a).

It is not amiss to remark that the rigorous methods we present in this book provide corrections, occasionally appreciable, to effective moduli results obtained earlier by other (approximate) methods.

Naturally enough, the material of this book is restricted, to some extent, by our own interests. There are many other topics in the field of composite materials, of much practical importance and in a very active state of development, that are not covered here but may be found discussed in detail in many other sources (Van Fo Fy 1971, Sendeckyj 1974, Obraztsov et al. 1977, Christensen 1979, Bolotin 1980, Malmeister et al. 1980, Cherepanov 1983, Pobedrya 1984, Vanin 1985, Tamužs and Protasov (Eds.) 1986, Vanin and Semenyuk 1987, and Vasil'ev 1988).

This book is offered in the hope that it may prove useful to scientists, teachers, and students in the field of composite materials as well as to practicing engineers engaged in the design, fabrication, and use of these materials in various fields.

Contents

1

Continuum Mechanics: Basic Concepts and Problem Formulations

A major problem in continuum mechanics (or mechanics of deformable media) consists of predicting the deformations and internal forces arising in a material body subject to a given set of external forces. In what may be called the classical formulation of the problem, only forces of a purely mechanical nature are considered. However, as far as composite materials and their behavior in various environments are concerned—particularly when dealing with piezoelectric composites—the effects of temperature and the presence of electromagnetic fields must be introduced and, in mathematical terms, a coupled system of equations considered (see Parton 1985, Parton and Kudryavtsev 1988).

The mechanics of composites is a subject of considerable current interest, both theoretically and experimentally. A wide variety of heterogeneous media have been studied and the development of fundamental methods for reliably predicting their mechanical behavior is the main objective of those involved in the field. This chapter discusses some of the methods commonly employed for the analysis of the macroscopic behavior of different types of heterogeneous media (or composites) and describes relationships between the effective properties of such media and characteristics of their individual constituent materials. We develop the subject primarily from the point of view of linear elasticity (thermoelasticity) theory, being perfectly aware, of course, that many aspects of mechanical behavior are left unaccounted for in this approach and referring the unsatisfied reader to monographs in which a level of higher generality is adopted.

As is customary in continuum mechanics studies, material properties are expressed in tensor form in this book. Another feature of our discussion is the thermodynamic expressions used for various combinations of the electric, magnetic, elastic, and thermal energies, which enable one to determine all the equilibrium properties by simple tensor derivatives of a number of thermodynamic potentials.

1. BASIC RELATIONS OF CONTINUUM MECHANICS

1.1. Kinematics of a moving material particle. The theory of deformation

Suppose a body at time $t = 0$ occupies a region V^0 (Figure 1.1). After a time t elapses, the body is deformed and occupies a region V. An arbitrary material point M within the body, located by the vector $\mathbf{x}^0(x_1^0, x_2^0, x_3^0)$, moves during the deformation to the spatial point M, located by the vector $\mathbf{x}(x_1, x_2, x_3)$, both the vectors being referred to rectangular axes x_i ($i = 1, 2, 3$) fixed in space. The coordinates x_i^0 ($i = 1, 2, 3$) specifying the point M^0 are called the material coordinates; the quantities x_i ($i = 1, 2, 3$) are the spatial coordinates of the point M, and the relations

$$\mathbf{x} = \mathbf{x}(\mathbf{x}^0, t) \tag{1.1}$$

or

$$x_i = x_i(x_1^0, x_2^0, x_3^0, t) \tag{1.2}$$

determine the motion of the material points of the body. The independent variables x_i^0 in Equation 1.1 usually are referred to as the Lagrange variables.

We may now invert Relation 1.1, assuming the coordinates x_i of point M in V to be unique functions of the time t and the coordinates x_i^0 of point M^0 in the reference state V^0. Clearly, the relations

$$\mathbf{x}^{(0)} = \mathbf{x}^{(0)}(\mathbf{x}, t) \tag{1.3}$$

or

$$x_i^{(0)} = x_i^{(0)}(x_1, x_2, x_3, t) \tag{1.4}$$

determine equally well the kinematics of the material body. The independent variables are in this case the spatial coordinates x_i (the Euler variables) corresponding to the coordinates of the particle that occupied point M^0 at the initial instant of time.

The strain tensor (describing the change in distance between various points in the deformed body) is conveniently defined by introducing the displacement vector

$$u_i = x_i - x_i^0 \tag{1.5}$$

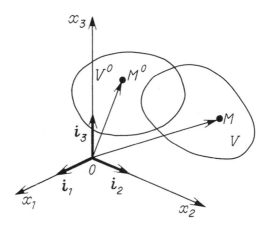

FIGURE 1.1.   Deformation of a material body V.

If two material particles were located initially at points x_i^0 and $x_i^0 + dx_i^0$, the squared distances between the particles will be determined by

$$ds_0^2 = dx_i^0 \, dx_i^0 \quad \text{and} \quad ds^2 = dx_i \, dx_i \qquad (1.6)$$

for respective instants 0 and t, and the deformed state of the body is characterized by the difference

$$ds^2 - ds_0^2$$

Working with the material variables x_i^0 and noting that

$$dx_i = \frac{\partial x_i}{\partial x_j^0} \, dx_j^0 \qquad (1.7)$$

we find

$$ds^2 - ds_0^2 = \frac{\partial x_i}{\partial x_j^0} \frac{\partial x_i}{\partial x_k^0} \, dx_j^0 \, dx_k^0 - dx_j^0 \, dx_j^0$$

$$= \left(\frac{\partial x_i}{\partial x_j^0} \frac{\partial x_i}{\partial x_k^0} - \delta_{jk} \right) dx_j^0 \, dx_k^0$$

$$= 2 \hat{\varepsilon}_{jk} \, dx_j^0 \, dx_k^0 \qquad (1.8)$$

where δ_{jk} is the Kronecker symbol and

$$\hat{\varepsilon}_{jk} = \frac{1}{2}\left(\frac{\partial x_i}{\partial x_j^0} \frac{\partial x_i}{\partial x_k^0} - \delta_{jk} \right)$$

is the strain tensor in Lagrange's representation.

Alternatively, we employ the spatial coordinates x_i as the independent variables and use the relation

$$dx_i^0 = \frac{\partial x_i^0}{\partial x_j} dx_j \tag{1.9}$$

to obtain

$$ds^2 - ds_0^2 = dx_j\, dx_j - \frac{\partial x_i^0}{\partial x_j} \frac{\partial x_i^0}{\partial x_k} dx_j\, dx_k$$

$$= \left(\delta_{jk} - \frac{\partial x_i^0}{\partial x_j} \frac{\partial x_i^0}{\partial x_k} \right) dx_j\, dx_k = 2\,\varepsilon_{jk}\, dx_j\, dx_k \tag{1.10}$$

where

$$\varepsilon_{jk} = \frac{1}{2}\left(\delta_{jk} - \frac{\partial x_i^0}{\partial x_j} \frac{\partial x_i^0}{\partial x_k} \right)$$

is the strain tensor in Euler's representation.

Using Equation 1.5, we now write

$$\hat{\varepsilon}_{jk} = \frac{1}{2}\left(\frac{\partial u_j}{\partial x_k^0} + \frac{\partial u_k}{\partial x_j^0} + \frac{\partial u_i}{\partial x_j^0} \frac{\partial u_i}{\partial x_k^0} \right) \tag{1.11}$$

and

$$\varepsilon_{jk} = \frac{1}{2}\left(\frac{\partial u_j}{\partial x_k} + \frac{\partial u_k}{\partial x_j} - \frac{\partial u_i}{\partial x_j} \frac{\partial u_i}{\partial x_k} \right) \tag{1.12}$$

for the finite strain tensor in, respectively, Lagrange's and Euler's representation; it is worth repeating here that both expressions refer to the same Cartesian coordinate system.

Because

$$\frac{\partial x_i}{\partial x_k^0}\frac{\partial x_k^0}{\partial x_j} = \delta_{ij} \tag{1.13}$$

the $\hat{\varepsilon}_{jk}$ and ε_{jk} tensors are readily found to be related by

$$\hat{\varepsilon}_{jk} = \varepsilon_{sn}\frac{\partial x_s}{\partial x_j^0}\frac{\partial x_n}{\partial x_k^0} \qquad \varepsilon_{jk} = \hat{\varepsilon}_{sn}\frac{\partial x_s^0}{\partial x_j}\frac{\partial x_n^0}{\partial x_k} \tag{1.14}$$

showing that if one of the tensors has all its components zero, so too does the other. (The case in point is the situation in which the material medium moves as a rigid body; we then have $ds = ds_0$ and $\hat{\varepsilon}_{jk} = \varepsilon_{jk} = 0$.)

In many practical applications it proves possible to neglect the products (or squares) of derivatives in Equations 1.11 and 1.12, and the well-known infinitesimal strain expression

$$\varepsilon_{ij} = \frac{1}{2}\left(\frac{\partial u_i}{\partial x_j} + \frac{\partial u_j}{\partial x_i}\right) \tag{1.15}$$

results.

It is often desirable that the components of the strain tensor be expressed in terms of some set of curvilinear coordinates θ^i ($i = 1, 2, 3$) defined by

$$x_i^0 = x_i^0(\theta^1, \theta^2, \theta^3) \tag{1.16}$$

Assuming that the $x_i^0(\theta^1, \theta^2, \theta^3)$ are single-valued continuous functions with continuous partial derivatives, the position of point M^0 in the nondeformed state is given by the vector

$$\mathbf{r}^0 = \mathbf{r}^0(\theta^1, \theta^2, \theta^3) \tag{1.17}$$

and the position this point assumes in the deformed state is determined by the vector

$$\mathbf{r} = \mathbf{r}(\theta^1, \theta^2, \theta^3) \quad \text{or} \quad x_i = x_i(\theta^1, \theta^2, \theta^3) \tag{1.18}$$

In the initial nondeformed state

$$ds_0^2 = g_{jk}(\theta^1, \theta^2, \theta^3)\,d\theta^j\,d\theta^k \tag{1.19}$$

where

$$g_{jk} = \mathbf{g}_j \cdot \mathbf{g}_k = \frac{\partial \mathbf{r}^0}{\partial \theta^j} \cdot \frac{\partial \mathbf{r}^0}{\partial \theta^k} = \frac{\partial x_i^0}{\partial \theta^j} \frac{\partial x_i^0}{\partial \theta^k}$$

is the metric tensor of the curvilinear coordinate system for the nonde-formed state, $\mathbf{g}_j$ being the basis vectors of this state.

During the deformation, the element ds_0 transforms into the element ds and

$$ds^2 = G_{jk}(\theta^1, \theta^2, \theta^3)\, d\theta^j\, d\theta^k \tag{1.20}$$

where

$$G_{jk}(\theta^1, \theta^2, \theta^3) = \mathbf{G}_j \cdot \mathbf{G}_k = \frac{\partial \mathbf{r}}{\partial \theta^j} \cdot \frac{\partial \mathbf{r}}{\partial \theta^k} = \frac{\partial x_i}{\partial \theta^j} \frac{\partial x_i}{\partial \theta^k}$$

is the metric tensor and $\mathbf{G}_j$ are the basis vectors of the deformed state.

The Lagrange strain tensor is now defined by

$$ds^2 - ds_0^2 = 2\hat{\varepsilon}_{jk}\, d\theta^j\, d\theta^k$$

and, by means of Equations 1.19 and 1.20,

$$\hat{\varepsilon}_{jk} = \tfrac{1}{2}(G_{jk} - g_{jk}) \tag{1.21}$$

Defining the displacement vector by

$$\mathbf{u} = u_n \mathbf{g}^n \tag{1.22}$$

and making use of the relations

$$\mathbf{r} = \mathbf{r}^0 + u_n \mathbf{g}^n$$

and

$$\mathbf{g}^n \mathbf{g}_k = \delta_k^n$$

we get, in terms of $g_{jk} = g_j g_k$,

$$\begin{aligned}
G_{jk} &= \frac{\partial \mathbf{r}}{\partial \theta^j} \cdot \frac{\partial \mathbf{r}}{\partial \theta^k} \\
&= \left[\frac{\partial \mathbf{r}^0}{\partial \theta^j} + (\nabla_j u_n)\mathbf{g}^n \right] \cdot \left[\frac{\partial \mathbf{r}^0}{\partial \theta^k} + (\nabla_k u_m)\mathbf{g}^m \right] \\
&= \left[\mathbf{g}_j + (\nabla_j u_n)\mathbf{g}^n \right] \cdot \left[\mathbf{g}_k + (\nabla_k u_m)\mathbf{g}^m \right] \\
&= g_{jk} + \nabla_j u_k + \nabla_k u_j + (\nabla_j u_n) \cdot (\nabla_k u^n)
\end{aligned} \tag{1.23}$$

where ∇_j denotes covariant differentiation involving the metric tensors g_{jk} and g^{jk}. From Equations 1.21 and 1.23, the finite strain tensor is given by

$$\hat{\varepsilon}_{jk} = \tfrac{1}{2}(\nabla_j u_k + \nabla_k u_j + \nabla_j u_n \cdot \nabla_k u^n) \tag{1.24}$$

An alternative representation of this tensor may be obtained by writing

$$\mathbf{u} = U_n \mathbf{G}^n \tag{1.25}$$

for the displacement vector. Noting that

$$\mathbf{r}^0 = \mathbf{r} - U_n \mathbf{G}^n \tag{1.26}$$

we then find

$$
\begin{aligned}
g_{jk} &= \frac{\partial \mathbf{r}^0}{\partial \theta^j} \cdot \frac{\partial \mathbf{r}^0}{\partial \theta^k} \\[2mm]
&= \left[\frac{\partial \mathbf{r}}{\partial \theta^j} - \left(\overline{\nabla}_j U_n \right) \mathbf{G}^n \right] \cdot \left[\frac{\partial \mathbf{r}}{\partial \theta^k} - \left(\overline{\nabla}_k U_m \right) \mathbf{G}^m \right] \\[2mm]
&= \left[\mathbf{G}_j - \left(\overline{\nabla}_j U_n \right) \mathbf{G}^n \right] \cdot \left[\mathbf{G}_k - \left(\overline{\nabla}_k U_m \right) \mathbf{G}^m \right] \\[2mm]
&= G_{jk} - \overline{\nabla}_j U_k - \overline{\nabla}_k U_j + \overline{\nabla}_j U_n \cdot \overline{\nabla}_k U^n
\end{aligned}
\tag{1.27}
$$

where $\overline{\nabla}_j$ denotes covariant differentiation involving the metric tensors G_{ij} and G^{ij}. We thus can write

$$\varepsilon_{jk} = \tfrac{1}{2}(G_{jk} - g_{jk}) = \tfrac{1}{2}\left(\overline{\nabla}_j U_k + \overline{\nabla}_k U_j - \overline{\nabla}_j U_n \cdot \overline{\nabla}_k U^n \right) \tag{1.28}$$

Employing the metric tensor g^{ij} of the nondeformed state or the metric tensor G^{ij} of the final (deformed) state, it is an easy matter to deduce a mixed-type tensor from the covariant strain tensor. In particular,

$$\hat{\varepsilon}_i^{\,k} = g^{ks}\hat{\varepsilon}_{si} = \tfrac{1}{2}\left(g^{ks}G_{si} - \delta_i^{\,k} \right)$$

For small deformations we have

$$\overline{\nabla}_i(\cdots) \cong \nabla_i(\cdots) \qquad u_k \cong U_k$$

and neglect of the terms quadratic in u reduces both Equation 1.24 and Equation 1.28 to

$$\varepsilon_{ik} = \tfrac{1}{2}(\nabla_i u_k + \nabla_k u_i) \tag{1.29}$$

where

$$\nabla_i U_k = \frac{\partial U_k}{\partial \theta^i} - \Gamma_{ki}^s U_s \tag{1.30}$$

and

$$\Gamma_{ki}^s = \frac{1}{2} g^{sp} \cdot \left(\frac{\partial g_{pk}}{\partial \theta^i} + \frac{\partial g_{pi}}{\partial \theta^k} - \frac{\partial g_{ik}}{\partial \theta^p} \right) \tag{1.31}$$

are the Christoffel symbols of the second kind.

Let us consider a system of orthogonal curvilinear coordinates with a fundamental quadratic form given by

$$ds^2 = H_i^2 (d\theta^i)^2 = H_1^2 (d\theta^1)^2 + H_2^2 (d\theta^2)^2 + H_3^2 (d\theta^3)^2 \tag{1.32}$$

The components of the metric tensor are in this case

$$g_{11} = H_1^2 \qquad g_{22} = H_2^2 \qquad g_{33} = H_3^2$$

$$g_{12} = g_{13} = g_{23} = 0$$

$$g^{11} = \frac{1}{H_1^2} \qquad g^{22} = \frac{1}{H_2^2} \qquad g^{33} = \frac{1}{H_3^2} \tag{1.33}$$

$$g^{12} = g^{13} = g^{23} = 0$$

and for the Christoffel symbols we find

$$\Gamma_{ki}^s = 0 \qquad (k \neq s \neq i)$$

$$\Gamma_{ki}^k = \Gamma_{ik}^k = \frac{1}{H_k} \frac{\partial H_k}{\partial \theta^i}$$

$$\Gamma_{kk}^k = \frac{1}{H_k} \frac{\partial H_k}{\partial \theta^k} \tag{1.34}$$

$$\Gamma_{kk}^s = -\frac{H_k}{H_s^2} \frac{\partial H_k}{\partial \theta^s}$$

If we denote by $u_{\theta 1}$, $u_{\theta 2}$, and $u_{\theta 3}$ the physical components of the displacement vector, then

$$u_1 = H_1 u_{\theta 1} \qquad u_2 = H_2 u_{\theta 2} \qquad u_3 = H_3 u_{\theta 3} \qquad (1.35)$$

and in accordance with Equation 1.29, the components of the strain tensor may be written (for small deformations) as

$$e_{11} = \frac{1}{H_1^2} \varepsilon_{11} = \frac{1}{H_1} \frac{\partial u_{\theta 1}}{\partial \theta^1} + \frac{1}{H_1 H_2} \frac{\partial H_1}{\partial \theta^2} u_{\theta 2} + \frac{1}{H_1 H_3} \frac{\partial H_1}{\partial \theta^3} u_{\theta 3}$$

$$e_{22} = \frac{1}{H_2^2} \varepsilon_{22} = \frac{1}{H_2} \frac{\partial u_{\theta 2}}{\partial \theta^2} + \frac{1}{H_2 H_1} \frac{\partial H_2}{\partial \theta^1} u_{\theta 1} + \frac{1}{H_2 H_3} \frac{\partial H_2}{\partial \theta^3} u_{\theta 3}$$

$$e_{33} = \frac{1}{H_3^2} \varepsilon_{33} = \frac{1}{H_3} \frac{\partial u_{\theta 3}}{\partial \theta^3} + \frac{1}{H_3 H_1} \frac{\partial H_3}{\partial \theta^1} u_{\theta 1} + \frac{1}{H_3 H_2} \frac{\partial H_3}{\partial \theta^2} u_{\theta 2}$$

$$e_{23} = \frac{1}{H_2 H_3} \varepsilon_{23} = \frac{1}{2}\left(\frac{1}{H_2} \frac{\partial u_{\theta 3}}{\partial \theta^2} + \frac{1}{H_3} \frac{\partial u_{\theta 2}}{\partial \theta^3} - \frac{1}{H_1 H_2} \frac{\partial H_3}{\partial \theta^2} u_{\theta 3} \right.$$
$$\left. - \frac{1}{H_2 H_3} \frac{\partial H_2}{\partial \theta^3} u_{\theta 2} \right)$$

$$e_{13} = \frac{1}{H_1 H_3} \varepsilon_{13} = \frac{1}{2}\left(\frac{1}{H_3} \frac{\partial u_{\theta 1}}{\partial \theta^3} + \frac{1}{H_1} \frac{\partial u_{\theta 3}}{\partial \theta^1} - \frac{1}{H_1 H_3} \frac{\partial H_1}{\partial \theta^3} u_{\theta 1} \right.$$
$$\left. - \frac{1}{H_3 H_1} \frac{\partial H_3}{\partial \theta^1} u_{\theta 3} \right)$$

$$e_{12} = \frac{1}{H_1 H_2} \varepsilon_{12} = \frac{1}{2}\left(\frac{1}{H_1} \frac{\partial u_{\theta 2}}{\partial \theta^1} + \frac{1}{H_2} \frac{\partial u_{\theta 1}}{\partial \theta^2} - \frac{1}{H_2 H_1} \frac{\partial H_2}{\partial \theta^1} u_{\theta 2} \right.$$
$$\left. - \frac{1}{H_1 H_2} \frac{\partial H_1}{\partial \theta^2} u_{\theta 1} \right) \qquad (1.36)$$

1.2. Stressed state at a point. Equilibrium equations

Select ΔS to be an elementary area within the deformed body V, and let $\Delta \mathbf{T}$ be the force that the part of the body on one side of ΔS exerts on the part on the other side. If ΔS is allowed to approach zero, the vector

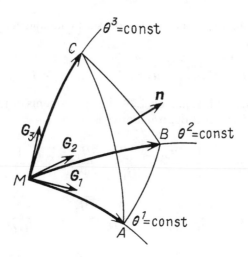

FIGURE 1.2.   Elemental volume *MABC* in a deformed body V_0.

$\Delta T / \Delta S$ becomes a vector **t** that describes the stresses arising in the area element ΔS and corresponds to a force acting on a unit area within the material.

Referring to Figure 1.2, let the small tetrahedron *MABC* have the representative point *M* as its apex and the edges *MA*, *MB*, and *MC* coincide with coordinate lines θ^i = constant. If a linear element along a coordinate line is defined by

$$d\mathbf{s}_i = \mathbf{G}_i \, d\theta^i \quad (i \text{ not summed}) \tag{1.37}$$

then the elemental area on the surface θ^i = constant will be given by

$$dS_i = \tfrac{1}{2}|d\mathbf{s}_j \times d\mathbf{s}_k| = \tfrac{1}{2}|\mathbf{G}_j \times \mathbf{G}_k| \, d\theta^j \, d\theta^k \quad (i \neq j \neq k)$$

which, by the definition of reciprocal coordinate vectors,

$$\mathbf{G}_i \times \mathbf{G}_j = \sqrt{G} \, G^k \quad (i \neq j \neq k)$$

$$G = \det G_{ij}$$

may be written

$$dS_i = \tfrac{1}{2}\sqrt{G G^{ii}} \, d\theta^j \, d\theta^k \quad (i \neq j \neq k, i \text{ not summed}) \tag{1.38}$$

or, in vector form,

$$dS_i = \frac{G^i \, dS_i}{\sqrt{G^{ii}}} \tag{1.39}$$

Denoting by dS the area of the face ABC and by $\mathbf{n}$ the unit normal to the face we may write

$$\mathbf{n} \, dS = \sum_{i=1}^{3} \frac{G^i \, dS_i}{\sqrt{G^{ii}}} \tag{1.40}$$

and setting

$$\mathbf{n} = n_i G^i \tag{1.41}$$

we obtain

$$n_i \sqrt{G^{ii}} \, dS = dS_i \quad (i \text{ not summed}) \tag{1.42}$$

Turning now to the derivation of equilibrium equations for the elemental tetrahedron $MABC$, it is expedient to introduce a stress vector $\mathbf{t}_i$ associated with elementary areas on the surfaces $\theta^i = $ constant. Clearly the vector $\mathbf{t}$ at point M of the deformed body is defined by

$$\mathbf{t} \, dS = \mathbf{t}_i \, dS_i \tag{1.43}$$

so that, by Equation 1.42,

$$\mathbf{t} = \sum_{i=1}^{3} n_i \mathbf{t}_i \sqrt{G^{ii}} \tag{1.44}$$

Because the vector $\mathbf{t}$ is invariant and the vector $\mathbf{n}$ covariant, it follows that the vector quantities $\mathbf{t}_i \sqrt{G^{ii}}$ (i not summed) may be represented in the form

$$\mathbf{t}_i \sqrt{G^{ii}} = \tau^{ij} \mathbf{G}_j \tag{1.45}$$

where τ^{ij} is the contravariant stress tensor.

The mixed and covariant stress tensors are defined by

$$\tau_j^i = G_{sj} \tau^{is} \qquad \tau_{ij} = G_{is} \tau_j^s \tag{1.46}$$

If the covariant and contravariant components of the stress vector **t** are defined by

$$\mathbf{t} = t_j \mathbf{G}^j \qquad \mathbf{t} = t^j \mathbf{G}_j \tag{1.47}$$

it follows from Equations 1.44 and 1.45 that

$$t^j = \tau^{ij} n_i \qquad t_j = n_i \tau_j^i \tag{1.48}$$

Rewriting Equation 1.45 in the form

$$\mathbf{t}_i = \sum_{j=1}^{3} \frac{\tau^{ij}}{\sqrt{G^{ii}}} \mathbf{G}_j = \sum_{j=1}^{3} \sqrt{\frac{G_{jj}}{G^{ii}}} \, \tau^{ij} \left(\frac{\mathbf{G}_j}{\sqrt{G_{jj}}} \right) = \sum_{j=1}^{3} \sigma_{ij} \left(\frac{\mathbf{G}_j}{\sqrt{G_{jj}}} \right) \tag{1.49}$$

and noting that the $\mathbf{G}_j / \sqrt{G_{jj}}$ are the unit vectors, we conclude that the quantities

$$\sigma_{ij} = \sqrt{\frac{G_{jj}}{G^{ii}}} \, \tau^{ij} \qquad (i, j \text{ not summed}) \tag{1.50}$$

are the physical components of the stress tensor at the elementary area whose normal coincides with the vector $\mathbf{G}_i$.

In a body occupying a volume V, bounded by a surface S, the equations of equilibrium of the volume are

$$\iint_S \mathbf{t} \, dS + \iiint_V \rho \mathbf{P} \, dV = 0 \tag{1.51}$$

and

$$\iint_S (\mathbf{r} \times \mathbf{t}) \, dS + \iiint_V (\mathbf{r} \times \rho \mathbf{P}) \, dV = 0 \tag{1.52}$$

where $\rho \mathbf{P}$ is the body force vector and ρ the mass density in the deformed state. Equations 1.51 and 1.52 express the requirements that, respectively, the forces acting on V and the momenta of these forces should sum to zero.

Applying the Green–Gauss theorem,

$$\iint_S F^i n_i \, dS = \iiint_V \bar{\nabla}_i F^i \, dV \qquad (\mathbf{n} = n_i \mathbf{G}^i)$$

and using Equations 1.47 and 1.48, the surface integrals in Equations 1.51 and 1.52 are converted to

$$\iint_S \mathbf{t}\, dS = \iint_S t^j \mathbf{G}_j\, dS$$

$$= \iint_S \left(\tau^{ij}\mathbf{G}_j\right) n_i\, dS$$

$$= \iiint_V \overline{\nabla}_i\left(\tau^{ij}\mathbf{G}_j\right) dV$$

$$\iint_S (\mathbf{r} \times \mathbf{t})\, dS = \iint_S \left(\mathbf{r} \times t^j \mathbf{G}_j\right) dS$$

$$= \iint_S \left(\mathbf{r} \times \tau^{ij}\mathbf{G}_j\right) n_i\, dS$$

$$= \iiint_V \overline{\nabla}_i\left(\mathbf{r} \times \tau^{ij}\mathbf{G}_j\right) dV$$

giving

$$\overline{\nabla}_i\left(\tau^{ij}\mathbf{G}_j\right) = \rho\mathbf{P} = 0 \tag{1.53}$$

$$\overline{\nabla}_i\left(\mathbf{r} \times \tau^{ij}\mathbf{G}_j\right) + (\mathbf{r} \times \rho\mathbf{P}) = 0 \tag{1.54}$$

in view of the arbitrariness of the volume being considered. Combining the last two equations gives

$$(\mathbf{G}_i \times \mathbf{G}_j)\tau^{ij} = \varepsilon_{ijs}\tau^{ij}\mathbf{G}^s$$

$$= (\tau^{23} - \tau^{32})\mathbf{G}^1 + (\tau^{31} - \tau^{13})\mathbf{G}^2 + (\tau^{12} - \tau^{21})\mathbf{G}^3$$

$$= 0 \tag{1.55}$$

showing the symmetry of the stress tensor: $\tau^{ij} = \tau^{ji}$.

Equation 1.53 represents the vector form of the differential equations of equilibrium and, if desired, may be rewritten as

$$\frac{1}{\sqrt{G}}\frac{\partial}{\partial\theta^i}\left(\sqrt{G}\,\tau^{ij}\mathbf{G}_j\right) + \rho\mathbf{P} = 0 \tag{1.56}$$

using the well-known result (see Sedov 1972)

$$\bar{\nabla}_i F^i = \frac{1}{\sqrt{G}} \frac{\partial}{\partial \theta^i} (\sqrt{G} F^i)$$

Instead of the deformed volume V, the nondeformed volume V^0 may be considered in Equation 1.51. Using

$$dV^0 = \sqrt{\frac{g}{G}} \, dV \quad (g = \det g_{ij})$$

the equilibrium equations then may be written in the form

$$\nabla_i \left(\tau^{ij} \mathbf{G}_j \sqrt{\frac{G}{g}} \right) + \rho^0 \mathbf{P}^0 = 0 \tag{1.57}$$

where $\mathbf{P}^0$ is the body force vector referred to the nondeformed volume.

To terms of the order of strain we have

$$G \cong g \qquad \nabla_i \cong \bar{\nabla}_i \qquad \rho^0 \mathbf{P}^0 \cong \rho \mathbf{P}$$

in Equations 1.53 and 1.57, and the equilibrium equations may be written

$$\nabla_i \left(\tau^{ij} \mathbf{G}_j \right) + \rho \mathbf{P} = 0 \tag{1.58}$$

or

$$\nabla_i \tau^{ij} + \rho P^i = 0 \tag{1.59}$$

where $\rho \mathbf{P} = \rho P^i \mathbf{G}_i$. The physical components of the stress tensor are given by

$$\sigma_{ij} = \tau^{ij} \sqrt{\frac{g_{jj}}{g^{ii}}} \tag{1.60}$$

For an orthogonal system of curvilinear coordinates with the quadratic form Equation 1.32, the physical components of the stress tensor are defined as

$$\sigma_{11} = \tau^{11} H_1^2 \qquad \sigma_{12} = \tau^{12} H_1 H_2 \qquad \sigma_{13} = \tau^{13} H_1 H_3$$
$$\sigma_{22} = \tau^{22} H_2^2 \qquad \sigma_{23} = \tau^{23} H_2 H_3 \qquad \sigma_{33} = \tau^{33} H_3^2 \tag{1.61}$$

and expanding Equation 1.59 in terms of these we have

$$\frac{1}{H_1 H_2 H_3}\left[\frac{\partial}{\partial\theta^1}(H_2 H_3 \sigma_{11}) + \frac{\partial}{\partial\theta^2}(H_3 H_1 \sigma_{12}) + \frac{\partial}{\partial\theta^3}(H_1 H_2 \sigma_{13})\right]$$

$$+ \frac{1}{H_1 H_2}\frac{\partial H_1}{\partial\theta^2}\sigma_{12} + \frac{1}{H_1 H_3}\frac{\partial H_1}{\partial\theta^3}\sigma_{13} - \frac{1}{H_1 H_2}\frac{\partial H_2}{\partial\theta^1}\sigma_{22}$$

$$- \frac{1}{H_1 H_3}\frac{\partial H_3}{\partial\theta^1}\sigma_{33} + \rho P_1 = 0$$

$$\frac{1}{H_1 H_2 H_3}\left[\frac{\partial}{\partial\theta^1}(H_2 H_3 \sigma_{12}) + \frac{\partial}{\partial\theta^2}(H_3 H_1 \sigma_{22})\right.$$

$$\left.+ \frac{\partial}{\partial\theta^3}(H_1 H_2 \sigma_{23})\right] + \frac{1}{H_2 H_3}\frac{\partial H_2}{\partial\theta^3}\sigma_{23}$$

$$+ \frac{1}{H_2 H_1}\frac{\partial H_2}{\partial\theta^1}\sigma_{12} - \frac{1}{H_3 H_2}\frac{\partial H_3}{\partial\theta^2}\sigma_{33}$$

$$- \frac{1}{H_1 H_2}\frac{\partial H_1}{\partial\theta^2}\sigma_{11} + \rho P_2 = 0$$

$$\frac{1}{H_1 H_2 H_3}\left[\frac{\partial}{\partial\theta^1}(H_2 H_3 \sigma_{13}) + \frac{\partial}{\partial\theta^2}(H_3 H_1 \sigma_{23})\right.$$

$$\left.+ \frac{\partial}{\partial\theta^3}(H_1 H_2 \sigma_{33})\right] + \frac{1}{H_3 H_1}\frac{\partial H_3}{\partial\theta^1}\sigma_{13}$$

$$+ \frac{1}{H_3 H_2}\frac{\partial H_3}{\partial\theta^2}\sigma_{23} - \frac{1}{H_1 H_3}\frac{\partial H_1}{\partial\theta^3}\sigma_{11}$$

$$- \frac{1}{H_2 H_3}\frac{\partial H_2}{\partial\theta^3}\sigma_{22} + \rho P_3 = 0 \tag{1.62}$$

where $P = P^i H_i$ (i not summed) are the physical components of the body force vector.

1.3. Elastic potential. Stress–strain relations in an elastic body

Let $\delta \mathbf{u}$ be the variation of the displacement vector $\mathbf{u}$. Multiplying the equation of motion (Equation 1.56) by $\delta \mathbf{u}$ and integrating over the deformed volume, because $dV = \sqrt{G}\, d\theta^1\, d\theta^2\, d\theta^3$, Equation 1.56 becomes

$$\iiint_V \left[\frac{\partial}{\partial \theta^i} \left(\sqrt{G}\, \tau^{ij} \mathbf{G}_j \right) \cdot \delta \mathbf{u} + \rho \sqrt{G}\, \mathbf{P} \cdot \delta \mathbf{u} \right] d\theta^1\, d\theta^2\, d\theta^3 = 0 \quad (1.63)$$

or

$$\iiint_V \left[\frac{\partial}{\partial \theta^i} \left(\sqrt{G}\, \tau^{ij} \mathbf{G}_j \cdot \delta \mathbf{u} \right) - \sqrt{G}\, \tau^{ij} \mathbf{G}_j \frac{\partial}{\partial \theta^i} (\delta \mathbf{u}) \right.$$

$$\left. + \rho \sqrt{G}\, \mathbf{P} \cdot \delta \mathbf{u} \right] d\theta^1\, d\theta^2\, d\theta^3 = 0 \quad (1.64)$$

After using the Green–Gauss theorem on the left-hand side of Equation 1.64 we obtain

$$\iint_S \mathbf{t} \cdot \delta \mathbf{u}\, dS + \iiint_V \rho \mathbf{P} \cdot \delta \mathbf{u}\, dV = \iiint_V \tau^{ij} \mathbf{G}_j \frac{\partial}{\partial \theta^i} (\delta \mathbf{u})\, dV \quad (1.65)$$

where we have also used the relation

$$n_i \tau^{ij} \mathbf{G}_j = \mathbf{t}$$

in which n_i is the unit vector normal to the bounding surface S. Because τ^{ij} is a symmetric tensor, as discussed in the previous section, the integrand of the right-hand side of Equation 1.65 can be written as

$$\tau^{ij} \mathbf{G}_j \cdot \frac{\partial}{\partial \theta^i} (\delta \mathbf{u}) = \frac{1}{2} \tau^{ij} \left[\mathbf{G}_j \cdot \frac{\partial}{\partial \theta^i} (\delta \mathbf{u}) + \mathbf{G}_i \cdot \frac{\partial}{\partial \theta^j} (\delta \mathbf{u}) \right] \quad (1.66)$$

This can be transformed by noting that the variations of the basis vectors $\mathbf{G}_j$ are

$$\delta \mathbf{G}_j = \delta \left(\mathbf{g}_j + \frac{\partial \mathbf{u}}{\partial \theta^j} \right) = \delta \left(\frac{\partial \mathbf{u}}{\partial \theta^j} \right) = \frac{\partial}{\partial \theta^j} (\delta \mathbf{u}) \quad (1.67)$$

where $\delta \mathbf{g}_j = 0$ because the vectors $\mathbf{g}_j$ are independent of $\mathbf{u}$. From Equation 1.28, then,

$$2\,\delta\varepsilon_{ij} = \delta G_{ij} = \delta(\mathbf{G}_i \cdot \mathbf{G}_j) = \delta\mathbf{G}_i \cdot \mathbf{G}_j + \mathbf{G}_i \cdot \delta\mathbf{G}_j$$

$$= \mathbf{G}_i \cdot \frac{\partial}{\partial\theta^j}(\delta\mathbf{u}) + \mathbf{G}_j \cdot \frac{\partial}{\partial\theta^i}(\delta\mathbf{u}) \tag{1.68}$$

and Equation 1.66 changes to

$$\tau^{ij}\mathbf{G}_j \cdot \frac{\partial}{\partial\theta^i}(\delta\mathbf{u}) = \tau^{ij}\,\delta\varepsilon_{ij} \tag{1.69}$$

Turning back to Equation 1.65, we are now in a position to write

$$\delta^*A = \delta U \tag{1.70}$$

where

$$\delta^*A = \iint_S \mathbf{t} \cdot \delta\mathbf{u}\, dS + \iiint_V \rho\mathbf{P} \cdot \delta\mathbf{u}\, dV$$

is the virtual work done by the surface and body forces along the permissible displacement $\delta\mathbf{u}$ and

$$\delta U = \delta \iiint_V \Phi\rho\, dV = \iiint_V \delta\Phi\rho\, dV$$

$$\delta\Phi = \left(\frac{\tau^{ij}}{\rho}\right)\delta\varepsilon_{ij} \tag{1.71}$$

For an elastic body the function Φ depends on strain

$$\Phi = \Phi(\varepsilon_{ij}) \qquad \delta\Phi = \frac{\partial\Phi}{\partial\varepsilon_{ij}}\,\delta\varepsilon_{ij}$$

so that

$$\tau^{ij} = \frac{1}{2}\rho\left(\frac{\partial\Phi}{\partial\varepsilon_{ij}} + \frac{\partial\Phi}{\partial\varepsilon_{ji}}\right) \tag{1.72}$$

The function Φ is usually referred to as the elastic potential (per unit mass of the material) and is meaningful for reversible isothermal and adiabatic processes. In the former case, Φ is in fact the free energy of the body and depends on strain and temperature; in the latter case Φ is the internal energy, a function of strain and (constant) entropy.

Introducing the elastic potential $W = \rho^0 \Phi$ for a unit volume of a nondeformed body and using $\rho^0 \sqrt{g} = \rho \sqrt{G}$, Equation 1.72 becomes

$$\sqrt{G}\,\tau^{ij} = \frac{1}{2}\sqrt{g}\left(\frac{\partial W}{\partial \varepsilon_{ij}} + \frac{\partial W}{\partial \varepsilon_{ji}}\right) \tag{1.73}$$

For infinitesimal deformations $G \cong g$, Equation 1.73 yields

$$\tau^{ij} = \frac{1}{2}\left(\frac{\partial W}{\partial \varepsilon_{ij}} + \frac{\partial W}{\partial \varepsilon_{ji}}\right) \tag{1.74}$$

If there are no stresses and no strains in the initial (virgin) state of the body, then, assuming W to be invariant, we may define it by

$$W = \tfrac{1}{2}C^{ijkl}\varepsilon_{ij}\varepsilon_{kl} \tag{1.75}$$

where C^{ijkl} is the tensor of elastic constants with the symmetry properties

$$C^{ijkl} = C^{jikl} = C^{klij} = C^{lkij} \tag{1.76}$$

From Equations 1.74 and 1.75, the small-deformation linear elasticity relations take the form

$$\tau^{ij} = C^{ijkl}\varepsilon_{kl} \tag{1.77}$$

Now, if we introduce the mixed tensor of elastic constants,

$$C^{ij}_{kl} = C^{ijps}g_{pk}g_{ls} \tag{1.78}$$

with symmetry properties

$$C^{ij}_{kl} = C^{ji}_{kl} = C^{ij}_{lk} = C^{jk}_{lk}$$

we see from Equation 1.77 that

$$\tau^{ij} = C^{ij}_{kl}\varepsilon^{kl} \qquad \tau_{ij} = C^{kl}_{ij}\varepsilon_{kl} \tag{1.79}$$

Because covariant and contravariant tensor components are the same in a rectangular coordinate system,

$$C_{ijkl} = C^{ijkl}$$

and the linear elasticity relations may be expressed as

$$\sigma_{ij} = C_{ijkl} e_{kl} \tag{1.80}$$

It is sometimes convenient to introduce a contracted notation and to rewrite Equation 1.80 in the compact matrix form

$$\sigma_\alpha = C_{\alpha\beta} e_\beta \qquad (\alpha, \beta = 1, 2, \ldots, 6) \tag{1.81}$$

where

$$\sigma_1 = \sigma_{11} \quad \sigma_2 = \sigma_{22} \quad \sigma_3 = \sigma_{33} \quad \sigma_4 = \sigma_{23} \quad \sigma_5 = \sigma_{13} \quad \sigma_6 = \sigma_{12}$$

$$e_1 = e_{11} \quad e_2 = e_{22} \quad e_3 = e_{33} \quad e_4 = 2e_{23} \quad e_5 = 2e_{13} \quad e_6 = 2e_{12}$$

and the two-index components $C_{\alpha\beta}$ are obtained from the components of C_{ijkl} by the following algorithm:

$$11 \to 1 \qquad 22 \to 2 \qquad 33 \to 3 \qquad 23 \to 4 \qquad 13 \to 5 \qquad 12 \to 6$$

In its most general anisotropic form the (symmetric) matrix of elastic moduli, $C_{\alpha\beta}$ ($\alpha, \beta = 1, 2, 3, 4, 5, 6$), has 21 independent components, but this number is reduced if there are some symmetry elements in the elastic properties of the material. Consider, for example, the case of symmetry with respect to three mutually orthogonal planes $x_1 x_2$, $x_1 x_3$, and $x_2 x_3$ in a Cartesian coordinate system. This class is known as *orthotropy* and there remain nine independent components of $C_{\alpha\beta}$ in this case,

$$C_{\alpha\beta} = \begin{pmatrix} C_{11} & C_{12} & C_{13} & 0 & 0 & 0 \\ C_{12} & C_{22} & C_{23} & 0 & 0 & 0 \\ C_{13} & C_{23} & C_{33} & 0 & 0 & 0 \\ 0 & 0 & 0 & C_{44} & 0 & 0 \\ 0 & 0 & 0 & 0 & C_{55} & 0 \\ 0 & 0 & 0 & 0 & 0 & C_{66} \end{pmatrix} \tag{1.82}$$

The *hexagonal* symmetry class occurs when the material is isotropic in one of the orthogonal planes. Letting x_3 be normal to the plane of isotropy, we have five independent components

$$C_{\alpha\beta} = \begin{pmatrix} C_{11} & C_{12} & C_{13} & 0 & 0 & 0 \\ C_{12} & C_{11} & C_{13} & 0 & 0 & 0 \\ C_{13} & C_{13} & C_{33} & 0 & 0 & 0 \\ 0 & 0 & 0 & C_{44} & 0 & 0 \\ 0 & 0 & 0 & 0 & C_{44} & 0 \\ 0 & 0 & 0 & 0 & 0 & C_{66} \end{pmatrix} \tag{1.83}$$

with

$$C_{66} = \tfrac{1}{2}(C_{11} - C_{22})$$

Finally, if the elastic medium exhibits hexagonal symmetry with respect to any two mutually orthogonal axes, we have complete isotropy, and the matrix of elastic constants reduces to

$$C_{\alpha\beta} = \begin{pmatrix} C_{11} & C_{12} & C_{12} & 0 & 0 & 0 \\ C_{12} & C_{11} & C_{12} & 0 & 0 & 0 \\ C_{12} & C_{12} & C_{11} & 0 & 0 & 0 \\ 0 & 0 & 0 & C_{44} & 0 & 0 \\ 0 & 0 & 0 & 0 & C_{44} & 0 \\ 0 & 0 & 0 & 0 & 0 & C_{44} \end{pmatrix} \tag{1.84}$$

where C_{11}, C_{12}, and C_{44} are defined to be

$$C_{11} = \lambda + 2G \qquad C_{12} = \lambda \qquad C_{44} = G = \tfrac{1}{2}(C_{11} - C_{12})$$

and λ and G are the Lamé constants related by

$$E = \frac{(2G + 3\lambda)G}{(\lambda + G)} \qquad \nu = \frac{\lambda}{2(\lambda + G)} \tag{1.85}$$

to Young's modulus E and Poisson's ratio ν. The general form for the Cartesian components of the tensor of elastic constants is thus

$$C_{ijkl} = \lambda \delta_{ij} \delta_{kl} + G(\delta_{ik} \delta_{jl} + \delta_{il} \delta_{jk}) \tag{1.86}$$

for an isotropic medium.

2. BASIC EQUATIONS OF THERMOELASTICITY AND ELECTROELASTICITY

2.1. Generalized equation of heat conduction; thermoelastic constitutive equations

It is a well-known fact of thermodynamics that irreversible heat-transfer processes produce entropy in a solid body. Based on the energy conservation law, the equation of entropy transfer may be written in local form as (see Nowacki 1970)

$$\frac{ds}{dt} = -\frac{1}{T}\frac{\partial q_i}{\partial x_i} + \frac{w}{T} \tag{2.1}$$

where q_i is the heat flow density, s is the entropy density, T is the absolute temperature, and w is the heat source function.

Let us apply Equation 2.1 to the determination of the rate of change of entropy during the process of heat conduction. Rewriting Equation 2.1 in the form

$$\frac{ds}{dt} = -\frac{\partial}{\partial x_i}\left(\frac{q_i}{T}\right) - \frac{1}{T^2}\frac{\partial T}{\partial x_i}q_i + \frac{w}{T} \tag{2.2}$$

integrating over a volume V and using the Green–Gauss theorem yields

$$\iiint_V \frac{ds}{dt}\,dV = -\iint_S \frac{1}{T}q_i n_i\,ds + \iiint_V \left(-\frac{1}{T^2}\frac{\partial T}{\partial x_i}q_i + \frac{w}{T}\right)dV \tag{2.3}$$

showing that the time rate of change of entropy is associated with two factors, the heat flow across the bounding surface S (the first term on the right-hand side of Equation 2.3) and the production of entropy within the volume by heat sources and heat transfer processes (the second term).

By the second law of thermodynamics, the right-hand side of Equation 2.3 is nonnegative and

$$\sigma = \left(-q_i F_i + \frac{w}{T}\right) = \left(-q_i\frac{1}{T^2}\frac{\partial T}{\partial x_i} + \frac{w}{T}\right) > 0 \tag{2.4}$$

The thermodynamic forces

$$F_i = -\frac{1}{T^2}\frac{\partial T}{\partial x_i} \tag{2.5}$$

we have just introduced are thought of as causing irreversible thermal processes in a body and must be related to the heat flow q_i in some way or other. In the simplest—linear—case,

$$q_i = \Lambda_{ij} F_j$$

where $\Lambda_{ij} = \Lambda_{ji}$ (Onsager's reciprocity relations), and the vector q_i is related to the temperature through

$$q_i = -\frac{\Lambda_{ij}}{T^2} \frac{\partial T}{\partial x_j} = -\lambda_{ij} \frac{\partial T}{\partial x_j} \qquad (2.6)$$

which is Fourier's law for an anisotropic medium.

To derive the constitutive relations of thermoelasticity theory, we employ the thermodynamic relation (Nowacki 1970)

$$dU = \sigma_{ij}\, de_{ij} + T\, ds \qquad (2.7)$$

which presents the total differential of the internal energy of the body as the sum of the increment of the work of deformation and the amount of heat introduced into the volume under study. Because the internal energy is a function of strain e_{ij} and entropy s,

$$dU = \left(\frac{\partial U}{\partial e_{ij}}\right)_s de_{ij} + \left(\frac{\partial U}{\partial s}\right)_e ds \qquad (2.8)$$

and hence, by comparison with Equation 2.7,

$$\sigma_{ij} = \left(\frac{\partial U}{\partial e_{ij}}\right)_s \qquad T = \left(\frac{\partial U}{\partial s}\right)_e \qquad (2.9)$$

The dependence of the stresses σ_{ij} on the strains e_{ij} and the temperature T is found by considering the free-energy function

$$F = U - TS \qquad (2.10)$$

which has just strains and temperature as its independent variables. From

$$dF = dU - T\, ds - s\, dT = \sigma_{ij}\, de_{ij} - s\, dT$$

$$dF = \left(\frac{\partial F}{\partial e_{ij}}\right)_T de_{ij} + \left(\frac{\partial F}{\partial T}\right)_e dT \qquad (2.11)$$

it is seen that

$$\sigma_{ij} = \left(\frac{\partial F}{\partial e_{ij}} \right)_T \qquad s = -\left(\frac{\partial F}{\partial T} \right)_e \qquad (2.12)$$

For small deformations and small temperature changes $\theta = T - T_0$ ($\theta / T_0 \ll 1$), we may expand $F(e_{ij}, T)$ in a power series of its arguments in the neighborhood of the virgin state and may retain only linear and quadratic terms in the expansion. Introducing the notation

$$C_{ijkl} = \frac{\partial^2 F(0, T_0)}{\partial e_{ij} \partial e_{kl}} \qquad \beta_{ij} = -\frac{\partial^2 F(0, T_0)}{\partial e_{ij} \partial T} \qquad m = \frac{\partial^2 F(0, T_0)}{\partial T^2}$$

and remembering that $e_{ij} = 0$ and $T = T_0$ in the virgin state, we get

$$F(e_{ij}, T) = \frac{1}{2} C_{ijkl} e_{ij} e_{kl} - \beta_{ij} e_{ij} \theta + \frac{m}{2} \theta^2 \qquad (2.13)$$

where use has been made of the relations

$$F(0, T_0) = 0 \qquad \frac{\partial F(0, T_0)}{\partial T} = 0 \qquad \frac{\partial F(0, T_0)}{\partial e_{ij}} = 0$$

which follow from the fact that s and σ_{ij} are initially zero.

Using Equation 2.13 in the first part of Equation 2.12 now yields the equation

$$\sigma_{ij} = C_{ijkl} e_{ij} - \beta_{ij} \theta \qquad (2.14)$$

which expresses the Duhamel–Neumann law for an anisotropic body. When solved for strains, Equation 2.14 gives

$$e_{ij} = s_{ijkl} \sigma_{kl} + \alpha_{ij}^T \theta \qquad (2.15)$$

where s_{ijkl} are the elastic compliances, and α_{ij}^T are the temperature coefficients of expansion and shear related to β_{ij} by

$$\beta_{ij} = \alpha_{kl}^T C_{ijkl} \qquad (2.16)$$

To determine the entropy as a function of strains e_{ij} and temperature T, consider the total differential of the function $s(e_{ij}, T)$

$$ds = \left(\frac{\partial s}{\partial e_{ij}}\right)_T de_{ij} + \left(\frac{\partial s}{\partial T}\right)_e dT \qquad (2.17)$$

Noting that

$$\left(\frac{\partial \sigma_{ij}}{\partial T}\right)_e = -\left(\frac{\partial s}{\partial e_{ij}}\right)_T$$

from Equations 2.12 and that

$$\frac{\partial \sigma_{ij}}{\partial T} = -\beta_{ij} \qquad T\left(\frac{\partial s}{\partial T}\right)_e = c_e$$

where c_e denotes the specific heat at constant strain, Equation 2.17 becomes

$$ds = \beta_{ij} de_{ij} + \frac{c_e}{T} dT \qquad (2.18)$$

which when integrated under the virgin state conditions $e_{ij} = 0$ and $T = T_0$ gives

$$s = \beta_{ij} e_{ij} + c_e \ln\left(1 + \frac{\theta}{T_0}\right) \qquad (2.19)$$

or, for small temperature changes ($\theta/T \ll 1$),

$$s = \beta_{ij} e_{ij} + \left(\frac{c_e}{T_0}\right)\theta \qquad (2.20)$$

Inserting Equations 2.20 and 2.6 into Equation 2.1 and linearizing with respect to θ yields the generalized heat conduction equation

$$\frac{\partial}{\partial x_i}\left(\lambda_{ij}\frac{\partial \theta}{\partial x_j}\right) - c_e\frac{\partial \theta}{\partial t} - T_0 \beta_{ij}\frac{\partial e_{ij}}{\partial t} = -w \qquad (2.21)$$

The third term on the left-hand side of Equation 2.21 couples the temperature and strain fields of the problem and makes it necessary that, simultaneously with Equation 2.21, the equations of motion (or equilibrium) be considered. In Cartesian coordinates these are

$$\frac{\partial \sigma_{ij}}{\partial x_j} + \rho P_i = \rho \frac{\partial^2 u_i}{\partial t^2} \tag{2.22}$$

Substituting from Equation 2.14 and noting that

$$e_{ij} = \frac{1}{2} \left(\frac{\partial u_i}{\partial x_j} + \frac{\partial u_j}{\partial x_i} \right)$$

the following system of equations is obtained:

$$C_{ijkl} \frac{\partial^2 u_k}{\partial x_l \, \partial x_j} + \rho P_i = \rho \frac{\partial^2 u_i}{\partial t^2} + \beta_{ij} \frac{\partial \theta}{\partial x_j} \tag{2.23}$$

The differential equations of Equations 2.21 and 2.23 make a complete system of equations of the coupled dynamic thermoelastic problem for an anisotropic body and enable one, in principle, to describe the deformations due to nonstationary thermal and mechanical influences and the temperature changes associated with the deformations.

Of course we have to adjoin appropriate boundary and initial conditions to these equations. At the initial instant of time, it is necessary to specify temperature, displacements, and velocities at all points within the volume V, that is, we must be able to write

$$\theta(M, 0) = \theta_0(M)$$

$$u_i(M, 0) = u_i^0(M) \tag{2.24}$$

$$\frac{\partial u_i(M, 0)}{\partial t} = v_i^0(M) \qquad (M \in V, t = 0)$$

with θ_0, u_i^0, and v_i^0 as known functions of point $M \in V$.

The most widely used boundary condition for the temperature function is the heat exchange condition of the third kind (known also as Newton's law), which states that the heat flow across the surface S of the body is proportional to the difference between the temperature of the surface,

$\theta(P, t)$ ($P \in S$), and the (known) temperature of the surrounding medium, θ_m. Mathematically this is expressed by writing

$$\lambda_{ij} \frac{\partial \theta(P, t)}{\partial x_j} n_i + \alpha_S [\theta(P, t) - \theta_c(P, t)] = 0 \qquad (P \in S) \quad (2.25)$$

where α_S is the coefficient of heat exchange at S, and n_i is the unit vector of the outward normal to S.

Alternatively, we may impose the boundary condition of the first kind

$$\theta(P, t) = \theta_0(P, t) \qquad (2.26)$$

or the second kind

$$-\lambda_{ij} \frac{\partial \theta(P, t)}{\partial x_j} n_i = q_0(P, t) \qquad (2.27)$$

with θ_0 and q_0 known functions of point $P \in S$.

Turning to the mechanical boundary conditions for thermoelastic problems, we must require that external loadings p_i or displacements $\tilde{u}_i$ be specified on the surface S as functions of spatial coordinates and time,

$$\sigma_{ij} n_j = p_i(P, t) \qquad u_i = \tilde{u}_i(P, t) \qquad (P \in S, t > 0) \quad (2.28)$$

If there are no mechanical influences in the problem and deformations are caused by a time-varying heating or cooling of the surface S of the body, we must set $P_0 = 0$ in Equation 2.23 and $p_i = 0$ in the boundary conditions (Equations 2.28).

The system of Equations 2.21 and 2.22 lends itself to a considerable simplification: Dropping the term $T_0 \beta_{ij}(\partial e_{ij}/\partial t)$ in the left-hand side of Equation 2.21, we obtain the uncoupled equations of dynamic thermoelasticity:

$$\frac{\partial}{\partial x_i} \left(\lambda_{ij} \frac{\partial \theta}{\partial x_j} \right) - c_e \frac{\partial \theta}{\partial t} = -w \qquad (2.29)$$

$$C_{ijkl} \frac{\partial^2 u_k}{\partial x_l \, \partial x_j} + \rho P_i = \rho \frac{\partial^2 u_i}{\partial t^2} + \beta_{ij} \frac{\partial \theta}{\partial x_j} \qquad (2.30)$$

which can be solved by first finding the temperature field θ and then calculating the displacements u_i from Equation 2.30. A further simplification consists of neglecting the inertial terms in Equation 2.30 with the result that a system of equations of the quasistatic thermoelastic problem is obtained.

The basic relations and thermoelastic equations we have developed in this section are general enough to fit any type of symmetry or, for that matter, the absence of symmetry. In an isotropic medium the equations are reduced to a simpler form by setting

$$\lambda_{ij} = \delta_{ij}\lambda$$

$$\alpha_{ij}^T = \delta_{ij}\alpha^T$$

$$\beta_{ij} = \delta_{ij}\beta$$

$$C_{ijkl} = \lambda\,\delta_{ij}\,\delta_{kl} + G(\delta_{ik}\delta_{ji} + \delta_{il}\delta_{jk}) \qquad (2.31)$$

where $\beta = \alpha^T(3\lambda + 2G)$ is the linear thermal-expansion coefficient of the material.

2.2. Electroelastic equations for a piezoelectric medium

The theory of thermoelasticity, a merger of classical elasticity theory and the theory of heat conduction, is a good illustrative example of a general theory of coupled fields that may be (and has been) developed in the framework of the mechanics of continuous media. Another very important aspect of the theory of coupled fields in such media is associated with electroelasticity and deals with phenomena caused by interactions between electric and mechanical fields. One of these phenomena is the so-called piezoelectric effect (see Parton and Kudryavtsev 1988). Discovered experimentally in 1880, the *direct* piezoelectric effect is concerned with the appearance of electric charges on the surface of a body subject to a mechanical loading. The effect is conveniently described by the polarization vector $\mathbf{P}_i$, which represents the electric moment per unit volume or the polarization charge per unit area and is related to the stress tensor components σ_{kl} by the linear expression

$$P_i = d_{ikl}\sigma_{kl} \qquad (2.32)$$

where d_{ikl} is the third-rank tensor of piezoelectric moduli.

A simple thermodynamic argument shows that the *inverse* piezoelectric effect must also exist. In this effect the electric field components and (small) strains e_{ij} are related by the linear formula

$$e_{ij} = d_{kij} E_k \tag{2.33}$$

where the tensor d_{kij} is symmetric with respect to interchange of the indices i and j because of the symmetry of the e_{ij} tensor. It should be noted that both the direct and inverse modes of the piezoelectric effect are only possible if there is no center of symmetry in the crystal; otherwise all components of the d_{kij} tensor turn to zero for reasons of symmetry.

In the pyroelectric effect, which also appears only for noncentrosymmetrical crystals, it is cooling or heating that changes the polarization of a material. For small temperature change ΔT, the change ΔP_i in polarization can be written

$$\Delta P_i = p_i \, \Delta T \tag{2.34}$$

where the p_i constants are known as pyroelectric coefficients.

The relations between the mechanical, electrical, and thermal properties of real piezoelectric media can be demonstrated most easily from equilibrium thermodynamics and appear as the result of applying the so-called Maxwell relations to particular forms of thermodynamic potentials. If the strains e_{ij}, the electric field E_i, and the temperature $T = T_0 + \theta$ are regarded as the independent variables, the dependent variables will be the stresses σ_{ij}, the electric displacement $D_i = E_i + \epsilon_0 P_i$ (ϵ_0 being the dielectric constant), and the entropy s. As shown in books dealing with thermodynamics (e.g., Zheludev 1968), the thermodynamic potential corresponding to this choice of variables is the electric Gibbs function G_e,

$$G_e = U - E_m D_m - sT \tag{2.35}$$

(U denoting the internal energy), whose differential form is

$$dG_e = \sigma_{ij} \, de_{ij} - D_m \, dE_m - s \, dT \tag{2.36}$$

From Equation 2.36 we obtain the relations

$$\sigma_{ij} = \left(\frac{\partial G_e}{\partial e_{ij}} \right)_{E,T} \qquad D_m = -\left(\frac{\partial G_e}{\partial E_m} \right)_{e,T} \qquad s = -\left(\frac{\partial G_e}{\partial T} \right)_{e,E} \tag{2.37}$$

in which the subscripts to the parentheses indicate the variables that are held constant during the differentiation. Differentiating Equations 2.37

now gives

$$\left(\frac{\partial \sigma_{ij}}{\partial E_m}\right)_{e,T} = -\left(\frac{\partial D_m}{\partial e_{ij}}\right)_{E,T}$$

$$\left(\frac{\partial \sigma_{ij}}{\partial T}\right)_{e,E} = -\left(\frac{\partial s}{\partial e_{ij}}\right)_{E,T}$$

$$\left(\frac{\partial D_m}{\partial T}\right)_{e,E} = \left(\frac{\partial s}{\partial E_m}\right)_{e,T} \tag{2.38}$$

and the perfect differentials of the dependent variables are found to be

$$d\sigma_{ij} = \left(\frac{\partial \sigma_{ij}}{\partial e_{kl}}\right)_{E,T} de_{kl} + \left(\frac{\partial \sigma_{ij}}{\partial E_m}\right)_{E,T} dE_m + \left(\frac{\partial \sigma_{ij}}{\partial T}\right)_{E,e} dT$$

$$dD_m = \left(\frac{\partial D_m}{\partial e_{ij}}\right)_{E,T} de_{ij} + \left(\frac{\partial D_m}{\partial E_k}\right)_{e,T} dE_k + \left(\frac{\partial D_m}{\partial T}\right)_{e,E} dT$$

$$ds = \left(\frac{\partial s}{\partial e_{ij}}\right)_{E,T} de_{ij} + \left(\frac{\partial s}{\partial E_k}\right)_{e,T} dE_k + \left(\frac{\partial s}{\partial T}\right)_{e,E} dT \tag{2.39}$$

The partial derivatives may be regarded as being constant and Equations 2.39 become integrable if the range of variables is assumed to be narrow enough. We then obtain the constitutive relations of a deformable piezoelectric medium (Berlincourt et al. 1964),

$$\sigma_{ij} = C_{ijkl} e_{kl} - e_{ijm} E_m - \beta_{ij} \theta$$

$$D_m = \varepsilon_{mk} E_k + e_{mij} e_{ij} + p_m \theta$$

$$\Delta s = \beta_{ij} e_{ij} + p_m E_m + \frac{\rho c}{T_0} \theta \tag{2.40}$$

where the partial derivatives have special names as follows:

$$C_{ijkl} = \left(\frac{\partial \sigma_{ij}}{\partial e_{kl}}\right)_{E,T} \qquad \text{elastic moduli}$$

$$e_{ijm} = -\left(\frac{\partial \sigma_{ij}}{\partial E_m}\right)_{e,T} \qquad \text{piezoelectric constants}$$

$$p_m = \left(\frac{\partial D_m}{\partial T}\right)_{e,\,T} \qquad \text{pyroelectric constants}$$

$$\varepsilon_{mk} = \left(\frac{\partial D_m}{\partial E_k}\right)_{e,\,T} \qquad \text{dielectric constants}$$

$$\beta_{ij} = -\left(\frac{\partial \sigma_{ij}}{\partial T}\right)_{e,\,E} \qquad \text{temperature-expansion coefficients}$$

$$\left(\frac{\rho C}{T_0}\right) = \left(\frac{\partial s}{\partial T}\right)_{e,\,E} \qquad \begin{array}{l}\text{specific heat per unit vol-}\\ \text{ume divided by the abso-}\\ \text{lute temperature}\end{array}$$

When dealing with specific electroelastic problems, it is convenient to rewrite Equation 2.40 in a matrix form. Using the notation of Equation 1.81 and the matrices

$$C_{\alpha\beta} = C_{ijkl} \quad \text{and} \quad e_{m\alpha} = e_{mij} \qquad (i,j,k,l = 1,2,3;\ \alpha,\beta = 1,2,\ldots,6)$$

we get

$$\sigma_\alpha = C_{\alpha\beta} e_\beta - e_{\alpha m} E_m - \beta_\alpha \theta$$

$$D_m = \varepsilon_{mk} E_k + e_{m\alpha} e_\alpha + p_m \theta$$

$$\Delta s = \beta_\alpha e_\alpha + p_m E_m + \frac{\rho c}{T_0} \theta \qquad (2.41)$$

where the temperature-expansion coefficients are redefined by

$$\beta_{ij} = \beta_\alpha \qquad (i,j = 1,2,3;\ \alpha = 1,2,\ldots,6)$$

Situations in which temperature effects may be neglected are of interest. In this case Equation 2.41 reduces to

$$\sigma_\alpha = C_{\alpha\beta} e_\beta - e_{\alpha m} E_m$$

$$D_m = \varepsilon_{mk} E_k + e_{m\alpha} e_\alpha \qquad (2.42)$$

A brief discussion should be given here of the number of independent coefficients that represent the elastic and electric properties of the medium in Equation 2.42. In the most general kind of anisotropic solid (triclinic system), Equation 2.42 contains 21 elastic constants, 18 piezoelectric constants, and 6 dielectric constants. Crystal symmetry eliminates some possible components of the material tensors. In particular, for polarized ceramic (6 mm symmetry class), directing the axis of x_3 along the polarization direction, we have

$$
C_{\alpha\beta} =
\begin{pmatrix}
C_{11} & C_{12} & C_{13} & 0 & 0 & 0 \\
C_{12} & C_{11} & C_{13} & 0 & 0 & 0 \\
C_{13} & C_{13} & C_{33} & 0 & 0 & 0 \\
0 & 0 & 0 & C_{44} & 0 & 0 \\
0 & 0 & 0 & 0 & C_{44} & 0 \\
0 & 0 & 0 & 0 & 0 & C_{66}
\end{pmatrix}
$$

$$
C_{66} = \tfrac{1}{2}(C_{11} - C_{12})
$$

$$
e_{\alpha m} =
\begin{pmatrix}
0 & 0 & e_{13} \\
0 & 0 & e_{13} \\
0 & 0 & e_{33} \\
0 & e_{15} & 0 \\
e_{15} & 0 & 0 \\
0 & 0 & 0
\end{pmatrix}
$$

$$
\varepsilon_{mk} =
\begin{pmatrix}
\varepsilon_{11} & 0 & 0 \\
0 & \varepsilon_{11} & 0 \\
0 & 0 & \varepsilon_{33}
\end{pmatrix}
\tag{2.43}
$$

3. MECHANICAL MODELS OF COMPOSITE MATERIALS

The behavior of composite materials (or of heterogeneous media, to take a broader view) is analyzed in this book primarily from the standpoint of continuum mechanics; thus, when dealing with a particular problem, we usually invoke an idealized geometric model of the heterogeneous system under study and proceed to obtain the theoretical predictions of the macroscopic properties of the system in terms of the geometrical and physical properties of the individual constituent materials. The model must necessarily contain piecewise-continuous material dependences in its governing relations, and the basic problem of the mechanics of composites is to determine the effective properties of some idealized homogeneous medium that, insofar as its response to external influences is concerned, is equivalent to the actual heterogeneous medium. In this book we limit ourselves to elastic, electric, and thermal influences (and properties); therefore it is from the point of view of the theories of thermoelasticity and electroelasticity that the subject of the behavior of composites is developed.

3.1. Effective mechanical characteristics of composite materials

We will consider a composite system as consisting of a number of regions (or phases or components), each described by governing elasticity relations of the form

$$\sigma_{ij} = C_{ijkl}(\mathbf{x})e_{kl} \tag{3.1}$$

with functions $C_{ijkl}(\mathbf{x})$ continuous in or, as a special case, independent of the spatial coordinate $\mathbf{x} = (x_1, x_2, x_3)$. We will denote by L the characteristic dimension of the composite system. One of the components of such a system is usually referred to as the *matrix*, whereas the others form what is called the *reinforcement* and are usually termed *inclusions*. Depending on the shape and distribution of inclusions we may distinguish, for example, laminates of fiber (filamentary) composites, each of the terms being self-explanatory, or particular composites, if all three dimensions of the inclusions are of the same order (and are much less than L). The choice—or design—of a composite material depends on its intended use, of course, and it must be kept in mind that combinations that enhance a particular property often involve the degradation of another property. Fiber materials, for example, provide good strength and stiffness properties; with rigid spherical inclusions, lower price and better dynamic response characteristics are achieved.

We follow here the averaging procedure described in Sendeckyj (1974) (see also Christensen 1979) in defining the effective stiffness of a composite material. Assuming that the characteristic dimension of inhomogeneity l is much less than L, the averaged stress and the averaged strain are defined by

$$\langle \sigma_{ij} \rangle = \frac{1}{V} \int_V \sigma_{ij}(x)\, dV$$

$$\langle e_{ij} \rangle = \frac{1}{V} \int_V e_{ij}(x)\, dV \tag{3.2}$$

where V is the volume of the representative element of the heterogeneous medium. The effective stiffness properties C_{ijkl} of the linearly elastic body may now be defined through their presence in the relation

$$\langle \sigma_{ij} \rangle = \tilde{C}_{ijkl}\langle e_{kl} \rangle \tag{3.3}$$

To perform these operations rigorously requires a knowledge of the stress and strain fields, $\sigma_{ij}(x)$ and $e_{ij}(x)$ in the heterogeneous medium, but in some special cases considerable simplifications can be made. Consider, for example, a two-material heterogeneous system, one material of which is continuous and the other in the form of a discrete inclusion, and both materials are isotropic. In this case (see Russel and Acrivos 1972)

$$\tilde{C}_{ijkl}\langle e_{kl}\rangle = \lambda_m\,\delta_{ij}\langle e_{kk}\rangle + 2G_M\langle e_{ij}\rangle + \frac{1}{V}\int_{V_I}$$

$$\times(\sigma_{ij} - \lambda_M\,\delta_{ij}e_{kk} - 2G_M e_{ij})\,dV \tag{3.4}$$

where λ_M and G_M are the matrix Lamé constants and V_I is the volume of the inclusion. Note that only the conditions within the inclusion are required for the evaluation of the C_{ijkl} tensor.

By specializing Equation 3.4 to the case of dilute suspension conditions, that is, by assuming a small value of the volume fraction of the inclusion(s), rather simple expressions for the effective shear modulus $\tilde{G}$ (or for the bulk modulus $\tilde{k}$) can be derived (Christensen 1979) in terms of the homogeneous strain in the inclusion(s). In doing this, a result due to Eshelby (1957) may be employed, that a single ellipsoidal inclusion embedded in an infinite elastic medium undergoes a homogeneous deformation proportional to that imposed at large distances from the inclusion.

Alternatively, we may define the effective properties C_{ijkl} through the equality of the strain energy stored in the heterogeneous medium to that stored in the equivalent homogeneous medium. We write, namely,

$$\frac{1}{2}\int_V \sigma_{ij}e_{ij}\,dV = \frac{1}{2}\tilde{C}_{ijkl}\langle e_{ij}\rangle\langle e_{kl}\rangle \tag{3.5}$$

This energy criterion may be used (again, see Christensen 1979) for determining the effective shear modulus in an elastic medium with a low concentration of spherical inclusions. Consider an isotropic medium in a state of simple shear deformation, with displacement components specified by

$$u_1 = \tau x_1 \qquad u_2 = -\tau x_2 \qquad u_3 = 0 \qquad (\tau\text{-constant}) \tag{3.6}$$

at large distances from a single spherical inclusion of radius a. We now calculate—and subsequently equate to each other—the deformation energies of two spherical volumes, a composite one containing the inclusion and an equivalent homogeneous one, of radius b. Assuming that $c =$

$(a/b)^3 \ll 1$ (small value of the volume fraction of the inclusions), this yields

$$\frac{\tilde{G}}{G_M} = 1 - \frac{15(1 - \nu_M)(1 - G_I/G_M)c}{7 - 5\nu_M + 2(4 - 5\nu_M)(G_I/G_M)} \tag{3.7}$$

where ν_M is the matrix Poisson ratio and G_I and G_M are the shear moduli of the inclusion and matrix materials, respectively.

One more method for determining the shear and bulk moduli of an elastic heterogeneous medium is due to Hashin (1962). In the composite spheres model he proposes that a system of spherical particles of different radii is embedded in a continuous unbounded matrix phase, and with each particular particle (of radius a) a spherical region (of radius b) of the matrix phase is associated. The composite particles fill up the entire volume being considered, and the ratio of radii a/b is taken to be a constant for each composite sphere, independent of its absolute size. To determine the effective bulk modulus within this model, one considers the deformation of the composite spherical particle of radius b and that of the equivalent homogeneous particle of the same radius, both particles undergoing a hydrostatic stress at $r = b$. Clearly, the strain states of these two particles will be identical in a volume-averaged sense if their radial displacements are equal at $r = b$. This condition yields the effective bulk modulus

$$\tilde{k} = k_M + \frac{(k_I - k_M)c}{1 + (1 - c)[(k_I - k_M)/(k_M + \tfrac{4}{3}G_M)]} \tag{3.8}$$

where k_I and k_M denote the inclusion and matrix bulk moduli, respectively.

Turning now to the effective shear modulus, it should be admitted that its determination is a rather difficult task in the composite spheres model because when simple shear-type displacement components are prescribed on the surface of a composite sphere, the resulting boundary stresses are not those corresponding to a state of simple shear stress. To obviate this difficulty, different approaches are needed for the shear modulus problem. In the so-called three-phase model, for example, a single composite spherical particle is considered. The particle is placed within an equivalent homogeneous medium whose effective characteristics G and k may be found (among other possibilities) from the condition that the energy stored in the composite equals that stored in an equivalent homogeneous medium (Christensen 1979).

A brief discussion of similar results for fiber composites is now in order. First we observe that a medium containing a system of parallel fibers has symmetrical properties in the plane normal to the fiber direction. Choosing this latter to coincide with the x_1 axis, the elastic relations of this *transversely isotropic* medium may be written as

$$\sigma_{11} = C_{11}e_{11} + C_{12}e_{22} + C_{12}e_{33}$$

$$\sigma_{22} = C_{12}e_{11} + C_{22}e_{22} + C_{23}e_{33}$$

$$\sigma_{33} = C_{12}e_{11} + C_{23}e_{22} + C_{22}e_{33}$$

$$\sigma_{23} = (C_{22} - C_{23})e_{23} \qquad \sigma_{13} = 2C_{66}e_{23} \qquad \sigma_{12} = 2C_{66}e_{12} \quad (3.9)$$

where we have introduced the conventional two-suffix notation for the five independent stiffness constants C_{ij} relevant to this particular symmetry. Note that in practical applications of Equation 3.9, the experimentally measurable "engineering properties," defined by

$$E_{11} = C_{11} - \frac{2C_{12}^2}{C_{22} + C_{23}} \qquad v_{12} = v_{13} = \frac{C_{12}}{C_{22} + C_{23}} \qquad k_{23} = \frac{1}{2}(C_{22} + C_{23})$$

$$G_{12} = G_{31} = C_{66} \quad \text{and} \quad G_{23} = \frac{1}{2}(C_{22} - C_{23}) \qquad (3.10)$$

are more convenient.

Analogous to the case of spherical inclusions, we may employ the *composite cylinders model* for the evaluation of the effective properties of fiber-reinforced composites (Hashin and Rosen 1964). We associate a certain outer cylinder with each particular fiber and assume the ratio a/b to be constant for all the fibers, a and b denoting the respective radii of the fiber and the cylinder. The stress–strain analysis of the resulting composite cylinder then yields (see Christensen 1979) the following expressions for four of the five constants involved in Equation 3.10:

$$E_{11} = cE_F + (1 - c)E_M$$

$$+ \frac{4c(1 - c)(v_F - v_M)^2 G_M}{(1 - c)G_M/(k_F + G_F/3) + cG_M(k_M + G_M/3)^{-1} + 1} \qquad (3.11)$$

$$\nu_{12} = (1 - c)\nu_M + c\nu_F$$

$$+ \frac{c(1 - c)(\nu_F - \nu_M)\left[G_M(k_M + G_M/3)^{-1} - G_M(k_F + G_F/3)\right]}{(1 - c)G_M(k_F + G_F/3)^{-1} + cG_M(k_M + G_M/3)^{-1} + 1}$$

$$(3.12)$$

$$k_{23} = k_M + \frac{1}{3}G_M$$

$$+ \frac{c}{[k_F - k_M + \frac{1}{3}(G_F - G_M)]^{-1} + (1 - c)/(k_M + \frac{4}{3}G_M)} \qquad (3.13)$$

$$\frac{G_{12}}{G_M} = \frac{G_F(1 + c) + G_M(1 - c)}{G_F(1 - c) + G_M(1 + c)} \qquad (3.14)$$

where $c = (a/b)^3$ is the volume fraction of fibers, and the subscripts M and F refer to the matrix and fiber materials, respectively.

To determine the shear modulus G_{23} in the plane of isotropy, again the three-phase model should be preferred—this time with cylindrical inclusions—in which all the composite cylinders except one are replaced by an equivalent homogeneous medium. For low values of $c = (a/b)^3$, we then obtain (Christensen 1979)

$$\frac{G_{23}}{G_M} = 1 + \frac{c}{G_M(G_F - G_M)^{-1} + (k_M + \frac{7}{3}G_M)(2k_M + \frac{8}{3}G_M)^{-1}} \qquad (3.15)$$

Up to this point we have been concerned with elastic characteristics of a composite material. The same methods may be applied to the calculation of other effective properties. In what follows, the thermal conductivity of a composite material will be evaluated assuming steady-state temperature fields and neglecting coupling between the mechanical and thermal variables. The effective thermal conductivities of the material will be defined by

$$\langle q_i \rangle = -\tilde{\lambda}_{ij} \frac{\partial \langle \theta \rangle}{\partial x_j} \qquad (3.16)$$

where $\langle q_i \rangle$ and $\langle \theta \rangle$ are the volume-averaged values of the heat flow and temperature, respectively. We consider the spherical inclusion case to fix our ideas, and we employ the three-phase model in our analysis, that is, we examine the temperature field in an unbounded region that contains a single composite spherical particle of radius b, consisting of a spherical inclusion of radius a plus a spherical region of matrix material; the surrounding medium is thought of as an isotropic material with an unknown effective thermal conductivity $\tilde{\lambda}$. Specifying the temperature gradient by

$$\theta \rightarrow \beta x_3 \qquad (3.17)$$

at large r and taking advantage of the symmetry with respect to the x_3 axis, the problem reduces to that of determining three harmonic functions of temperature that solve the equations

$$\nabla^2 \theta_I = 0 \qquad (0 \le r \le a)$$

$$\nabla^2 \theta_M = 0 \qquad (a \le r \le b)$$

$$\nabla^2 \theta = 0 \qquad (b \le r < \infty) \qquad (3.18)$$

subjected to perfect thermal contact conditions of the form

$$\theta_I = \theta_M \qquad \lambda_I \frac{\partial \theta_I}{\partial r} = \lambda_M \frac{\partial \theta_M}{\partial r} \qquad (r = a)$$

$$\theta_M = \theta \qquad \lambda_M \frac{\partial \theta_M}{\partial r} = \tilde{\lambda} \frac{\partial \theta}{\partial r} \qquad (r = b) \qquad (3.19)$$

where λ_I and λ_M are the thermal conductivities of the inclusion and matrix materials, respectively. Note also that

$$\theta \rightarrow \beta r \cos \theta \quad \text{as } r \rightarrow \infty \qquad (3.20)$$

in accordance with Equation 3.17. Changing now to spherical coordinates with axial symmetry relative x_3 and noting that

$$\nabla^2 = \frac{1}{r} \frac{\partial}{\partial r}\left(r^2 \frac{\partial}{\partial r}\right) + \frac{1}{r^2 \sin \theta} \frac{\partial}{\partial \theta}\left(\sin \theta \frac{\partial}{\partial \theta}\right)$$

the solution to Equation 3.18 is found to be

$$\theta_I = A_1 r \cos \theta \qquad (0 \leq r \leq a)$$

$$\theta_M = \left(A_2 r + \frac{B_2}{r^2} \right) \cos \theta \qquad (a \leq r \leq b)$$

$$\theta = \left(Ar + \frac{B}{r^2} \right) \cos \theta \qquad (b \leq r < \infty)$$

(3.21)

where $A = \beta$ and the constants A_1, A_2, B_2, and B are to be determined from Condition 3.19. We obtain, as a result,

$$\tilde{\lambda} = \lambda_M \left[1 + \frac{c}{\lambda_M / (\lambda_I - \lambda_M) + (1 - c)/3} \right]$$

(3.22)

where as always, c is the volume fraction of the (spherical) inclusions. It is of interest to point out here that Equation 3.22 is also applicable mutatis mutandis, for evaluating the dielectric permittivity of the spherical inclusion model (Kerner 1956).

In summary, then, any heterogeneous medium or composite described by governing relations of the form of Equation 3.1 may, for purposes of analysis, be replaced by or associated with a certain idealized homogeneous medium, describable by the same relations but with different (effective) characteristics. An important point to be made about this homogeneous medium is that its symmetry properties differ from those of the individual constituent materials so that it may be anisotropic even when these latter are perfectly isotropic. The theory based on the use of effective characteristics is often spoken of as being an effective modulus theory (Pobedrya 1984) and has good predictive capability in many composite materials mechanics problems where the distribution of stress, strain, displacement, and temperature fields in individual components is of no consequence. Clearly, to take account of these microcharacteristics, more sophisticated approaches are needed.

3.2. Effective moduli from the Hashin–Shtrikman variational principle

Understandably, it is often desirable to have a lower and an upper bound for the effective characteristics of a composite material. One possible estimate for the effective bulk modulus k and the effective shear modulus G is given by the so-called Vougt–Reuss bounds which may be derived using two minimum-energy theorems, the theorem of minimum potential energy and the theorem of minimum complementary energy. For a macroscopically isotropic N-phase nonhomogeneous medium, denoting by subscript i the partial moduli and volume fractions, we have (Christensen 1979, Pobedrya 1984)

$$\left(\sum_{i=1}^{N} \frac{c_i}{k_i} \right)^{-1} \leq \tilde{k} \leq \left(\sum_{i=1}^{N} c_i k_i \right)$$

$$\left(\sum_{i=1}^{N} \frac{c_i}{G_i} \right)^{-1} \leq \tilde{G} \leq \left(\sum_{i=1}^{N} c_i G_i \right)$$

(3.23)

which in the case $N = 2$ reduces to

$$\left[\frac{\gamma}{k_1} + \frac{1 - \gamma}{k_2} \right]^{-1} \leq \tilde{k} \leq \gamma k_1 + (1 - \gamma)k_2$$

$$\left[\frac{\gamma}{G_1} + \frac{1 - \gamma}{G_2} \right]^{-1} \leq \tilde{G} \leq \gamma G_1 + (1 - \gamma)G_2 \qquad (3.24)$$

with the obvious notation $c_1 = \gamma$ and $c_2 = 1 - \gamma$. This estimate is, however, too rough to be of any practical significance (Christensen 1979). We will discuss here a more accurate approach, based on the variational principle due to Hashin and Shtrikman (1962). We will follow the line of argument adopted by Pobedrya (1984) and will limit ourselves to elastic composite materials.

Let the heterogeneous body occupy a volume V and be bounded by a surface S. We consider for this body the static elasticity problem

$$\frac{\partial \sigma_{ij}}{\partial x_j} = 0 \qquad \sigma_{ij} = C_{ijkl}(\mathbf{x})e_{kl} \qquad e_{kl} = \frac{1}{2} \left(\frac{\partial u_k}{\partial x_l} + \frac{\partial u_l}{\partial x_k} \right) \qquad (3.25)$$

$$u_k|_S = v_k^{(0)} \qquad (3.26)$$

where

$$\sigma_{ij} = \frac{\partial U}{\partial e_{ij}} \qquad U(e) = \frac{1}{2}c_{ijkl}e_{ij}e_{kl} \qquad (3.27)$$

and we consider also the elasticity problem

$$\frac{\partial \sigma_{ij}^{(c)}}{\partial x_j} = 0 \qquad \sigma_{ij}^{(c)} = c_{ijkl}^{(c)} e_{kl}^{(c)} \qquad e_{kl}^{(c)} = \frac{1}{2}\left(\frac{\partial u_k^{(c)}}{\partial x_l} + \frac{\partial u_l^{(c)}}{\partial x_k}\right) \qquad (3.28)$$

$$u_k^{(c)}\big|_S = v_k^{(0)} \qquad (3.29)$$

$$\sigma_{ij}^{(c)} = \frac{\partial U^{(c)}}{\partial e_{ij}^{(c)}} \qquad U^{(c)}(e^{(c)}) = \frac{1}{2}C_{ijkl}^{(c)} e_{kl}^{(c)} e_{ij}^{(c)} \qquad (3.30)$$

for the homogeneous body imagined to occupy the same volume and referred to as the *comparison* body.

If the solution to Equations 3.25 and 3.26 is taken in the form

$$u_k = u_k^{(c)} + u_k' \qquad (3.31)$$

then

$$e_{kl} = e_{kl}^{(c)} + e_{kl}'$$

We introduce the notation

$$U^{(p)}(e) = U(e) - U^{(c)}(e) = \left(c_{ijkl}(\mathbf{x}) - c_{ijkl}^{(c)}\right)e_{ij}e_{kl} \qquad (3.32)$$

and define the symmetrical second-rank polarization tensor by

$$P_{ij} = \frac{\partial U^{(p)}}{\partial e_{ij}} = \sigma_{ij} - c_{ijkl}^{(c)} e_{kl} = \left(c_{ijkl}(\mathbf{x}) - c_{ijkl}^{(c)}\right)e_{kl} \qquad (3.33)$$

Inverting Equation 3.3 yields

$$e_{ij} = \frac{\partial W(p)}{\partial p_{ij}} = \left(c_{ijkl}(\mathbf{x}) - c_{ijkl}^{(c)}\right)^{-1} P_{kl} = s_{ijkl} P_{kl} \qquad (3.34)$$

where

$$W(p) = \tfrac{1}{2}s_{ijkl} P_{kl} P_{ij} = \tfrac{1}{2} P_{ij} e_{ij} \qquad (3.35)$$

and combining Equations 3.32 and 3.34 we find

$$U(e) - \tfrac{1}{2}c_{ijkl}^{(c)} e_{ij} e_{kl} + W(p) = P_{ij} e_{ij} \qquad (3.36)$$

Now, from Equations 3.31 and 3.33,

$$\sigma_{ij} = p_{ij} + c_{ijkl}^{(c)} e_{kl} = p_{ij} + c_{ijkl}^{(c)} e_{kl}^{(c)} + C_{ijkl}^{(c)} e_{kl}' \tag{3.37}$$

which when inserted into Equations 3.25, 3.26, 3.28, and 3.29 yields the following problem for the displacement perturbation u_k':

$$\frac{\partial p_{ij}}{\partial x_j} + c_{ijkl}^{(c)} \frac{\partial^2 u_k'}{\partial x_l \, \partial x_k} = 0 \tag{3.38}$$

$$u_k'|_S = 0 \tag{3.39}$$

Using Condition 3.39, it can be shown that

$$\int_V \sigma_{ij} e_{ij}' \, dV = \int_V \left[\frac{\partial}{\partial x_j} (\sigma_{ij} u_i') - \frac{\partial \sigma_{ij}}{\partial x_j} u_i' \right] dV$$

$$= \int_S \sigma_{ij} u_i' n_i \, dS = 0 \tag{3.40}$$

so that the Lagrangian of the static elasticity problem of Equations 3.25 and 3.26 becomes

$$L = \int_V U(e) \, dV = \int_V \left[U(e) - \tfrac{1}{2} \sigma_{ij} e_{ij} + \tfrac{1}{2} (\sigma_{ij} e_{ij} - \sigma_{ij} e_{ij}') \right] dV \tag{3.41}$$

The last term in the integrand of Equation 3.41 is now rewritten as

$$\sigma_{ij}(e_{ij} - e_{ij}') = \sigma_{ij} e_{ij}^{(c)}$$

$$= p_{ij} e_{ij}^{(c)} + c_{ijkl}^{(c)} e_{kl} e_{ij}^{(c)}$$

$$= c_{ijkl}^{(c)} e_{ij}^{(c)} e_{kl} + 2 p_{ij} e_{ij}^{(c)} + p_{ij} e_{ij}' - p_{ij} e_{ij} \tag{3.42}$$

to give

$$L = \int_V \left[U(e) - \tfrac{1}{2} c_{ijkl}^{(c)} e_{ij} e_{kl} - \tfrac{1}{2} p_{ij} e_{ij} \right.$$

$$\left. + \tfrac{1}{2} \left(c_{ijkl}^{(c)} e_{kl} e_{ij}^{(c)} + 2 p_{ij} e_{ij}^{(c)} + p_{ij} e_{ij}' - p_{ij} e_{ij} \right) \right] dV \tag{3.43}$$

or, substituting from Equation 3.36,

$$L = \tfrac{1}{2}\int_V \left[c_{ijkl}^{(c)} e_{ij}^{(c)} e_{kl} + 2 p_{ij} e_{ij}^{(c)} + p_{ij} e_{ij}' - 2W(p) \right] dV$$

$$= \tfrac{1}{2}\int_V \left[c_{ijkl}^{(c)} e_{ij}^{(c)} e_{kl} + p_{ij} e_{ij}^{(c)} \right] dV \tag{3.44}$$

where use also has been made of the relation

$$2W(p) = p_{ij} e_{ij} = p_{ij} e_{ij}^{(c)} + p_{ij} e_{ij}'$$

If we assume that the tensor p_{ij} is independent of σ_{ij} and $c_{ijkl}^{(c)} e_{kl}$, the term $\partial p_{ij}/\partial x_j$ may be interpreted as a body force in Equation 3.38 and we show now that these latter express the stationarity of the functional in Equation 3.44. Taking p_{ij} and e_{ij}' as variables in this functional we obtain, using Equation 3.33,

$$\delta L = \frac{1}{2}\int_V \left[2\,\delta p_{ij} e_{ij}^{(c)} + \delta p_{ij} e_{ij}' + p_{ij}\,\delta e_{ij}' - 2\frac{\partial W}{\partial p_{ij}}\,\delta p_{ij} \right] dV$$

$$= \frac{1}{2}\int_V (\delta e_{ij}' p_{ij} - \delta p_{ij} e_{ij}')\, dV \tag{3.45}$$

We have to show now that the integral in Equation 3.45 is zero. If we transform Equation 3.38 by multiplying by $\delta u_i'$ ($\delta u_i'|_S = 0$) and integrating over the volume V we find that

$$\int_V \left(\frac{\partial p_{ij}}{\partial x_j}\,\delta u_i' + c_{ijkl}^{(c)} \frac{\partial e_{kl}'}{\partial x_j}\,\delta u_i' \right) dV$$

$$= -\int_V \left(p_{ij}\,\delta e_{ij}' + c_{ijkl}^{(c)} e_{kl}'\,\delta e_{ij}' \right) dV = 0 \tag{3.46}$$

In a similar manner, integrating over V the product of

$$\frac{\partial}{\partial x_j}\left(\delta p_{ij} + c_{ijkl}^{(c)}\,\delta e_{kl}' \right) = 0$$

and u_i' ($u_i'|_S = 0$) gives

$$\int_V \left(\frac{\partial \delta p_{ij}}{\partial x_j} u_i' + c_{ijkl}^{(c)} \frac{\partial \delta e_{kl}'}{\partial x_j} u_i' \right) dV = -\int_V \left(\delta p_{ij} e_{ij}' + c_{ijkl}^{(c)} e_{ij}' \delta e_{kl}' \right) dV = 0$$

(3.47)

Now, using the symmetry $c_{ijkl}^{(c)} = c_{klij}^{(c)}$ in Equation 3.47 results in

$$\delta L = \tfrac{1}{2} \int_V (\delta e_{ij}' p_{ij} - \delta p_{ij} e_{ij}') \, dV = 0$$

(3.48)

showing that the stationarity of the functional in Equation 3.44 leads to Equation 3.38 and Condition 3.39, provided (we recall) the tensor p_{ij} is independent of σ_{ij} and $c_{ijkl}^{(c)} e_{kl}$.

Two statements concerning the functional in Equation 3.44 are of importance here:

1. The stationary point of the functional in Equation 3.44 is a maximum if $c_{ijkl}^{(c)} < c_{ijkl}(\mathbf{x})$.
2. The stationary point of the functional in Equation 3.44 is a maximum if $c_{ijkl}^{(c)} > c_{ijkl}(\mathbf{x})$.

To prove statement 1, consider the second variation of the functional in Equation 3.44,

$$\delta^2 L = \int_V \left[\delta p_{ij} \, \delta e_{ij}' - \frac{\partial^2 W(p)}{\partial p_{ij} \, \partial p_{kl}} - \delta p_{kl} \, \delta p_{ij} \right] dV$$

$$= \int_V [\, \delta p_{ij} \, \delta e_{ij}' - s_{ijkl} \, \delta p_{kl} \, \delta p_{ij}] \, dV$$

(3.49)

Noting that

$$\int_V \left[\delta p_{ij} \, \delta e_{ij}' + c_{ijkl}^{(c)} \, \delta e_{kl}' \, \delta e_{ij}' \right] dV = 0$$

(3.50)

(which follows from the equation $\partial p_{ij}/\partial x_j + c_{ijkl}^{(c)}(\partial \delta e_{kl}'/\partial x_j) = 0$) and using the condition $\delta u_i'|_S = 0$, Equation 3.49 can be expressed as

$$\delta^2 L = -\int_V \left[c_{ijkl}^{(c)} \, \delta e_{kl} \, \delta e_{ij} + s_{ijkl} \, \delta p_{kl} \, \delta p_{ij} \right] dV$$

(3.51)

Now, because

$$s_{ijkl}\, \delta p_{kl} = \left(c_{ijkl} - c_{ijkl}^{(c)}\right)^{-1}\left(c_{mnkl} - c_{mnkl}^{(c)}\right) \delta e_{mm} = \delta e_{ij}$$

Equation 3.51 may be written

$$\delta^2 L = -\int_V \left[c_{ijkl}^{(c)}\, \delta e_{kl}\, \delta e_{ij} + \left(c_{ijkl} - c_{ijkl}^{(c)}\right) \delta e_{ij}\, \delta e_{kl} \right] dV \quad (3.52)$$

indicating that

$$\delta^2 L < 0 \quad \text{if } c_{ijkl}^{(c)} < c_{ijkl} \quad (3.53)$$

thus completing the proof of statement 1. The proof of statement 2 involves more general forms of the governing relations of continuum mechanics and is somewhat too cumbersome to be reproduced here. The interested reader may consult Pobedrya (1984) for details.

We wish to show now that the foregoing variational principle provides a more accurate estimate for composite material properties than do the Voigh–Reuss bounds of Equation 3.23. To this end we rewrite the functional in Equation 3.44 in the form

$$A \equiv \frac{2L}{V} = A_1 + A_2 + A_3 + A_4 \quad (3.54)$$

where we define

$$A_1 = \langle c_{ijkl}^{(c)} e_{ij}^{(c)} e_{kl} \rangle \qquad A_2 = 2\langle p_{ij} e_{ij}^{(c)} \rangle$$

$$A_3 = -2\langle W(p) \rangle \qquad A_4 = \langle p_{ij} e_{ij}' \rangle \quad (3.55)$$

The quantities A_1, A_2, and A_3 considered as functions of the tensor p_{ij} are determined from the problem given in Equations 3.28 and 3.29 for a homogeneous medium; from the problem given in Equations 3.38 and 3.39, the quantities e_{ij}' as functions of the same argument are found. With these results we shall be able to calculate the extremum of the functional $A(p)$, which will be a maximum $A_{max}(p)$ or a minimum $A_{min}(p)$ depending, as discussed earlier, on whether $c_{ijkl}^{(c)}$ is, respectively, greater or less than c_{ijkl}. We may then write

$$A_{min}(p) < A^*(p) < A_{max}(p) \quad (3.56)$$

where $A^*(p)$ is the value of the functional as calculated from the effective modulus theory.

It will be convenient for later purposes to assume that the tensor p_{ij} is constant within each of the composite phases and that the boundary conditions given by Equation 3.29 for Problem 3.28 are of the form

$$u_k^{(c)}\big|_S = e_{kj}^{(0)} x_j \tag{3.57}$$

Then

$$\langle p_{ij} \rangle = \sum_{\alpha=1}^{N} c_\alpha p_{ij}^{(\alpha)} \qquad \left(p_{ij}^{(\alpha)} = \text{constant} \right) \tag{3.58}$$

so that the functional A will depend on constant quantities $p_{ij}^{(\alpha)}$.

Omitting all the intermediate steps of derivation, the extremum value of A for an elastic isotropic composite is given by

$$A = \left[k^{(c)} + \frac{1}{9} \frac{m}{(1 + a_c m)} \right] (e_{kk}^{(0)})^2$$

$$+ \left[2G^{(c)} + \frac{n}{1 + b_c n} \right] e_{ij}^{(0)} e_{ij}^{(0)} \tag{3.59}$$

where

$$a_c = -\frac{1}{3} (3k^{(c)} + 4G^{(c)})^{-1}$$

$$b_c = -\frac{3}{5} \frac{(k^{(c)} + 2G^{(c)})}{G^{(c)}(3k^{(c)} + 4G^{(c)})}$$

$$m = \sum_{\alpha=1}^{N} c_\alpha \left[\frac{1}{9(k_\alpha - k^{(c)})} - a_c \right]^{-1}$$

$$n = \sum_{\alpha=1}^{N} c_\alpha \left[\frac{1}{2(G_\alpha - G^{(c)})} - b_c \right]^{-1}$$

On the other hand, we have

$$A^* = k^*(e_{kk}^{(0)})^2 + 2G^* e_{ij}^{(0)} e_{ij}^{(0)} \tag{3.60}$$

for the same composite. Now if $k^{(c)} = k_{\min}$ and $G^{(c)} = G_{\min}$, where $k_{\min}$ and $G_{\min}$ are the least of all the partial moduli of the composite, then

$c_{ijkl}^{(c)} < c_{ijkl}(x)$ and we have

$$A_{max} = \left[k_{min} + \frac{1}{9} \frac{m}{(1 + a_c m)} \right](e_{kk}^{(0)})^2$$

$$+ \left[2G_{min} + \frac{n}{(1 + b_c n)} \right] e_{ij}^{(0)} e_{ij}^{(0)} \qquad (3.61)$$

By the same token, setting $k^{(c)} = k_{max}$ and $G^{(c)} = G_{max}$, where k_{max} and G_{max} are the greatest of all the partial moduli, we write

$$A_{min} = \left[k_{max} + \frac{1}{9} \frac{m}{1 + a_c m} \right](e_{kk}^{(0)})^2$$

$$+ \left[2G_{max} + \frac{n}{1 + b_c n} \right] e_{ij}^{(0)} e_{ij}^{(0)} \qquad (3.62)$$

Turning now to Inequalities 3.56 and comparing Expressions 3.60–3.62 yields the so-called Hashin–Shtrikman estimate,

$$k' < k^* < k'' \quad \text{and} \quad G' < G^* < G'' \qquad (3.63)$$

where

$$k' = k_{max} + \frac{1}{9} \frac{m}{(1 + a_c m)}$$

$$k'' = k_{min} + \frac{1}{9} \frac{m}{(1 + a_c m)}$$

$$G' = G_{max} + \frac{1}{2} \frac{n}{(1 + b_c n)}$$

$$G'' = G_{min} + \frac{1}{2} \frac{n}{(1 + b_c n)} \qquad (3.64)$$

3.3. Effective properties of a periodic composite structure

A heterogeneous medium, or composite, is said to have a regular periodic structure if its mechanical behavior is described by constitutional relations of the form

$$\sigma_{ij} = c_{ijkl}(\mathbf{x}) e_{kl} \qquad (3.65)$$

where $c_{ijkl}(\mathbf{x})$ is a periodic function of the spatial coordinate $\mathbf{x} = (x_1, x_2, x_3)$ in the sense that

$$c_{ijkl}(\mathbf{x} + n_p \mathbf{a}_p) = c_{ijkl}(\mathbf{x}) \tag{3.66}$$

where the n_p are arbitrary integer numbers and the constant vectors $\mathbf{a}_p$ determine the period of the structure.

We will start our analysis by considering an unbounded isotropic medium (the matrix) containing periodically distributed inclusions of a dissimilar material. We consider the system to be composed of equal-size parallelepipedal cells Y and assume that there is only one inclusion in each cell, the region occupied by the inclusion being I.

If the (constant) stresses acting on the matrix are related to the strains by

$$\sigma_{ij}^{(0)} = c_{ijkl}^{(M)} e_{kl}^{(0)} \tag{3.67}$$

where

$$c_{ijkl}^{(M)} = \lambda_M \delta_{ij} \delta_{kl} + G_M (\delta_{ik} \delta_{jl} + \delta_{il} \delta_{jk}) \tag{3.68}$$

then the periodically distributed inclusions with elastic moduli

$$c_{ijkl}^{(I)} = \lambda_I \delta_{ij} \delta_{kl} + G_I (\delta_{ik} \delta_{jl} + \delta_{il} \delta_{jk}) \tag{3.69}$$

perturb the stress and strain fields in such a manner that, for given strains e_{kl}, the cell-averaged stresses

$$\langle \sigma_{ij} \rangle = \frac{1}{V_Y} \int_{V_Y} \sigma_{ij} \, dV$$

(V_Y being the cell volume) will be defined by the relation

$$\langle \sigma_{ij} \rangle = \tilde{c}_{ijkl} e_{kl}^{(0)} \tag{3.70}$$

where $\tilde{c}_{ijkl}$ are the effective elastic moduli.

Following Nemat-Nasser et al. (1982) we designate by u_i and e_{ij} the displacement and strain perturbations due to inclusion in the cell Y and we note that

$$e_{ij} = \frac{1}{2} \left(\frac{\partial u_i}{\partial x_j} + \frac{\partial u_j}{\partial x_i} \right)$$

and that $u_i = 0$ on ∂Y.

The full stress field in the cell Y will then be given by

$$\sigma_{ij}^{(M)} = c_{ijkl}^{(M)}(e_{kl}^{(0)} + e_{kl}) \quad \text{in } Y/I \tag{3.71}$$

$$\sigma_{ij}^{(I)} = c_{ijkl}^{(I)}(e_{kl}^{(0)} + e_{kl}) \quad \text{in } I \tag{3.72}$$

and it follows from the equation of equilibrium that

$$\frac{\partial \sigma_{ij}^{(M)}}{\partial x_j} = 0 \quad \text{in } Y/I$$

$$\frac{\partial \sigma_{ij}^{(I)}}{\partial x_j} = 0 \quad \text{in } I \tag{3.73}$$

The boundary conditions to be satisfied at the surface of an inclusion are as follows:

$$\left[c_{ijkl}^{(M)}(e_{kl}^{(0)} + e_{kl})^+ - c_{ijkl}^{(I)}(e_{kl}^{(0)} + e_{kl})^- \right] n_j = 0 \tag{3.74}$$

$$u_i^+ - u_i^- = 0 \quad \text{on } \partial I \tag{3.75}$$

when **n** is the unit vector in the outward normal direction to the surface ∂I and the superscripts "$-$" and "$+$" designate the value the quantity assumes in the immediate vicinity of ∂I, respectively, inside and outside the inclusion.

Equation 3.73 is conveniently solved by introducing a so-called transformation strain e_{kl}^* such that (Nemat-Nasser et al. 1982)

$$c_{ijkl}^{(M)}(e_{kl}^{(0)} + e_{kl} - e_{kl}^*) = c_{ijkl}^{(I)}(e_{kl} + e_{kl}^{(0)}) \quad \text{in } I \tag{3.76}$$

and

$$e_{kl}^* = 0 \quad \text{in } Y/I \tag{3.77}$$

With this definition, the equation

$$\frac{\partial \sigma_{ij}^{(I)}}{\partial x_j} = 0 \quad \text{in } I$$

becomes

$$\frac{\partial}{\partial x_j}\left[c_{ijkl}^{(M)}(e_{kl}^{(0)} + e_{kl} - e_{kl}^*)\right] = 0 \quad \text{in } I \tag{3.78}$$

or, equivalently,

$$c_{sjkl}^{(M)}\frac{\partial^2 u_k}{\partial x_l\, \partial x_s} = c_{pjmn}^{(M)}\frac{\partial e_{mn}^*}{\partial x_p} \tag{3.79}$$

In view of the spatial periodicity of the displacement components u_k, it is expedient that both the solution of and the quantities e_{mn} in Equation 3.79 be expanded in the Fourier series

$$u_k(\mathbf{x}) = \sum_{n_p = -\infty}^{\infty} \hat{u}_k(\boldsymbol{\xi})e^{i\boldsymbol{\xi}\cdot\mathbf{x}} \tag{3.80}$$

$$e_{mn}^* = \sum_{n_p = -\infty}^{\infty} \hat{e}_{mn}^*(\boldsymbol{\xi}) \cdot e^{i\boldsymbol{\xi}\cdot\mathbf{x}} \tag{3.81}$$

where

$$\boldsymbol{\xi} = (\xi_1, \xi_2, \xi_3)$$

$$\xi_p = \frac{2\pi n_p}{\Lambda_p} \quad (p = 1, 2, 3)$$

$$\boldsymbol{\xi}\cdot\mathbf{x} = \xi_1 x_1 + \xi_2 x_2 + \xi_3 x_3$$

$$\sum_{n_p = -\infty}^{\infty} = \sum_{n_1 = -\infty}^{\infty}\sum_{n_2 = -\infty}^{\infty}\sum_{n_3 = -\infty}^{\infty} \quad (i = \sqrt{-1})$$

Upon substituting into Equation 3.79 we then obtain

$$\hat{u}_k(\boldsymbol{\xi}) = -iN_{kj}(\boldsymbol{\xi})c_{pjmn}^{(M)}\hat{e}_{mn}^*(\boldsymbol{\xi})\xi_p \tag{3.82}$$

$$u_k(\mathbf{x}) = -i\sum_{n_p = -\infty}^{\infty} N_{kj}(\boldsymbol{\xi})c_{pjmn}^{(M)}\hat{e}_{mn}^*(\boldsymbol{\xi})\xi_p e^{i\boldsymbol{\xi}\cdot\mathbf{x}} \tag{3.83}$$

where

$$N_{kj}(\boldsymbol{\xi}) = (c_{sjkl}\xi_l\xi_s)^{-1} \tag{3.84}$$

From Equation 3.83 the perturbed strain components can be shown to be given by (Nemat-Nasser et al. 1982)

$$e_{jk} = \sum_{n_p = -\infty}^{\infty}{}' g_{jkmn}(\boldsymbol{\xi}) \hat{e}_{mn}^*(\boldsymbol{\xi}) e^{i\boldsymbol{\xi} \cdot \mathbf{x}} \tag{3.85}$$

where the prime indicates that the term with $n = \sqrt{n_0 n_p} = 0$ is excluded from the sum, and, in terms of $\xi^2 = \xi_k \xi_k$,

$$g_{jkmn}(\boldsymbol{\xi}) = \frac{1}{2}\left(N_{ks}(\boldsymbol{\xi})\xi_j + N_{js}(\boldsymbol{\xi})\xi_k\right) c_{psmn}^{(M)} \xi_p$$

$$= \frac{1}{2\xi^2}\left\{\xi_k(\delta_{jn}\xi_m + \delta_{jm}\xi_n) + \xi_j(\delta_{kn}\xi_m + \delta_{km}\xi_n)\right\}$$

$$- \frac{1}{(1 - \nu_M)}\frac{\xi_j \xi_k \xi_m \xi_n}{\xi^4} + \frac{\nu_M}{1 - \nu_M}\frac{\xi_j \xi_k}{\xi^2}\delta_{mn} \tag{3.86}$$

where ν_M is the Poisson ratio of matrix material.

Now the Fourier coefficients in Equation 3.81 are given by

$$\hat{e}_{mn}^*(\boldsymbol{\xi}) = \frac{1}{V_Y}\int_Y e_{mn}^*(\mathbf{x}') e^{-i\boldsymbol{\xi} \cdot \mathbf{x}'}\, d\mathbf{x}' \tag{3.87}$$

which when substituted into Equation 3.85 yields

$$e_{jk}(\mathbf{x}) = \frac{1}{V_Y}\sum_{n_p = -\infty}^{\infty}{}' g_{jkmn}(\boldsymbol{\xi})\int_I e_{mn}^*(\mathbf{x}') e^{i\boldsymbol{\xi} \cdot (\mathbf{x} - \mathbf{x}')}\, d\mathbf{x} \tag{3.88}$$

At this point use is made of the relation

$$e_{ij}^{(0)} = A_{ijkl}e_{kl}^*(\mathbf{x}) - e_{ij}(\mathbf{x}) \tag{3.89}$$

which follows from Equation 3.76 and in which

$$A_{ijkl} = \left(c_{ijmn}^{(M)} - c_{ijmn}^{(I)}\right)c_{mnkl}^{(M)}$$

$$= \frac{G_M}{2(G_M - G_I)}(\delta_{ik}\delta_{jl} + \delta_{il}\delta_{jk})$$

$$+ \frac{(G_M \lambda_I - G_I \lambda_M)\delta_{ij}\delta_{kl}}{(G_M - G_I)[3(\lambda_\mu - \lambda_I) - 2(G_M - G_I)]} \tag{3.90}$$

Substituting Equation 3.88 into Equation 3.89 yields the following integral equation for the quantities $e_{mn}^*(\mathbf{x})$:

$$e_{jk}^{(0)} = A_{jksl}e_{sl}^*(\mathbf{x})$$

$$-\frac{1}{V_Y}\sum_{n_p=-\infty}^{\infty}{}' \, g_{jkmn}(\boldsymbol{\xi}) \cdot \int_I e_{mn}^*(\mathbf{x}')e^{i\xi(\mathbf{x}-\mathbf{x}')}\,d\mathbf{x}' \qquad (3.91)$$

Deferring until later the analysis of methods for treating Equation 3.91, we show here that to determine the effective properties of a periodic composite, all that is needed is to calculate the volume average of the transformation strain over the inclusion,

$$\langle e_{ij}^* \rangle = \frac{1}{V_I}\int_I e_{ij}^*(\mathbf{x})\,d\mathbf{x} \qquad (3.92)$$

To this end we note that for the periodicity cell Y subjected to a uniform strain $e_{ij}^{(0)}$ and thereby to a uniform stress $\sigma_{ij}^{(0)} = c_{ijkl}^{(M)}e_{kl}^{(0)}$, the full strain energy is given by

$$W_0 = \tfrac{1}{2}\int_Y \sigma_{ij}^{(0)}e_{ij}^{(0)}\,d\mathbf{x} = \tfrac{1}{2}V_Y\sigma_{ij}^{(0)}e_{ij}^{(0)} \qquad (3.93)$$

in the absence of an inclusion, and

$$W_e = \tfrac{1}{2}\int_Y c_{ijkl}^{(M)}(e_{kl}^{(0)} + e_{kl} - e_{kl}^*)\left(e_{ij}^{(0)} + e_{ij}\right)d\mathbf{x} \qquad (3.94)$$

in the presence of an inclusion. Recalling that $e_{kl}^* = 0$ in Y/I and that $u_i = 0$ on ∂Y, Equation 3.94 can be rewritten as (Nemat-Nasser et al. 1982)

$$W_e = W_0 - \tfrac{1}{2}\gamma\sigma_{ij}^{(0)}\langle e_{ij}^* \rangle V_Y \qquad (3.95)$$

where $\gamma = V_I/V_Y$ is evidently the volume fraction of the inclusion. On the other hand, for a homogeneous cell with effective properties c_{ijkl},

$$W_e = \tfrac{1}{2}V_Y\tilde{c}_{ijkl}e_{kl}^{(0)}e_{ij}^{(0)} \qquad (3.96)$$

By comparing Equations 3.95 and 3.96, the equation for determining c_{ijkl} is found to be

$$\tilde{c}_{ijkl}e_{kl}^{(0)}e_{ij}^{(0)} = c_{ijkl}^{(M)}e_{kl}^{(0)}e_{ij}^{(0)} - \gamma c_{ijkl}^{(M)}e_{kl}^{(0)}\langle e_{ij}^* \rangle \quad (3.97)$$

where only the (average) quantity given by Equation 3.92 is involved, which proves the assertion.

Returning now to Equation 3.91, we easily reduce this integral equation to an infinite system of linear algebraic equations for the variable

$$E_{jk}^*(\boldsymbol{\xi}) = \frac{1}{V_I}\int_I e_{jk}^*(\mathbf{x}')e^{-i\boldsymbol{\xi}\cdot\mathbf{x}'}\,d\mathbf{x}' \quad (3.98)$$

if we multiply Equation 3.91 by $\exp(-i\boldsymbol{\xi}\cdot\mathbf{x})$ and integrate over the volume *I*. We obtain

$$e_{ij}^{(0)}Q(\boldsymbol{\eta}) = A_{ijkl}E_{kl}^*(\boldsymbol{\eta}) - \gamma \sum_{n_p=0}^{\infty} g_{ijkl}(\boldsymbol{\xi})Q(\boldsymbol{\eta}-\boldsymbol{\xi})E_{kl}^*(\boldsymbol{\xi}) \quad (3.99)$$

where

$$Q(\boldsymbol{\eta}) = \frac{g_0(-\boldsymbol{\eta})}{V} \qquad g_0(-\boldsymbol{\eta}) = \int_I e^{-i\boldsymbol{\eta}\cdot\mathbf{x}}\,d\mathbf{x}$$

$$\eta_i = \frac{2\pi m_i}{\Lambda_i} \qquad (i = 1,2,3)$$

The solution to Equation 3.99 having been found, Equation 3.98 gives

$$\langle e_{ij}^* \rangle = E_{ij}^*(0) \quad (3.100)$$

An alternative method for estimating the quantity $\langle e_{ij}\rangle$ is based on the replacement of the quantity $e_{ij}^*(\mathbf{x}')$ in Equation 3.91 by its average value in Equation 3.92 (Nemat-Nasser et al. 1982). Integrating Equation 3.91 over *I* then yields

$$e_{ij}^{(0)} = (A_{ijkl} - S_{ijkl})e_{kl}^* \quad (3.101)$$

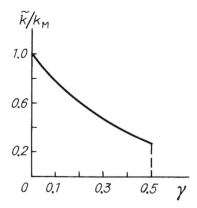

FIGURE 3.1. Effective bulk modulus versus the volume fraction of a spherical cavity in a composite unit cell.

where

$$S_{ijkl} = \sum_{n_p = -\infty}^{\infty}{}' P(\boldsymbol{\xi})g_{ijkl}(\boldsymbol{\xi})$$

(3.102)

$$P(\boldsymbol{\xi}) = \frac{1}{V_I V_Y} g_0(\boldsymbol{\xi})g_0(-\boldsymbol{\xi})$$

Equation 3.101 is now inverted, and e_{kl}^* (expressed in terms of $e_{ij}^{(0)}$) is substituted into Equation 3.97 to give

$$\left[\tilde{c}_{ijkl} - c_{ijkl}^{(M)} + \gamma c_{ijmn}^{(M)}(A_{mnkl} - S_{mnkl})^{-1}\right]e_{kl}^{(0)}e_{ij}^{(0)} = 0$$

(3.103)

By making use of the arbitrariness of $e_{ij}^{(0)}$ we conclude that

$$\tilde{c}_{ijkl} = c_{ijkl}^{(M)} - \gamma c_{ijmn}^{(M)}(A_{mnkl} - S_{mnkl})^{-1}$$

(3.104)

Nemat-Nasser et al. (1982) used this formula for estimating the effective bulk modulus $\tilde{k}$ and the effective shear modulus $\tilde{G}$ of an isotropic elastic medium with a periodic array of spherical cavities. Figures 3.1 and 3.2 reproduced from this work show the ratios $\tilde{k}/k_M$ and $\tilde{G}/G_M$ as functions of the parameter γ for a cubical periodicity cell, the quantities $\tilde{k}$ and $\tilde{G}$ being defined by respective relations $\tilde{k} = (1/9)c_{iijj}$ and $\tilde{G} = c_{2323} = c_{3131}$.

Numan and Keller (1984) discussed methods for calculating effective elastic moduli in an isotropic medium with a periodic array of rigid

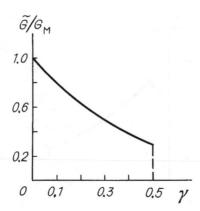

FIGURE 3.2. Effective shear modulus versus the volume fraction of a spherical cavity in a composite unit cell.

spherical inclusions. Using Equation 3.4 and the fundamental periodic solution of the equations of the theory of elasticity, Numan and Keller obtained the following expression for the effective elastic constants:

$$\tilde{c}_{ijkl} = (\lambda_M + G_M f)\delta_{ij}\delta_{kl} + G_M(1 + \beta)(\delta_{ik}\delta_{jl} + \delta_{il}\delta_{jk})$$

$$+ \, G_M(\alpha - \beta)\delta_{ijkl}$$

where λ_M and G_M are the Lamé constants of the matrix material; $\delta_{ijkl} = 1$ if $i = j = k = l$ and $\delta_{ijkl} = 0$ if $i \neq j \neq k \neq l$; and α, β, and f are functions of inclusion concentration, the matrix Poisson ratio, and the parameters of the (periodic) arrangement.

2

Asymptotic Homogenization of Regular Structures

Admittedly, the methods discussed in Section 3 (Chapter 1) may be regarded as of academic rather than practical interest. The assumptions and hypotheses that are the basis of these methods fail to give the boundaries of applicability of the results obtained; nor, as a rule, are the latter susceptible to any improvement in terms of accuracy within the same mathematical framework. To get more useful results, the general practice in the field of composite materials is to develop rigorous mathematical models utilizing as much as possible the specific geometry of a particular problem. This is especially the case with composites having a regular structure—the class of composites most widely used at the present time. The property of periodicity often makes a problem readily amenable to theoretical treatment, and physical processes occurring in such composites therefore have become a subject of extensive study in the last two decades or so.

In the approach developed by Kuz'menko et al. (1982) and Kozhevnikova and Kuz'menko (1984), the regularly nonuniform nonlinear elastic problem is reduced by symmetry arguments to a boundary value problem for a unit cell of the system and is then solved by various numerical methods, such as the finite-element scheme or iteration technique. The results obtained by the application of this method to filamentary composites include the elastic and nonlinearly elastic characteristics, the distribution of local strains for different average values of strain, and the effects of temperature on the effective elastic moduli of such materials. It has been shown (Manevich et al. 1983, Oshmyan et al. 1984, Knunyants et al. 1986) that the state of stress and strain in a regular composite can be found by considering a boundary value problem for a unit cell; the numerical solution of this problem yields the effective properties of the material and the local mechanical fields necessary for strength analysis. Mention should also be made of the work of Sgubini et al. (1983), Podlipenets (1984), and Shul'ga (1984), where the properties of differential equations with periodic coefficients are used in the treatment of wave propagation processes in elastic composites with a regular structure.

Processes occurring in composite materials can be discussed by means of differential equations with rapidly oscillating coefficients. These latter are periodic functions of spatial coordinates in composites with a regular structure, and because the period is usually much smaller than the characteristic dimensions of the problem, it proves possible to apply to such composites the asymptotic method developed for the homogenization of equations with rapidly oscillating periodic coefficients (Babuška 1976a, 1976b, 1977, Tartar 1978, Bensoussan et al. 1978, Sanchez-Palencia 1980, Lions 1981, Bakhvalov and Panasenko 1984). In this method, the fist step is to obtain average (or effective) characteristics of the material from a local problem formulated for the unit cell; at the second step, a boundary value problem for a (quasi-)homogeneous material described by these characteristics is considered.

4. HOMOGENIZATION TECHNIQUES FOR PERIODIC STRUCTURES

The physical behavior of a heterogeneous medium (or composite) with a regular structure is governed by differential equations with rapidly oscillating coefficients dependent on the material properties of the individual components. Because of these coefficients, the relevant boundary value problem is extremely difficult to solve as it stands, even if a high-speed computer is available. To get around this difficulty, it is necessary to model the nonhomogeneous medium by a set of simpler equations in which some averaged coefficients replace the exact ones; it is implicit, of course, that the boundary problem based on these averaged equations give predictions differing as little as possible from those of the original problem. A mathematical framework from which to predict the mechanical behavior of regularly nonhomogeneous media has been developed under the assumption that there is an ordered microstructure in such media, describable by a characteristic nonhomogeneity dimension (Bensoussan et al. 1978, Sanchez-Palencia 1980, Bakhvalov and Panasenko 1984, Pobedrya 1984).

If we denote this (relative) dimension by ε, the partial differential equations of the problem have (rapidly oscillating) coefficients of the form $a(x/\varepsilon)$, $a(y)$ being a periodic function of its argument; the corresponding boundary value problem may be treated by asymptotically expanding the solution in powers of the small parameter ε with the help of the so-called two-scale expansion method, known from the theory of ordinary differential equations. According to this method, two scales of spatial variables are

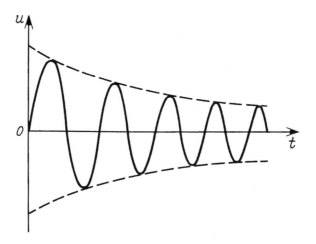

FIGURE 4.1. Motion of a pendulum in a weakly resistant medium.

considered, one for the microscopic description and the other for the macroscopic description of the process under study; correspondingly, the expansion coefficients are made to depend both on the (slow) macroscopic variables x and the (rapid) microscopic variables $y = x/\varepsilon$. The former enable one to describe the global changes in the physical fields of interest, whereas the latter refer to the changes that take place on a local level.

The two-scale technique has already proved useful in the analysis of weakly perturbed periodical processes in the theory of vibrations. If we consider the simplest example of this kind, the motion of a pendulum in a weakly resisting medium, we see (referring to Figure 4.1) that although two neighboring "periods" are almost equal in length, the difference gradually increases in time as the attenuation effect builds up. The usual way of treating this type of problem is by introducing two time scales, the rapid time $t^* = t$ and the slow time $\tau = \varepsilon t$ (ε being the small parameter), and by applying to the solution the asymptotic process

$$u_\varepsilon(t) = u_0(t,\tau) + \varepsilon u_1(t,\tau) + \cdots$$

with $\varepsilon \to 0$.

The sample principle can be extended to processes occurring in composite materials with a regular structure. Assigning to a composite material a coordinate system $x = (x_1, x_2, x_3)$ in space R^3 and assuming periodicity in mechanical, thermal, and other physical properties, let us separate in this composite a region Ω, which will be regarded as a collection of parallelepiped unit cells of identical dimensions εY_1, εY_2, εY_3. In a local coordinate system $y = x/\varepsilon$, $y = (y_1, y_2, y_3)$, the sides of the

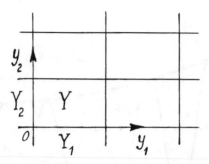

FIGURE 4.2. Unit cell of a composite material of regular structure.

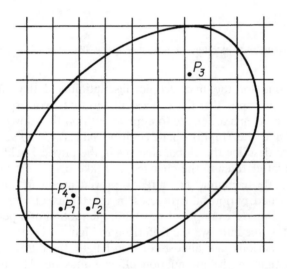

FIGURE 4.3. Characteristic points in a composite material of periodic structure.

parallelepiped will be Y_1, Y_2, and Y_3 as shown in Figure 4.2. It seems more or less self-evident that functions determining the material behavior of the composite should be expanded as

$$u_\varepsilon(x) = u_0(x, y) + \varepsilon u_1(x, y) + \cdots$$

where $y = x/\varepsilon$, and the functions $u_0(x, y)$, $u_1(x, y), \ldots$ are smooth with respect to x and Y-periodic in y.

To gain some insight into the nature of the local periodicity of the functions $u_i(x, y)$, let us compare the values of $u_1(x, y)$ at congruent points P_1 and P_2 in two neighboring cells, as shown in Figure 4.3 (see Sanchez-Palencia 1987). Because of the assumed periodicity in y and the short distance between P_1 and P_2, the function $u_1(x, y)$ has nearly equal values at these points. On the other hand, although y-dependences are the same in points P_3 and P_1, x-dependences are different because of the two points being rather far apart. Finally, if we compare the values of $u_i(x, y)$ at points P_1 and P_2 within the same cell, it is easy to see that y-dependences are different in this case whereas the x-dependences are virtually the same.

A very simple problem involving the solution of the divergent elliptic equation will serve to illustrate some general features of the two-scale homogenization procedure (Bensoussan et al. 1978, Sanchez-Palencia 1980). This type of equation provides a basis for studying many physical processes, of which the (steady-state) heat conduction in a nonhomogeneous solid with a regular structure may be cited as an example. We thus consider an unbounded medium with a periodic structure in a region Ω in space R^3 and note that the material properties in the parallelepiped unit cell Y are determined by a symmetrical matrix $a_{ij}^*(x) = a_{ij}(y)$, where $y = x/\varepsilon$, $x = (x_1, x_2, x_3)$, and the functions a_{ij} are periodic in spatial variables $y = (y_1, y_2, y_3)$. The boundary value problem to be dealt with is

$$A^\varepsilon u = f \text{ in } \Omega \qquad (4.1)$$

$$u_\varepsilon = 0 \quad \text{on } \partial\Omega \qquad (4.2)$$

where the function f is defined in Ω and

$$A^\varepsilon = -\frac{\partial}{\partial x_i}\left(a_{ij}\left(\frac{x}{\varepsilon}\right)\frac{\partial}{\partial x_j}\right) \qquad (4.3)$$

is the elliptical operator.

To begin, we represent the solution to the system of Equations 4.1 and 4.2 as a two-scale asymptotic expansion of the form

$$u_\varepsilon(x) = u_0(x, y) + \varepsilon u_1(x, y) + \varepsilon^2 u_2(x, y) + \cdots \qquad (4.4)$$

where the functions $u_j(x, y)$ are Y-periodic in y ($\forall\ x \in \Omega$), and we recall the rule of indirect differentiation to write (Bensoussan et al. 1978)

$$A^\varepsilon = \varepsilon^{-2}A_1 + \varepsilon^{-1}A_2 + \varepsilon^0 A_3 \qquad (4.5)$$

where

$$A_1 = -\frac{\partial}{\partial y_i}\left(a_{ij}(y)\frac{\partial}{\partial y_j}\right) \qquad A_3 = -\frac{\partial}{\partial x_i}\left(a_{ij}(y)\frac{\partial}{\partial x_j}\right)$$

$$A_2 = -\frac{\partial}{\partial y_i}\left(a_{ij}(y)\frac{\partial}{\partial x_j}\right) - \frac{\partial}{\partial x_i}\left(a_{ij}(y)\frac{\partial}{\partial y_j}\right)$$

Using Equations 4.4 and 4.5, the left-hand side of Equation 4.1 can also be expanded in powers of ε; collecting terms of equal powers of ε, we find in the usual manner

$$A_1 u_0 = 0 \tag{4.6}$$

$$A_1 u_1 + A_2 u_0 = 0 \tag{4.7}$$

$$A_1 u_2 + A_2 u_1 + A_3 u_0 = f \tag{4.8}$$

$$\vdots$$

If x and y are considered as independent variables, Equations 4.6 through 4.8 form a recursive system of differential equations for the functions $u_0, u_1, \ldots$ parametrized by x. Before proceeding to the analysis of this system, it will be useful to note that the equation

$$A_1 u = F \quad \text{in } Y \tag{4.9}$$

for a Y-periodic function u has a unique solution if (Bakhvalov and Panasenko 1984)

$$\langle F \rangle = \frac{1}{|Y|}\int_Y F\, dy = 0 \tag{4.10}$$

$|Y|$ denoting the volume of the unit cell.

It immediately follows from Equation 4.6 that

$$u_0 = u(x) \tag{4.11}$$

and upon substituting into Equation 4.7 we find

$$A_1 u_1 = \frac{\partial a_{ij}(y)}{\partial y_i}\frac{\partial u(x)}{\partial x_j} \tag{4.12}$$

Noting the separation of variables on the right-hand side of Equation 4.12, the solution of this equation may be represented in the form

$$u_1(x, y) = U_j(y) \frac{\partial u(x)}{\partial x_j} + \bar{u}_1(x) \tag{4.13}$$

where $U_j(y)$ is the Y-periodic solution of the local equation

$$A_1 U_j(y) = \frac{\partial a_{ij}(y)}{\partial y_i} \quad \text{in } Y \tag{4.14}$$

We next turn to the problem of solving Equation 4.8 for u_2, taking x as a parameter. It follows from Condition 4.10 that Equation 4.8 will have a unique solution if

$$\frac{1}{|Y|} \int_Y (A_2 u_1 + A_3 u_0) \, dy = f \tag{4.15}$$

which when combined with Equation 4.13 yields the following homogenized (macroscopic) equation for $u(x)$:

$$-\tilde{a}_{ij} \frac{\partial^2 u}{\partial x_i \, \partial x_j} = f \tag{4.16}$$

where the quantities

$$\tilde{a}_{ij} = \frac{1}{|Y|} \int_Y \left(a_{ij}(y) + a_{ik}(y) \frac{\partial U_j}{\partial y_k} \right) dy \tag{4.17}$$

are the effective coefficients of the homogenized operator

$$A = -\tilde{a}_{ij} \frac{\partial^2}{\partial x_i \, \partial x_j}$$

We have thus demonstrated that if we keep only two terms in the asymptotic expansion Equation 4.4, the homogenization of Equation 4.1 splits into two problems: The determination of the local functions $U_j(y)$ from Equation 4.14 is solved on the unit cell, whereas the other is concerned with one solution of the homogenized equation (Equation 4.16) set in Ω with the boundary condition $u = 0$ on $\partial\Omega$ (the function $u_1(x)$ in Equation 4.13 may be considered to be arbitrary and may be set equal to zero). If we go to higher terms in the asymptotic series Equation 4.4, we can follow the foregoing procedure in deriving a homogenized equation of the type of Equation 4.16 for $u_1(x)$; in this case, one more local problem will have to be solved on the unit cell (Bensoussan et al. 1978). There is, however, a very wide class of problems in mechanics in which very realistic predictions can be made even within the framework of the two-term technique, that is, by the use of Equations 4.14 and 4.16 and Relations 4.17. This approach is sometimes referred to as the zeroth-order approximation theory (Pobedrya 1984).

It should be noted that the asymptotic process for Equation 4.1 is amenable to the important extension to the case of piecewise-smooth (periodic) coefficients $a_{ij}(y)$. We may demand, for example, that the function $a_{ij}(y)$ assume constant values $a_{ij}^{(1)}$ and $a_{ij}^{(2)}$ in two regions $Y^{(1)}$ and $Y^{(2)}$ comprising the unit cell and having a smooth boundary Γ between them. It turns out that both the expansion given by Equation 4.4 and the Expressions 4.17 for the averaged coefficients retain their forms in this case, but the local problem given by Equation 4.14 for the Y-periodic functions $U_j(y)$ is replaced by the system

$$-\frac{\partial}{\partial y_i}\left(a_{ij}(y) + a_{ik}(y)\frac{\partial U_j(y)}{\partial y_k}\right) = 0 \quad \text{in } Y^{(1)}, Y^{(2)} \qquad (4.18)$$

subject to the additional matching conditions

$$[U_j] = 0 \quad \text{and} \quad \left[\left(a_{ij} + a_{ik}\frac{\partial U_j}{\partial y_k}\right)n_i\right] = 0 \qquad (4.19)$$

across the boundary Γ, to which n_i is the unit normal.

The topic we consider next is the application of the preceding homogenization scheme to so-called perforated media. (We refer the reader to Lions [1987] for a comprehensive account of the work on the subject.)

Let a region Ω in R^n ($n = 2, 3$) contain in it a system $\mathcal{O}_\varepsilon$ of periodically arranged holes, and let Ω_ε designate the region excluding the holes,

$$\Omega_\varepsilon = \Omega \setminus \mathcal{O}_\varepsilon$$

whose boundary $\partial\Omega_\varepsilon$ is specified to include the hole boundaries $\partial\mathcal{O}_\varepsilon$. A unit-side cube is taken as the unit cell of the problem (Y). Within the cube a region $\mathcal{O}$ bounded by a surface S with an outward unit normal ν_i is chosen as shown in Figure 4.4; let $\mathcal{Y} = Y \setminus \mathcal{O}$. The (elementary) boundary

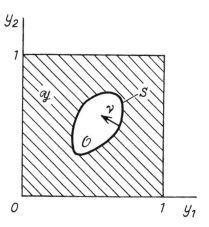

FIGURE 4.4. Unit cell with a hole $\mathcal{O}$.

value problem we consider is

$$-\frac{\partial^2 u_\varepsilon}{\partial x_i\, \partial x_i} = f \quad \text{in } \Omega_\varepsilon \tag{4.20}$$

$$u_\varepsilon = 0 \quad \text{on } \Gamma_\varepsilon \tag{4.21}$$

$$\frac{\partial u_\varepsilon}{\partial \nu} = 0 \quad \text{on } S_\varepsilon \tag{4.22}$$

where $f = f(x)$.

If we take the solution of this problem in the form

$$u_\varepsilon = u_0(x, y) + \varepsilon u_1(x, y) + \cdots \tag{4.23}$$

where $u_k(x, y)$ is Y-periodic in $y = x/\varepsilon$, we find

$$\varepsilon^{-2} \frac{\partial^2 u_0}{\partial y_i\, \partial y_i} + \varepsilon^{-1}\left(\frac{\partial^2 u_1}{\partial y_i\, \partial y_i} + 2\frac{\partial^2 u_0}{\partial x_i\, \partial y_i} \right)$$

$$+ \varepsilon^0 \left(\frac{\partial^2 u_2}{\partial y_i\, \partial y_i} + 2\frac{\partial^2 u_1}{\partial x_i\, \partial y_i} + \frac{\partial^2 u_0}{\partial x_i\, \partial x_i} \right) + \cdots = f \quad \text{in } \Omega_\varepsilon \tag{4.24}$$

$$\varepsilon^{-1} \nu_i \frac{\partial u_0}{\partial y_i} + \varepsilon^0 \left(\frac{\partial u_0}{\partial x_i}\nu_i + \frac{\partial u_1}{\partial y_i}\nu_i \right)$$

$$+ \varepsilon\left(\frac{\partial u_2}{\partial y_i}\nu_i + \frac{\partial u_1}{\partial x_i}\nu_i \right) + \cdots = \quad \text{on } S_\varepsilon \tag{4.25}$$

from Equations 4.20 and 4.22. Comparing terms in equal powers of ε, the following two problems result:

$$-\frac{\partial^2 u_0}{\partial y_i \, \partial y_i} = 0 \quad \text{and} \quad \frac{\partial u_0}{\partial y_i} \nu_i = 0 \tag{4.26}$$

This gives us $u_0 = u_0(x)$; hence the problem we are concerned with can be rewritten as

$$-\frac{\partial^2 u_1}{\partial y_i \, \partial y_i} = 0 \quad \text{in } \mathcal{Y} \tag{4.27}$$

$$\frac{\partial u_1}{\partial y_i} \nu_i + \frac{\partial u_0}{\partial x_i} \nu_i = 0 \quad \text{on } S \tag{4.28}$$

and

$$\frac{\partial^2 u_2}{\partial y_i \, \partial y_i} + 2\frac{\partial^2 u_1}{\partial x_i \, \partial y_i} + \frac{\partial^2 u_0}{\partial x_i \, \partial x_i} = -f \quad \text{in } \mathcal{Y} \tag{4.29}$$

$$\frac{\partial u_2}{\partial y_i} \nu_i + \frac{\partial u_1}{\partial x_i} \nu_i = 0 \quad \text{on } S \tag{4.30}$$

Setting

$$u_1(x, y) = -\frac{\partial u_0(x)}{\partial x_k} U_k(y) \tag{4.31}$$

with functions $U_k(y)$ 1-periodic ($x \in \Omega$), Equations 4.27 and 4.28 now lead to the local problem

$$\frac{\partial^2 U_k(y)}{\partial y_i \, \partial y_i} = 0 \quad \text{in } \mathcal{Y} \tag{4.32}$$

$$\frac{\partial U_k(y)}{\partial y_i} \nu_i = \nu_k \quad \text{on } S \tag{4.33}$$

and Equations 4.29 and 4.30 assume the respective forms

$$\frac{\partial^2 u_2}{\partial y_i \, \partial y_i} + \frac{\partial^2 u_0(x)}{\partial x_k \, \partial x_i}\left(\delta_{ik} - 2\frac{\partial U_k(y)}{\partial y_i}\right) = -f(x) \quad \text{in } \mathcal{Y} \quad (4.34)$$

$$\frac{\partial u_2}{\partial y_i} \nu_i - \frac{\partial^2 u_0}{\partial x_i \, \partial x_k} - U_k(y)\nu_i = 0 \quad \text{on } S \quad (4.35)$$

where Equation 4.21 has been used.

The condition for the problem given by Equations 4.34 and 4.35 to be solvable is

$$\frac{1}{|\mathcal{Y}|}\int_S \frac{\partial u_2}{\partial y_i}\nu_i \, dy = \frac{1}{|\mathcal{Y}|}\int_{\mathcal{Y}}\left(\delta_{ik} - 2\frac{\partial U_k(y)}{\partial y_i}\right)dy \, \frac{\partial^2 u_0}{\partial x_k \, \partial x_i} - f(x) \quad (4.36)$$

and is now combined with Equation 4.35 to give the homogenized equation

$$-q_{ik}\frac{\partial^2 u_0}{\partial x_k \, \partial x_i} = f \quad \text{in } \Omega \quad (4.37)$$

where the effective coefficients q_{ik} are defined by

$$q_{ik} = \frac{1}{|\mathcal{Y}|}\int_{\mathcal{Y}}\left(\delta_{ik} - \frac{\partial U_k(y)}{\partial y_i}\right)dy \quad (4.38)$$

$|\mathcal{Y}|$ denoting the volume of the region $\mathcal{Y}$. We thus conclude that the homogenized problem for a perforated medium reduces to the solution of Equation 4.37 under the condition that u_0 should vanish at the boundary Γ of the region Ω.

It can be shown (Lions 1987) that the approach just discussed applies with equal force to the more general boundary value problem

$$\frac{\partial}{\partial x_i}\left(a_{ij}\left(\frac{x}{\varepsilon}\right)\frac{\partial u_\varepsilon}{\partial x_j}\right) = -f \quad \text{in } \Omega_\varepsilon \quad (4.39)$$

$$u_\varepsilon = 0 \quad \text{on } \Gamma_\varepsilon \quad (4.40)$$

$$\nu_i a_{ij}\left(\frac{x}{\varepsilon}\right)\frac{\partial u_\varepsilon}{\partial x_j} = 0 \quad \text{in } S_\varepsilon \quad (4.41)$$

where the quantities a_{ij} may or may not be symmetric in their indices and are assumed to satisfy the condition

$$a_{ij}(y)\xi_i\xi_j \geq \alpha\xi_i\xi_j \quad (\alpha > 0)$$

If only two terms are retained in the asymptotic expansion

$$u_\varepsilon = u_0(x) + \varepsilon u_1(x, y) + \cdots \tag{4.42}$$

the function $u_0(x)$ turns out to solve the homogenized problem

$$-q_{ik}\frac{\partial^2 u_0(x)}{\partial x_i \, \partial x_k} = f \quad \text{in } \Omega \tag{4.43}$$

$$u_0(x) = 0 \quad \text{on } \Gamma \tag{4.44}$$

with

$$q_{ik} = \frac{1}{|\mathscr{Y}|}\int_{\mathscr{Y}}\left(a_{ik}(y) - a_{ij}\frac{\partial U_k(y)}{\partial y_j}\right)dy \tag{4.45}$$

and the 1-periodic functions $U_k(y)$ must be determined from the local problem

$$\frac{\partial}{\partial y_i}\left(a_{ij}(y)\frac{\partial U_k(y)}{\partial y_j}\right) = \frac{\partial a_{ik}(y)}{\partial y_i} \quad \text{in } \mathscr{Y} \tag{4.46}$$

$$v_i a_{ij}(y)\frac{\partial U_k(y)}{\partial y_j} = v_i a_{ik}(y) \quad \text{on } S \tag{4.47}$$

and are defined by (cf. Equation 4.31)

$$u_1(x, y) = -\frac{\partial u_0(x)}{\partial x_k}U_k(y) \tag{4.48}$$

An important result that seems worth mentioning here concerns the spectral analysis of the perforated medium problem (Cioranescu and Paulin 1979, Kesavan 1979a, 1979b; see also Lions 1985),

$$-\frac{\partial}{\partial x_i}\left(a_{ij}\left(\frac{x}{\varepsilon}\right)\frac{\partial u_\varepsilon}{\partial x_j}\right) = \lambda(\varepsilon)u_\varepsilon \quad \text{in } \Omega_\varepsilon \tag{4.49}$$

$$u_\varepsilon = 0 \quad \text{on } \Gamma_\varepsilon \tag{4.50}$$

$$v_i a_{ij}\left(\frac{x}{\varepsilon}\right)\frac{\partial u}{\partial x_j} = 0 \quad \text{on } S_\varepsilon \tag{4.51}$$

and states that the eigenvalues $\lambda_m(\varepsilon)$ of this problem $(m = 1, 2, \ldots;$ $0 \le \lambda_1(\varepsilon) \le \lambda_2(\varepsilon) \le \cdots)$ have the property

$$\lambda_m(\varepsilon) \to \lambda_m \quad \text{as } \varepsilon \to 0 \tag{4.52}$$

where $\{\lambda_m\}$ is the spectrum of the Dirichlet problem

$$-q_{ij}\frac{\partial u}{\partial x_i\, \partial x_j} = \lambda u \quad \text{in } \Omega \tag{4.53}$$

$$u = 0 \quad \text{on } \Gamma \tag{4.54}$$

with coefficients q_{ij} defined by Equation 4.45.

It should be noted at this point that an application of the two-scale asymptotic process in the version outlined previously may present difficulties in satisfying the boundary condition of a homogenized problem. We can illustrate this by considering the problem of homogenization for a perforated region Ω_ε with Dirichlet conditions imposed at the boundaries of the holes (or cavities) (Lions 1985), that is, we write,

$$-\varepsilon^2 \frac{\partial^2 u_\varepsilon}{\partial x_i\, \partial x_i} = f \quad \text{in } \Omega_\varepsilon \tag{4.55}$$

$$u_\varepsilon = 0 \quad \text{on } \Gamma_\varepsilon \tag{4.56}$$

As a solution of this problem we assume

$$u_\varepsilon = u_0(x, y) + \varepsilon u_1(x, y) + \cdots \tag{4.57}$$

where the functions u_0 and u_1 are 1-periodic in y.

When Equation 4.57 is substituted into Equation 4.55 the result is

$$-\frac{\partial^2 u_0(x, y)}{\partial y_i\, \partial y_i} = f(x) \tag{4.58}$$

$$-\frac{\partial^2 u_1(x, y)}{\partial y_i\, \partial y_i} - 2\frac{\partial^2 u_0(x, y)}{\partial x_i\, \partial y_i} = 0 \tag{4.59}$$

$$-\frac{\partial^2 u_2(x, y)}{\partial y_i\, \partial y_i} - 2\frac{\partial^2 u_1(x, y)}{\partial x_i\, \partial y_i} = 0 \tag{4.60}$$

$$\vdots$$

Now, if we write

$$u_0(x, y) = W_0(y)f(x) \tag{4.61}$$

$$u_1(x, y) = W_k(y)\frac{\partial f(x)}{\partial x_k} \tag{4.62}$$

$$\vdots$$

then the functions W_0 and W_k will have to be determined from the following local problems:

$$-\frac{\partial^2 W_0(y)}{\partial y_i \, \partial y_i} = 1 \quad \text{in } \mathcal{Y} \tag{4.63}$$

$$W_0 = 0 \quad \text{on } S \tag{4.64}$$

and

$$-\frac{\partial^2 W_k(y)}{\partial y_i \, \partial y_i} = 2\frac{\partial W_0(y)}{\partial y_k} \quad \text{in } \mathcal{Y} \tag{4.65}$$

$$W_k = 0 \qquad \text{on } S \tag{4.66}$$

We see that if we take the solution of Equation 4.55 in the form

$$u_\varepsilon = W_0(y)f(x) + \varepsilon W_k(y)\frac{\partial f(x)}{\partial x_k} + \cdots \tag{4.67}$$

the boundary condition given by Equation 4.56 will be satisfied if and only if

$$f(x) = \frac{\partial f(x)}{\partial x_k} = 0 \quad \text{on } \Gamma \tag{4.68}$$

Otherwise boundary-layer-type solutions will have to be introduced into asymptotic processes, the construction of which is discussed in the works of Panasenko (1979), Bakhvalov and Panasenko (1984), and Lions (1985); the paper by Sanchez-Palencia (1987) is also relevant.

Clearly, the solutions provided by the locally periodic expansions like Equation 6.4 are only adequate at points remote both from the boundary

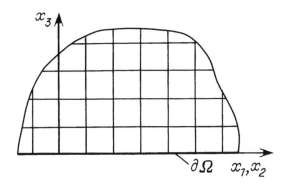

FIGURE 4.5. The boundary of a body made of a composite material of periodic structure.

of the region Ω and from those regions within Ω where nonperiodic effects due to cracks (discontinuities), inclusions, and similar factors may be of importance. In the immediate vicinity of $\partial\Omega$, boundary layer solutions must be considered. If the boundary is taken to coincide with the plane $x_3 = 0$, as in Figure 4.5, these solutions will have the form

$$u_\varepsilon(x) = u_0^{(n)}(x, y) + \varepsilon u_1^{(n)}(x, y) + \cdots \qquad (4.69)$$

where the functions $u_i^{(n)}(x, y)$ are periodic in y_1 and y_2 (with periods Y_1 and Y_2, respectively) but nonperiodic in x_1, x_2, y_3. Note that the boundary layer solution given by Equation 4.69 is supposed to satisfy the true boundary conditions at $y_3 = 0$ and must be made consistent with Solutions 4.4 in the limit as $x_3 \to 0$, $y_3 \to \infty$. Because the tangential variables x_1, x_2, y_1, and y_2 come simply as parameters in this case, one can write (Sanchez-Palencia 1987)

$$u_\varepsilon(x) = u_0(x_3) + \varepsilon u_1(x_3) + \cdots \qquad (4.70)$$

for the external (bulk) expansion, and

$$u_\varepsilon(x) = u_0^{(n)}(y_3) + \varepsilon u_1^{(n)}(y_3) + \cdots \qquad (4.71)$$

for the internal (boundary layer) expansion. It should be emphasized that the external form depends only on the external variable x_3 and the internal form depends only on $y_3 = x_3/\varepsilon$.

It is common practice in performing the matching procedure for an asymptotic expansion to define the external (internal) limit of the function $u_\varepsilon(x)$ as the value it attains in the limit as ε tends to zero at fixed external (internal) variable x_3 (y_3). Referring to Nayfeh (1972) and

Sanchez-Palencia (1987), the following simple rule expresses the required consistency between the external and internal solutions in the transition region:

> The internal limit of the external expansion is equal to the external limit of the internal expansion.

According to this rule,

$$u_0(0) = u_0^{(n)}(\infty) \tag{4.72}$$

A general discussion of the matching procedure for an n-term asymptotic expansion is given in Nayfeh (1972).

5. HOMOGENIZATION METHOD FOR REGIONS WITH A WAVY BOUNDARY

The boundary value problems with rapidly oscillating coefficients are by no means the only area of application for the two-scale expansion method. The method has been successfully employed, for example, in the treatment of partial differential equations in regions bounded by small-period wavy surfaces. We consider as an illustration a two-dimensional heat conduction problem for the half-plane Ω_ε with a wavy boundary $\partial\Omega$ as shown in Figure 5.1, taking the ambient temperature to be zero and assuming that a heat exchange condition is specified at $\partial\Omega$. A general outline of the homogenization analysis of this problem has been given in the monograph by Sanchez-Palencia (1980). Here, a more detailed discussion seems to be appropriate.

Let the boundary $\partial\Omega$ of the region $\Omega_\varepsilon = \{x_1, x_2 : x_2 < \varepsilon F(x/\varepsilon), |x_1| < \infty\}$ be specified by the equation $x_2 = \varepsilon F(x_1/\varepsilon)$, where ε is a small positive parameter and $F(y_1)$ a smooth 1-periodic function. Assuming the thermal and physical characteristics of the medium to be constant and denoting $x = (x_1, x_2)$, we write Fourier's law as

$$q_\alpha^{(\varepsilon)}(x, t) = -\lambda_{\alpha\beta}\frac{\partial T^{(\varepsilon)}(x, t)}{\partial x_\beta} \tag{5.1}$$

and the heat balance condition as

$$-f(x) + c\rho\frac{\partial T^{(\varepsilon)}(x, t)}{\partial t} = -\frac{\partial q_\alpha^{(\varepsilon)}(x, t)}{\partial x_\alpha} \quad \text{in } \Omega_\varepsilon \tag{5.2}$$

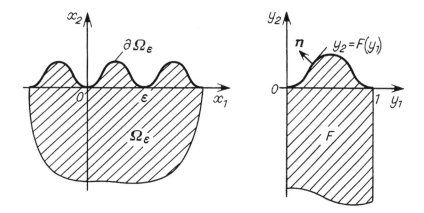

FIGURE 5.1. A region with a wavy boundary of periodic structure.

where the temperature $T^{(\varepsilon)}(x,t)$ is the unknown quantity of the problem, $q_\alpha^{(\varepsilon)}$ is the heat flow vector, $\lambda_{\alpha\beta}$ the (constant) thermal conductivities, c the specific heat, ρ the mass density, and $f(x)$ the (known) heat source density function; the subscripts α and β assume values 1 and 2.

Let the solution of Equations 5.1 and 5.2 satisfy the initial condition

$$T^{(\varepsilon)}(x,0) = 0 \tag{5.3}$$

and boundary conditions of the form

$$q_\alpha^{(\varepsilon)}(x,t)n_\alpha = \alpha T^{(\varepsilon)}(x,t) \quad \text{on } \partial\Omega_\varepsilon$$

$$T^{(\varepsilon)}(x,t) \to 0 \qquad \text{as } x_2 \to -\infty \tag{5.4}$$

where

$$n_1 = -\frac{dF}{dy_1}\left[1 + \left(\frac{dF}{dy_1}\right)^2\right]^{-1/2}$$

$$n_2 = \left[1 + \left(\frac{dF}{dy_1}\right)^2\right]^{-1/2}$$

and α is the heat-transfer coefficient.

The asymptotic expansion postulated for the problem by Sanchez-Palencia (1980) reads

$$T^{(\varepsilon)}(x,t) = T^{(0)}(x,t) + \varepsilon T^{(1)}(x,y,t) + \cdots \tag{5.5}$$

where $y = x/\varepsilon$, $T^{(1)}(x, y, t)$ is 1-periodic in y, and

$$\frac{\partial T^{(1)}}{\partial y_2} \to 0 \quad \text{as } y_2 \to -\infty \qquad (5.6)$$

Substituting Equation 5.5 into Equations 5.1 and 5.2 we find

$$\frac{\partial q_\alpha^{(0)}(x, y, t)}{\partial y_\alpha} = 0 \qquad (5.7)$$

$$-f(x) + c\rho \frac{\partial T^{(0)}(x, t)}{\partial t} = -\frac{\partial q_\alpha^{(0)}(x, y, t)}{\partial x_\alpha} - \frac{\partial q_\alpha^{(1)}(x, y, t)}{\partial y_\alpha} \qquad (5.8)$$

where

$$q_\alpha^{(n)}(x, y, t) = -\lambda_{\alpha\beta} \left(\frac{\partial T^{(n)}(x, y, t)}{\partial x_\beta} + \frac{\partial T^{(n+1)}(x, y, t)}{\partial y_\beta} \right) \qquad (5.9)$$

Now Equation 5.7 corresponding to the local problem set on the unit cell $F = \{y_1, y_2 : y_1 \in [0, 1], y_2 < F(y_1)\}$ takes the form

$$\lambda_{\alpha\beta} \frac{\partial^2 T^{(1)}(x, y, t)}{\partial y_\alpha \partial y_\beta} = 0 \quad \text{in } F \qquad (5.10)$$

and from the boundary condition given by Equation 5.4 we have

$$-\lambda_{\alpha\beta} \frac{\partial T^{(1)}(x, y, t)}{\partial y_\beta} \bigg|_{y_2 = F(y_1)} n_\alpha(y_1)$$

$$= \lambda_{\alpha\beta} \frac{\partial T^{(0)}(x, y, t)}{\partial x_\beta} \bigg|_{x_2 = \varepsilon F(x_1/\varepsilon)} n_\alpha(y_1) + \alpha T^{(0)}(x, t)|_{x_2 = \varepsilon F(x_1/\varepsilon)}$$

$$(5.11)$$

Note that $T^{(1)}(x, y, t)$ viewed as a function of y_1 must satisfy Condition 5.6 and must be 1-periodic in y. It is also important to recall here the uniqueness condition for the (periodic) function $T^{(1)}$ solving the problem given by Equations 5.10, 5.11, and 5.6. This condition is obtained by integrating Equation 5.10 over the region F, and in the limiting case of $\varepsilon \to 0$ it becomes

$$\lambda_{\alpha\beta} \frac{\partial T^{(0)}(x,t)}{\partial x_\beta} + \alpha l_F T^{(0)}(x,t) = 0 \quad \text{as } x_2 = \varepsilon F\left(\frac{x_1}{\varepsilon}\right) \to 0 \quad (5.12)$$

where l_F is the length of the arc of the curve $y_2 = F(y_1)$ $(0 \le y_1 \le 1)$.

With Equation 5.12, the boundary condition of the local problem Equation 5.7, as given by Equation 5.11, can be expressed as

$$-\lambda_{\alpha\beta} \left. \frac{\partial T^{(1)}(x,y,t)}{\partial y_\beta} \right|_{y_2=F(y_1)} n_\alpha(y_1)$$

$$= \left(\lambda_{\alpha\beta} n_\alpha(y_1) - \frac{\lambda_{2\beta}}{l_F} \right) \left. \frac{\partial T^{(0)}(x,t)}{\partial x_\beta} \right|_{x_2 \to 0} \quad (5.13)$$

and if we write

$$T^{(1)}(x,y,t) = U_\gamma(y) \frac{\partial T^{(0)}(x,t)}{\partial x_\gamma} \quad (5.14)$$

then the function $U_\gamma(y)$ (periodic in y_1) will have to be found from the problem

$$\lambda_{\alpha\beta} \frac{\partial^2 u_\gamma}{\partial y_\alpha \partial y_\beta} = 0 \quad \text{in } F \quad (5.15)$$

$$-\lambda_{\alpha\beta} \left. \frac{\partial U_\gamma(y)}{\partial y_\beta} \right|_{y_2=F(y_1)} n_\alpha(y_1) = \lambda_{\alpha\gamma} n_\alpha(y_1) - \frac{\lambda_{2\gamma}}{l_F} \quad (5.16)$$

$$\frac{\partial u_\gamma(y)}{\partial y_\beta} \to 0 \quad \text{as } y_2 \to -\infty \ (\gamma = 1,2) \quad (5.17)$$

The macroscopic heat conduction equation is readily obtained by taking the volume average of Equation 5.8 over the periodicity cell F. Because the region F is unbounded in the y_2 direction $(-\infty < y_2 < F(y_1))$, this is

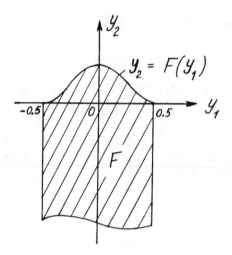

FIGURE 5.2.   Unit cell of a region with a wavy boundary.

effected by first averaging over the bounded region $F_0 = \{y_1, y_2: -h < y_2 < F(y_1), \ 0 < y_1 < 1\}$ and then passing to the limit as $h \to \infty$. Using Condition 5.17 and recalling the periodicity of $U_\gamma(y)$ in y_1, we obtain

$$-f(x) + c\rho\frac{\partial T^{(0)}(x,t)}{\partial t} = -\lambda_{\alpha\gamma}\frac{\partial^2 T^{(0)}(x,t)}{\partial x_\alpha \, \partial x_\gamma} \tag{5.18}$$

in the region $\Omega_0 = \{x_1, x_2: x_2 < 0, |x_1| < \infty\}$.

With the expansion given by Equation 5.5 taken to represent the solution, we thus see that we must start by determining its first term from the limiting problem for Equations 3.18 in the half-plane $x_2 < 0$ under the boundary condition given by Equation 5.12. This done, the function $T^{(1)}$ can be found up to a constant. A proof of the uniqueness of $T^{(1)}$ is given in Sanchez-Palencia (1980).

It is of interest to examine the solution of the local problem given by Equations 5.15 through 5.17 in the special case where $\lambda_{\alpha\beta} = \lambda\delta_{\alpha\beta}$ and the function $F(y_1)$, the shape of the cell boundary, is symmetric on the interval $(-\frac{1}{2} < y_1 < \frac{1}{2})$ (see Figure 5.2). The problem given by Equations 5.15 through 5.17 reduces then to that of finding functions $U_\gamma(y_1, y_2)$ $(\gamma = 1, 2)$ periodic in y_1 and satisfying Laplace's equation

$$\frac{\partial^2 u_\gamma}{\partial y_1} + \frac{\partial^2 u_\gamma}{\partial y_2} = 0 \quad \text{in } F \tag{5.19}$$

subjected to the conditions

$$
\left(\frac{\partial U_\gamma}{\partial y_1} n_1 + \frac{\partial U_\gamma}{\partial y_2} n_2 \right)\bigg|_{y_2 = F(y_1)} = \begin{cases} -n_1 & (\gamma = 1) \\ -n_2 + \dfrac{1}{l_F} & (\gamma = 2) \end{cases} \tag{5.20}
$$

$$
\frac{\partial u_\gamma}{\partial y_2} \to 0 \quad \text{as } y_2 \to -\infty \tag{5.21}
$$

We satisfy Condition 5.21 and take account of the periodicity of $U_\gamma(y_1, y_2)$ in y_1 if we assume the solution to be of the form

$$
U_1(y_1, y_2) = \sum_{k=1}^{\infty} \frac{1}{k} A_k e^{2\pi k y_2} \sin(2\pi k y_1) \tag{5.22}
$$

$$
U_2(y_1, y_2) = \sum_{k=1}^{\infty} \frac{1}{k} B_k e^{2\pi k y_2} \cos(2\pi k y_1) \tag{5.23}
$$

with constants A_k and B_k to be determined from the boundary conditions given by Equation 5.20.

Using Equations 5.22 and 5.23 in Equation 5.20 results in the pair of equations

$$
2\pi \sum_{k=1}^{\infty} A_k e^{2\pi k F(y_1)} (F'(y_1)\cos(2\pi k y_1) - \sin(2\pi k y_1))
$$

$$
= F'(y_1) \tag{5.24}
$$

$$
2\pi \sum_{k=1}^{\infty} B_k e^{2\pi k F(y_1)} (-F'(y_1)\sin(2\pi k y_1) - \cos(2\pi k y_1))
$$

$$
= 1 - \frac{1}{l_F}\left(1 + (F'(y_1))^2\right)^{1/2} \qquad \left(|y_1| < \frac{1}{2}\right) \tag{5.25}
$$

which when multiplied by, respectively, $\sin(2\pi p y_1)$ and $\cos(2\pi p y_1)$ ($p = 1, 2, \dots$) and integrated with respect to y_1 from 0 to $\frac{1}{2}$ lead to the following infinite system of algebraic equations for determining the constants A_k

and B_k:

$$\sum_{k=1}^{\infty} A_k \alpha_{kp} = a_p$$

$$\sum_{k=1}^{\infty} B_k \beta_{kp} = -b_p \qquad (p = 1, 2, \dots) \tag{5.26}$$

where

$$\alpha_{kp} = 2\pi \int_0^{1/2} e^{2\pi k F(y_1)} (F'(y_1)\cos(2\pi ky_1) - \sin(2\pi ky_1))$$

$$\times \sin(2\pi py_1)\, dy_1 \tag{5.27}$$

$$\beta_{kp} = 2\pi \int_0^{1/2} e^{2\pi k F(y_1)} (F'(y_1)\sin(2\pi ky_1) + \cos(2\pi ky_1))$$

$$\times \cos(2\pi py_1)\, dy_1 \tag{5.28}$$

$$a_p = \int_0^{1/2} F'(y_1)\sin(2\pi py_1)\, dy_1 \tag{5.29}$$

$$b_p = \int_0^{1/2} \left[1 - \frac{1}{l_F} \left(1 + (F'(y_1))^2 \right)^{1/2} \right] \cos(2\pi py_1)\, dy_1 \tag{5.30}$$

To illustrate the use of this result, we consider the temperature distribution in an infinite cylinder with a regular wavy surface Σ (Figure 5.3), making the following assumptions:

1. T_0, the initial temperature of the cylinder, is constant (over the cylinder).
2. T_m, the ambient temperature, is constant (both in space and time).
3. The heat exchange across Σ is governed by

$$-\lambda \frac{\partial T}{\partial \mathbf{n}} \bigg|_\Sigma = \alpha(T_m - T|_\Sigma) \tag{5.31}$$

where $\mathbf{n}$ is the outward normal to Σ, and $\lambda_{\alpha\beta} = \lambda \delta_{\alpha\beta}$.

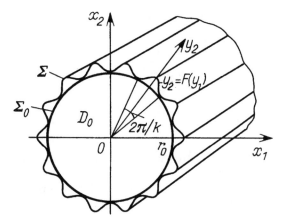

FIGURE 5.3. Cylindrical body with a regular wavy boundary.

To describe the wavy shape of the surface Σ, it is convenient to introduce the local coordinates

$$y_1 = \frac{r_0 \varphi}{\varepsilon} \quad \text{and} \quad y_2 = \frac{r - r_0}{\varepsilon} = F(y_1) \qquad (5.32)$$

where $\varepsilon = 2\pi r_0/k$, k being the number of the sectors that comprise the circle D_0: $r < r_0$ in Figure 5.3.

Using the asymptotic expansion

$$T(r, \varphi, t) = T^{(0)}(r, t) + \varepsilon T^{(1)}(r, y_1, y_2, t) + \cdots \qquad (5.33)$$

we obtain the following limiting problem for $T^{(0)}(r, t)$:

$$\frac{\partial^2 T^{(0)}}{\partial r^2} + \frac{1}{r} \frac{\partial T^{(0)}}{\partial r} - \frac{1}{\kappa^2} \frac{\partial T^{(0)}}{\partial t} = 0 \qquad (0 < r < r_0, t > 0) \quad (5.34)$$

$$T^{(0)}(r, 0) = T_0 \qquad (5.35)$$

$$-\lambda \frac{\partial T^{(0)}}{\partial r} \bigg|_{r=r_0} = l_F \alpha \left(T_m - T^{(0)} \big|_{r=r_0} \right) \qquad (5.36)$$

where

$$l_F = \frac{k}{2\pi r_0} \int_0^1 \sqrt{1 + (F'(y_1))^2} \, dy_1 \quad \text{and} \quad \kappa^2 = \frac{\lambda}{c\rho}$$

The function $T^{(1)}$ is written in the form

$$T^{(1)}(r, y_1, y_2, t) = U_\gamma(y) \frac{\partial T^{(0)}(r, t)}{\partial x_\gamma}$$

$$= [U_1(y_1, y_2)\cos \varphi + U_2(y_1, y_2)\sin \varphi] \frac{\partial T^{(0)}(r, t)}{\partial r} \quad (5.37)$$

and the functions $U_\gamma(y_1, y_2)$ are periodic in y_1 and solve the local problem given by Equations 5.19 through 5.21.

For the region D_0: $r < r_0$, the solution of Equations 5.34 through 5.36 is known to be given by

$$T^{(0)}(r, t) = T_m + 2(T_m - T_0)\frac{l_F \alpha}{\lambda} r_0$$

$$\times \sum_{n=1}^{\infty} A_n \exp\left(-\frac{\kappa^2 \mu_n^2 t}{r_0^2}\right) J_0\left(\frac{\mu_n r}{r_0}\right) \quad (5.38)$$

where

$$A_n = \frac{1}{J_0(\mu_n)\left(\mu_n^2 + l_F^2(\alpha^2/\lambda^2)r_0^2\right)} \quad (5.39)$$

in which μ_n are the roots of the equation.

6. LOCAL PROBLEMS AND EFFECTIVE COEFFICIENTS

It seems important to give some illustrations of how local problems arising from the use of asymptotic expansions of the form of Equation 4.4 can be treated. Consider, to start with, the two-dimensional, steady-state heat conduction problem for a rectangular region of a laminated material with lamina thickness ε. We will assume, referring to Figure 6.1, that the boundaries $x_2 = 0$ and $x_2 = l_2$ of the region D are thermally insulated and that at $x_1 = 0$ and $x_1 = l$ heat flows are prescribed. Under these circumstances the thermal and physical characteristics depend on x_1 alone (with period ε) and the problem of finding the temperature field $T^\varepsilon(x_1, x_2)$ reduces to the solution of the equation

$$\frac{\partial}{\partial x_\alpha}\left(\lambda_{\alpha\beta}\left(\frac{x_1}{\varepsilon}\right)\frac{\partial T^\varepsilon}{\partial x_\beta}\right) = 0 \quad \text{in } D \quad (6.1)$$

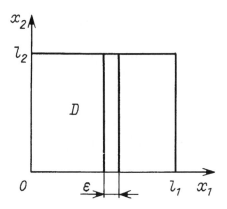

FIGURE 6.1. Laminate composite.

subject to the conditions

$$\lambda_{2\beta} \frac{\partial T^{\varepsilon}}{\partial x_{\beta}} = 0 \qquad \text{tor } x_2 = 0, \, x_2 = l_2 \tag{6.2}$$

$$\pm \lambda_{1\beta} \frac{\partial T^{\varepsilon}}{\partial x_{\beta}} = \mp q(x_2) \quad \text{for } x_1 = 0, \, x_1 = l_1 \tag{6.3}$$

At large distances from region D, Solution 6.1 takes the form

$$T^{\varepsilon}(x) = T_0(x) + \varepsilon T_1(x, y_1) + \cdots \qquad \left(y_1 = \frac{x_1}{\varepsilon} \right) \tag{6.4}$$

where, in accordance with Equation 4.13,

$$T_1(x, y) = U_{\gamma}(y_1) \frac{\partial T_0(x)}{\partial x_{\gamma}} \qquad (\gamma = 1, 2) \tag{6.5}$$

and the functions $U_{\gamma}(y_1)$ are 1-periodic solutions of the one-dimensional local problem (cf. Equation 4.14)

$$-\frac{d}{dy_1} \left(\lambda_{11}(y_1) \frac{dU_{\gamma}(y_1)}{dy_1} \right) = \frac{d\lambda_{1\gamma}}{dy_1} \qquad (0 < y_1 < 1) \tag{6.6}$$

Integrating Equation 6.6 we get

$$U_{\gamma}(y_1) = -\int_0^{y_1} \lambda_{1\gamma}(\xi) \lambda_{11}^{-1}(\xi) \, d\xi + C_{\gamma} \int_0^{y_1} \lambda_{11}^{-1}(\xi) \, d\xi \tag{6.7}$$

where the constants C_γ are to be determined from

$$U_\gamma(0) = U_\gamma(1) \tag{6.8}$$

It is easily verified that

$$C_\gamma = \int_0^1 \lambda_{1\gamma}(\xi)\lambda_{11}^{-1}(\xi)\,d_x \left(\int_0^1 \lambda_{11}^{-1}(\xi)\,d\xi\right)^{-1} \tag{6.9}$$

giving the solution

$$U_1(y_1) = \int_0^{y_1} \lambda_{11}^{-1}(\xi)\,d\xi \left(\int_0^1 \lambda_{11}^{-1}(\xi)\,d\xi\right)^{-1} - y_1 \tag{6.10}$$

$$U_2(y_1) = \left(\int_0^1 \lambda_{12}(\xi)\lambda_{11}^{-1}(\xi)\,d\xi\right)\left(\int_0^1 \lambda_{11}^{-1}(\xi)\,d\xi\right)^{-1}$$

$$\times \int_0^{y_1} \lambda_{11}^{-1}(\xi)\,d\xi - \int_0^{y_1} \lambda_{12}(\xi)\lambda_{11}^{-1}(\xi)\,d\xi \tag{6.11}$$

so that the effective heat conductivities are found to be

$$\tilde{\lambda}_{11} = \left(\int_0^1 \lambda_{11}^{-1}(\xi)\,d\xi\right)^{-1}$$

$$\tilde{\lambda}_{12} = \left(\int_0^1 \lambda_{12}(\xi)\lambda_{11}^{-1}(\xi)\,d\xi\right)\left(\int_0^1 \lambda_{11}^{-1}(\xi)\,d\xi\right)^{-1}$$

$$\tag{6.12}$$

$$\tilde{\lambda}_{22} = \int_0^1 \lambda_{22}(\xi)\,d\xi + \left(\int_0^1 \lambda_{12}(\xi)\lambda_{11}^{-1}(\xi)\,d\xi\right)\left(\int_0^1 \lambda_{11}^{-1}(\xi)\,d\xi\right)^{-1}$$

$$- \int_0^1 \lambda_{12}^2(\xi)\lambda_{11}^{-1}(\xi)\,d\xi$$

using Relations 4.17. Averaging Equation 6.1 for the homogenized medium now yields

$$\tilde{\lambda}_{11}\frac{\partial^2 T_0}{\partial x_1^2} + 2\tilde{\lambda}_{12}\frac{\partial^2 T_0}{\partial x_1\,\partial x_2} + \tilde{\lambda}_{22}\frac{\partial^2 T_0}{\partial x_2^2} = 0 \tag{6.13}$$

and by using Equations 6.4 and 6.5 the boundary equations can be written as

$$\left(\lambda_{2\beta}(y_1) + \lambda_{21}(y_1)\frac{dU_\beta}{dy_1}\right)\frac{\partial T_0}{\partial x_\beta} = 0 \qquad \text{for } x_2 = 0, x_2 = l_2 \quad (6.14)$$

$$\mp \left(\lambda_{1\beta}(y_1) + \lambda_{11}(y_1)\frac{dU_\beta}{dy_1}\right)\frac{\partial T_0}{\partial x_\beta} = \pm q(x_2) \quad \text{for } x_1 = 0, x_1 = l_1 \quad (6.15)$$

giving

$$\tilde{\lambda}_{2\beta}\frac{\partial T_0}{\partial x_\beta} = 0 \qquad \text{for } x_2 = 0, x_2 = l_2 \qquad (6.16)$$

$$\mp \tilde{\lambda}_{1\beta}\frac{\partial T_0}{\partial x_\beta} = \pm q(x_2) \quad \text{for } x_1 = 0, x_1 = l_1 \qquad (6.17)$$

after the application of the averaging procedure.

It will be understood that the employment of averaged conditions at the boundaries $x_2 = 0$ and $x_2 = l_2$ in the homogenized problem given by Equations 6.13, 6.16, and 6.17 makes it necessary to invoke boundary layer type solutions in these vicinities. We consider the case $x_2 = 0$ in what follows.

We introduce an additional term into Equation 6.4 by writing

$$T^\varepsilon(x) = T_0(x) + \varepsilon(T_1(x, y_1) + T_1^{(n)}(x, y)) + \cdots$$

$$\left[y = \frac{x}{\varepsilon}, y = (y_1, y_2)\right] \qquad (6.18)$$

where $T_1(x, y_1)$ is determined from Equations 6.5 and 6.6, and $T_1^{(n)}(x, y)$ may be represented as

$$T_1^{(n)}(x, y) = U_\beta^{(n)}(y)\frac{\partial T_0(x)}{\partial x_\beta} \qquad (6.19)$$

It can be shown that the function $U^{(n)}(y)$ solves the equation

$$\frac{\partial}{\partial y_\gamma}\left(\lambda_{\gamma\alpha}(y_1)\frac{\partial U_\beta^{(n)}(y)}{\partial y_\alpha}\right) = 0 \qquad (6.20)$$

in the strip $S = \{(y_1, y_2): y_1 \in [0, 1], y_2 \in [0, \infty]\}$ and is 1-periodic in y_1.
Note also that

$$\frac{\partial U_\beta^{(n)}(y)}{\partial y_\alpha} \to 0 \quad \text{as } y_2 \to \infty \tag{6.21}$$

which when combined with Equation 6.20 gives

$$\int_0^1 \lambda_{2\alpha}(y_1) \frac{\partial U_\beta^{(n)}(y)}{\partial y_\alpha}\bigg|_{y_2 = 0} dy_1 = 0 \tag{6.22}$$

using the periodicity property just mentioned. Now if the expansion given
by Equation 6.18 is introduced into Equation 6.21 for $x_2 = 0$ and Condition 6.16 is used, the condition to be satisfied by the boundary layer type
solution may be expressed as

$$\lambda_{2\gamma}(y_1) \frac{\partial U_\beta^{(n)}(y)}{\partial y_\gamma}\bigg|_{y_2 = 0} = \tilde{\lambda}_{2\beta} - \lambda_{2\beta}(y_1) - \lambda_{21}(y_1)\frac{dU_\beta(y_1)}{dy_1} \tag{6.23}$$

The problem of finding $U_\beta^{(n)}(y)$ thus reduces to that of solving Equation
6.20 for a function satisfying Conditions 6.21 and 6.23 and 1-periodic in y_1.
In particular, if $\lambda_{\alpha\beta}(y_1) = \lambda(y_1)\delta_{\alpha\beta}$, the preceding problem reads

$$\frac{\partial}{\partial y_\alpha}\left(\lambda(y_1)\frac{\partial U_2^{(n)}(y)}{\partial y_\alpha}\right) = 0 \quad \text{in } S \tag{6.24}$$

$$\lambda(y_1)\frac{\partial U_2^{(n)}}{\partial y_2}\bigg|_{y_2 = 0} = \tilde{\lambda}_{22} - \lambda(y_1) \tag{6.25}$$

$$\frac{\partial U_2^{(n)}}{\partial y_\alpha} \to 0 \quad \text{as } y_2 \to \infty \tag{6.26}$$

We note that in this case $U_1^{(n)} = 0$ and the set of nonzero effective
coefficients is

$$\tilde{\lambda}_{11} = \left(\int_0^1 \lambda^{-1}(\xi)\, d\xi\right)^{-1}$$

$$\tag{6.27}$$

$$\tilde{\lambda}_{22} = \int_0^1 \lambda(\xi)\, d\xi$$

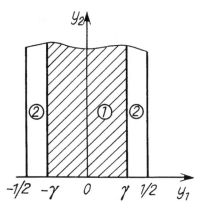

FIGURE 6.2. Unit cell of a two-component laminate composite.

We further particularize by assuming the unit cell to be composed of two laminae with constant but different conductivities, $\lambda(y_1) = \lambda_1$, for $|y_1| < \gamma$ and $\lambda(y_1) = \lambda_2$ for $\gamma < |y_1| < \frac{1}{2}$ (Figure 6.2). The function $U_2^{(n)}(y)$ will then satisfy Laplace's equation

$$\frac{\partial^2 U_2^{(n)}}{\partial y_1^2} + \frac{\partial^2 U_2^{(n)}}{\partial y_2^2} = 0 \qquad (6.28)$$

in regions

$$S_1 = \{(y_1, y_2): -\gamma < y_1 < \gamma, 0 < y_2 < \infty\}$$

and

$$S_2 = \{(y_1, y_2): \gamma < |y_1| < \tfrac{1}{2}, 0 < y_2 < \infty\}$$

and the interfacial contact conditions

$$U_2^{(n)}(-\gamma + 0, y_2) = U_2^{(n)}(-\gamma - 0, y_2)$$

$$U_2^{(n)}(\gamma - 0, y_2) = U_2^{(n)}(\gamma + 0, y_2) \qquad (6.29)$$

$$\lambda_1 \frac{\partial U_2^{(n)}}{\partial y_1}\bigg|_{y_1 = -\gamma + 0} = \lambda_2 \frac{\partial U_2^{(n)}}{\partial y_1}\bigg|_{y_1 = \gamma - 0}$$

$$\lambda_1 \frac{\partial U_2^{(n)}}{\partial y_1}\bigg|_{y_1 = \gamma - 0} = \lambda_2 \frac{\partial U_2^{(n)}}{\partial y_1}\bigg|_{y_1 = \gamma + 0} \qquad (6.30)$$

Because U_2 is an even function of y_1, it suffices to consider the second equations in Equations 6.29 and 6.30. Substituting Equations 6.27 into the boundary condition given by Equation 6.25, the latter becomes

$$
\left. \frac{\partial U_2^{(n)}}{\partial y_2} \right|_{y_1=0} =
\begin{cases}
-(1 - 2\gamma)\left(1 - \dfrac{\lambda_2}{\lambda_1}\right) & (|y_1| < \gamma) \\[12pt]
2\gamma \dfrac{\lambda_1}{\lambda_2}\left(1 - \dfrac{\lambda_2}{\lambda_1}\right) & \left(\gamma < |y_1| < \dfrac{1}{2}\right)
\end{cases}
\tag{6.31}
$$

In the appendix a special system of functions is developed in terms of which the solution to Equation 6.28 can be conveniently expanded. The functions are continuous and mutually orthogonal on the segment $[-\frac{1}{2}, \frac{1}{2}]$ and undergo a jump of prescribed magnitude in their first derivatives at $y_1 = \pm\gamma$. Remembering Equation 6.26, the solution even with respect to y_1 may thus be represented as

$$
U_2^{(n)}(y) =
\begin{cases}
\dfrac{1}{2}A_0 + \displaystyle\sum_{n=1}^{\infty} A_n e^{-p_n y_2} \cos(p_n y_1) & (0 \le y_1 < \gamma) \\[16pt]
\dfrac{1}{2}A_0 + \displaystyle\sum_{n=1}^{\infty} A_n e^{-p_n y_2} \\[12pt]
\quad\times \dfrac{\cos(p_n \gamma)}{\cos(p_n(\frac{1}{2} - \gamma))} \cos\left(p_n\left(\dfrac{1}{2} - y_1\right)\right) & \left(\gamma < y_1 < \dfrac{1}{2}\right)
\end{cases}
$$

$$\tag{6.32}$$

where A_n are constants (of which A_0 may be set equal to zero) and p_n are the roots of the equation

$$
\sin\left(p_n\left(\frac{1}{2} - \gamma\right)\right)\cos(p_n\gamma) + \frac{\lambda_1}{\lambda_2}\cos\left(p_n\left(\frac{1}{2} - \gamma\right)\right)\sin(p_n\gamma) = 0 \tag{6.33}
$$

It is left as an exercise for the reader to show by substitution that Solution 6.32 satisfies the conditions

$$
U_2^{(n)}(\gamma - 0, y_2) = U_2^{(n)}(\gamma + 0, y_2)
$$

$$
\left. \lambda_1 \frac{\partial U_2^{(n)}}{\partial y_1} \right|_{y_1 = \gamma - 0} = \left. \lambda_2 \frac{\partial U_2^{(n)}}{\partial y_1} \right|_{y_1 = \gamma + 0}
$$

To determine the constants A_n, substitute Equation 6.32 into Equation 6.31, giving

$$\sum_{n=1}^{\infty} P_n A_n \begin{cases} \cos(p_n y_1) & (0 \le y_1 < \gamma) \\ \dfrac{\cos(p_n \gamma)}{\cos(p_n(\frac{1}{2} - \gamma))} \cos\left(p_n\left(\frac{1}{2} - y_1\right)\right) & \gamma < y_1 < \frac{1}{2} \end{cases}$$

$$= \begin{cases} (1 - 2\gamma)\left(1 - \dfrac{\lambda_2}{\lambda_1}\right) & (0 \le y_1 < \gamma) \\ -2\gamma\dfrac{\lambda_1}{\lambda_2}\left(1 - \dfrac{\lambda_2}{\lambda_1}\right) & \left(\gamma < y_1 < \dfrac{1}{2}\right) \end{cases} \tag{6.34}$$

Using Formula A.20 of the appendix now yields, for $\delta = \lambda_1/\lambda_2$ and $\mu = 1$,

$$A_n = -\frac{2}{p_n^2 C_n}\left(1 - \frac{\lambda_2}{\lambda_1}\right)\left[(1 - 2\gamma)C_n^{(1)}\sin(p_n\gamma)\right.$$

$$\left. + 2\gamma\frac{\lambda_1}{\lambda_2}\sin\left(p_n\left(\frac{1}{2} - \gamma\right)\right)\right] \tag{6.35}$$

where

$$C_n^{(1)} = \sin\left(p_n\left(\frac{1}{2} - \gamma\right)\right)\sin(p_n\gamma)$$

$$- \frac{\lambda_1}{\lambda_2}\cos\left(p_n\left(\frac{1}{2} - \gamma\right)\right)\cos(p_n\gamma) \tag{6.36}$$

$$C_n = \left(\frac{1}{2} - \gamma + \frac{\lambda_1}{\lambda_2}\gamma\right)\cos\left(p_n\left(\frac{1}{2} - \gamma\right)\right)\cos(p_n\gamma)$$

$$- \left(\gamma + \frac{\lambda_1}{2\lambda_2} - \frac{\lambda_1}{\lambda_2}\gamma\right)\sin\left(p_n\left(\frac{1}{2} - \gamma\right)\right)\sin(p_n\gamma) \tag{6.37}$$

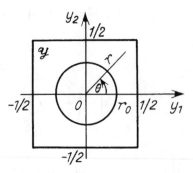

FIGURE 6.3. Unit cell containing a circular hole.

We now proceed to apply the two-dimensional local analysis to the problem given by Equations 4.32 and 4.33 for a unit cell taken in the form of a square with a circular hole of radius $r_0 < 1$ as in Figure 6.3. The task at hand is to solve the equation

$$\frac{\partial^2 U_\gamma}{\partial y_1^2} + \frac{\partial^2 U_\gamma}{\partial y_2^2} = 0 \quad \text{in } \mathcal{Y} \tag{6.38}$$

subject to the condition

$$\left. \frac{\partial U_\gamma}{\partial r} \right|_{r=r_0} = \begin{cases} \cos\theta & (\gamma = 1) \\ \sin\theta & (\gamma = 2) \end{cases} \tag{6.39}$$

where r and θ are polar coordinates with origin at $y_1 = 0$, $y_2 = 0$, as shown in the figure. Clearly the functions $U_\gamma(y_1, y_2)$ must be 1-periodic with respect to both of its arguments.

Introducing a complex variable $z = y_1 + iy_2$, the (doubly periodic) harmonic function $u_1(y_1, y_2)$ can be expanded in a series of the type

$$U_1 = \text{Re}\left\{ A_0\left[z - \frac{1}{\pi}\zeta(z) \right] + \sum_{k=0}^{\infty} A_{2k+4} r_0^{2k+4} \frac{\wp^{(2k+1)}(z)}{(2k+3)!} \right\} \tag{6.40}$$

where $\wp(z)$ is Weierstrass' elliptic function, $\zeta(z)$ is Weierstrass' zeta function, satisfying the quasiperiodicity conduction

$$\zeta(\mathcal{Z}+1) - \zeta(\mathcal{Z}) = \delta_1 \qquad \zeta(\mathcal{Z}+i) - \zeta(\mathcal{Z}) = \delta_2$$

the constants A_0 and A_{2k+4} are real, and the constants δ_1 and δ_2 are related by

$$i\delta_1 - \delta_2 = 2\pi i$$

(Legendre's relation) and

$$i\delta_1 + \delta_2 = 0$$

so that $\delta_1 = +\pi$ and $\delta_2 = -\pi i$. The previously mentioned periodicity property of $U_1(y_1, y_2)$ is readily deduced from the quasiperiodicity of $\zeta(z)$ and the double periodicity of Weierstrass's elliptic function and its derivatives.

From Equation 6.40, using $\zeta'(z) = -\wp(z)$,

$$\frac{\partial U_1}{\partial r} = \mathrm{Re}\left\{ A_0\left[1 - \frac{1}{\pi}\wp(\mathscr{Z})\right]e^{i\theta} + e^{i\theta}\sum_{k=0}^{\infty} A_{2k+4}r_0^{2k+4}\frac{\wp^{(2k+2)}(\mathscr{Z})}{(2k+3)!} \right\}$$

$$(6.41)$$

Now use will be made of the expansions (Grigolyuk and Fil'shtinskii 1970)

$$\wp(\mathscr{Z}) = \frac{1}{\mathscr{Z}^2} + \sum_{j=0}^{\infty} r_{j,0}\mathscr{Z}^{2j} \qquad (6.42)$$

$$\frac{\wp^{(2k+2)}(\mathscr{Z})}{(2k+3)!} = \frac{1}{\mathscr{Z}^{2k+4}} + \sum_{j=0}^{\infty} r_{j,k+1}\mathscr{Z}^{2j} \qquad (6.43)$$

where

$$r_{j,k} = \frac{(2j + 2k + 1)!\, g_{j+k+1}}{(2j)!(2k + 1)!\, 2^{2j+2k+2}}$$

$$g_{j+k+1} = \sideset{}{'}\sum_{m,n} \frac{1}{T_{m,n}^{2j+2k+2}}$$

$$T_{mn} = \frac{1}{2}(m + in)$$

a prime on Σ indicating that $m = n = 0$ is excluded in the summation. Substituting Equations 6.42 and 6.43 into Equation 6.41 and setting $z = r \exp(i\theta)$ we find

$$
\left.\frac{\partial U_1}{\partial r}\right|_{r=r_0} = \left[A_0\left(1 + \frac{1}{\pi r_0^2}\right) + \sum_{k=0}^{\infty} A_{2k+4} r_0^{2k+4} r_{0,\,k+1} \right]
$$

$$
\times \cos\theta + \sum_{j=1}^{\infty} \left[A_0 \frac{1}{\pi} r_0^{2j} r_{j,0} \right.
$$

$$
\left. + A_{2j+2} + \sum_{k=0}^{\infty} A_{2k+4} r_0^{2k+2j+4} r_{j,\,k+1} \right]
$$

$$
\times \cos(2j + 1)\theta \tag{6.44}
$$

which when combined with the boundary condition given by Equation 6.39, gives an infinite system of algebraic equations for constants A_0 and A_{2k},

$$
A_0\left(1 + \frac{1}{\pi r_0^2}\right) + \sum_{k=1}^{\infty} A_{2k+2} r_0^{2k+2} r_{0,\,k} = 1 \tag{6.45}
$$

$$
A_0 \frac{1}{\pi r_0^2} r_0^{2j+2} r_{j,0} + A_{2j+2} + \sum_{k=1}^{\infty} A_{2k+2} r_0^{2k+2+2j} r_{j,\,k} = 0 \tag{6.46}
$$

Eliminating A_0 from Equations 6.45 and 6.46 results in the following system of equations for A_{2k+2} $(k = 1, 2, 3, \ldots)$:

$$
A_{2j+2} + \sum_{k=1}^{\infty} A_{2k+2} r_0^{2k+2j+2} \left(r_{j,\,k} - \frac{r_0^2}{\pi r_0^2 + 1} r_{j,0} r_{0,\,k} \right)
$$

$$
= -\frac{r_0^{2j+2} r_{j,0}}{(\pi r_0^2 + 1)} \qquad (j = 1, 2, \ldots) \tag{6.47}
$$

Using Equation 6.43 and the well-known result (Grigolyuk and Fil'shtinskii 1970)

$$
\zeta(\mathscr{Z}) = \frac{1}{\mathscr{Z}} - \sum_{j=1}^{\infty} \frac{r_{j,0}}{(2j + 1)} \mathscr{Z}^{2j+1} \tag{6.48}
$$

Equation 6.40 becomes

$$
U_1 = \mathrm{Re}\left\{\left[A_0 + \sum_{k=1}^{\infty} A_{2k+2} r_0^{2k+2} \frac{r_{1,k-1}}{k(2k+1)}\right] \mathscr{Z}\right.
$$

$$
+ \sum_{j=1}^{\infty}\left[\frac{A_0}{\pi}\frac{r_{j,0}}{(2j+1)} + (2j+2)\sum_{k=1}^{\infty} A_{2k+2}\frac{r_0^{2k+2} r_{j+1,k-1}}{2k(2k+1)}\right]\mathscr{Z}^{2j+1}
$$

$$
\left. -A_0\frac{1}{\pi\mathscr{Z}} - \sum_{k=1}^{\infty} A_{2k+2}\frac{r_0^{2k+2}}{(2k+1)}\frac{1}{\mathscr{Z}^{2k+1}}\right\} \tag{6.49}
$$

The function $U_2(y_1, y_2)$, doubly periodic in its arguments, now is defined by analogy with Equation 6.40 as

$$
U_2 = -\mathrm{Re}\left\{A_0\left[i\mathscr{Z} - \frac{1}{\pi}\zeta(i\mathscr{Z})\right] + \sum_{k=0}^{\infty} A_{2k+4}\frac{r_0^{2k+4}\wp^{(2k+1)}(i\mathscr{Z})}{(2k+3)!}\right\}
$$

$$
\tag{6.50}
$$

where again A_0 and A_{2k+4} are real constants.

Taking the derivative $\partial U_2/\partial r$ and making use of the expansions given in Equations 6.42 and 6.43 we obtain, for $z = r_0 \exp(i\theta)$,

$$
\frac{\partial U_2}{\partial r}\bigg|_{r=r_0} = \left[A_0\left(1 + \frac{1}{\pi r_0^2}\right) + \sum_{k=1}^{\infty} A_{2k+2} r_0^{2k+2} r_{0,k}\right]\sin\theta
$$

$$
+ \sum_{j=1}^{\infty}(-1)^j\left[A_0\frac{1}{\pi}r_0^{2j}r_{j,0} + A_{2j+2}\right.
$$

$$
\left. + \sum_{k=1}^{\infty} A_{2k+2} r_0^{2k+2} r_{j,k}\right]\sin(2j+1)\theta \tag{6.51}
$$

This is now substituted into the boundary condition given in Equation 6.39 to give the system given by Equations 6.45 and 6.46 for determining the constants A_0 and A_{2k+2}; thus

$$
U_2 = -\text{Re}\left\{ i\left[A_0 + \sum_{k=1}^{\infty} A_{2k+2} r_0^{2k+2} \frac{r_{1,k-1}}{k(2k+1)} \right] \mathcal{Z} \right.
$$

$$
+ i \sum_{k=1}^{\infty} (-1)^j \left[\frac{A_0 r_{j,0}}{\pi(2j+1)} + (2j+2) \sum_{k=1}^{\infty} A_{2k+2} \frac{r_0^{2k+2} r_{j+1,k-1}}{2k(2k+1)} \right] \mathcal{Z}^{2j+1}
$$

$$
\left. + A_0 \frac{i}{\pi \mathcal{Z}} + i \sum_{k=1}^{\infty} A_{2k+2} \frac{r_0^{2k+2} (-1)^k}{(2k+1)} \frac{1}{\mathcal{Z}^{2k+1}} \right\} \tag{6.52}
$$

Using the expansions given in Equations 6.49 and 6.52 along with Formula 4.38, the effective coefficients involved in the averaged Equation 4.37 are

$$
q_{11} = \frac{1}{(1 - \pi r_0^2)} \int_{\mathcal{Y}} \left(1 - \frac{\partial U_1}{\partial y_1} \right) dy_1 \, dy_2
$$

$$
= 1 + \frac{r_0}{(1 - \pi r_0^2)} \int_0^{2\pi} U_1(r_0, \theta) \cos \theta \, d\theta
$$

$$
q_{22} = \frac{1}{(1 - \pi r_0^2)} \int_{\mathcal{Y}} \left(1 - \frac{\partial U_2}{\partial y_2} \right) dy_1 \, dy_2 \tag{6.53}
$$

$$
= 1 + \frac{r_0}{(1 - \pi r_0^2)} \int_0^{2\pi} U_2(r_0, \theta) \sin \theta \, d\theta
$$

$$
q_{12} = q_{21} = 0
$$

or, substituting Equations 6.49 and 6.52 into Equations 6.53,

$$
q_{11} = q_{22} = 1 - A_0 + \frac{\pi r_0^2}{(1 - \pi r_0^2)} \sum_{k=1}^{\infty} A_{2k+2} r_0^{2k+2} \frac{r_{1,k-1}}{k(2k+1)} \tag{6.54}
$$

$$
q_{12} = q_{21} = 0
$$

The equality $q_{11} = q_{22}$ is a clear consequence of the symmetry of the unit cell with respect to the y_1 and y_2 axes.

3

Elasticity of Regular Composite Structures

The asymptotic methods developed for boundary value problems described by partial differential equations with rapidly oscillating coefficients are relevant to the study of the elastic behavior of composite materials with a regular structure.

The method of homogenization outlined in Section 6 (Chapter 2) was introduced into classical linear elasticity theory in the 1970s by the pioneering work of Duvaut and co-workers (Duvaut 1976, 1977, Duvaut and Metellus 1976, Artola and Duvaut 1977) and Pobedrya and Gorbachev (1977). It has been shown (Duvaut 1976, Sanchez-Palencia 1980) that for a highly heterogeneous (composite) medium with a periodic structure, the solution of the original linear elasticity problem converges to that of the homogenized problem in the limit $\varepsilon \to 0$, ε being the (relative) dimension of the periodicity cell of the system. Useful estimates for the remainder of the asymptotic ε-expansions of the theory of elasticity have been given by Bakhvalov and Panasenko (1984); further work in this direction has been reviewed by Kalamkarov et al. (1987a) and Bakhvalov et al. (1988).

7. HOMOGENIZATION OF THE LINEAR ELASTICITY PROBLEM

Consider an anisotropic elastic body of regular structure occupying a bounded region Ω in space R^3 with a smooth boundary

$$\partial \Omega = \partial_1 \Omega \cup \partial_2 \Omega \qquad (\partial_1 \Omega \cap \partial_2 \Omega = 0)$$

We assume that the region Ω is made up by the periodical repetition of the unit cell Y in the form of a parallelepiped with dimensions εY_i

91

($i = 1, 2, 3$). The equations of motion and the boundary conditions of the linear elasticity problem of interest may be written in the form

$$\frac{\partial \sigma_{ij}^{(\varepsilon)}}{\partial x_j} + P_i = \rho^{(\varepsilon)} \frac{\partial^2 u_i}{\partial t^2} \quad \text{in } \Omega \tag{7.1}$$

$$u_i^{(\varepsilon)} = 0 \quad \text{on } \partial_1 \Omega$$
$$\sigma_{ij}^{(\varepsilon)} n_j = 0 \quad \text{on } \partial_2 \Omega \tag{7.2}$$

$$u_i^{(\varepsilon)} = \frac{\partial u_i^{(\varepsilon)}}{\partial t} = 0 \quad \text{at } t = 0 \tag{7.3}$$

In Equations 7.1 through 7.3,

$$\sigma_{ij}^{(\varepsilon)} = c_{ijkl}\left(\frac{x}{\varepsilon}\right) e_{kl}^{(\varepsilon)}$$

$$e_{kl}^{(\varepsilon)} = \frac{1}{2}\left(\frac{\partial u_k^{(\varepsilon)}}{\partial x_l} + \frac{\partial u_l^{(\varepsilon)}}{\partial x_k}\right)$$

the quantities P_i are independent of ε and represent the imposed body forces and $\rho^{(\varepsilon)} = \rho(y)$ is the mass density function of the material. The coefficients $c_{ijkl}(y)$ satisfying the symmetry conditions

$$c_{ijkl}(y) = c_{klij}(y) = c_{jikl}(y) \tag{7.4}$$

are positive-definite Y-periodic functions of $y = x/\varepsilon$ that are defined on the unit cell and may be piecewise-constant in practical applications. This is illustrated in Figure 7.1, where the unit cell consists of two dissimilar materials (one for the matrix and the other for fibers) and the functions $c_{ijkl}(y)$ are defined by

$$c_{ijkl}(y) = \begin{cases} c_{ijkl}^{(F)} & \text{in } Y_F \\ c_{ijkl}^{(M)} & \text{in } Y_m \end{cases} \tag{7.5}$$

with c_{ijkl}^F and c_{ijkl}^M constant. Note also that at the interface between the matrix and fiber materials, the conditions

$$\left[\sigma_{ij}^{(\varepsilon)} n_j\right] = 0 \quad \text{and} \quad \left[u_i^{(\varepsilon)}\right] = 0 \tag{7.6}$$

must be satisfied, where $[A]$ means the jump in the value of A.

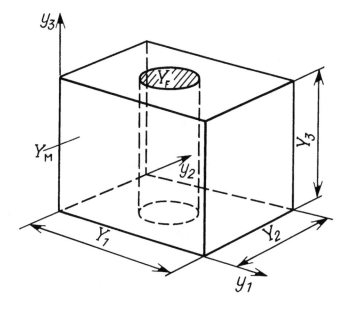

FIGURE 7.1. Unit cell of a fiber composite.

In the spirit of the two-scale expansion method and following Sanchez-Palencia (1980) and Bakhvalov and Panasenko (1984), we begin by writing the series representations

$$u_i^{(\varepsilon)} = u_i^{(0)}(x,t) + \varepsilon u_i^{(1)}(x,y,t) + \cdots \tag{7.7}$$

$$e_{ij}^{(\varepsilon)} = e_{ij}^{(0)}(x,y,t) + \varepsilon e_{ij}^{(1)}(x,y,t) + \cdots \tag{7.8}$$

$$\sigma_{ij}^{(\varepsilon)} = \sigma_{ij}^{(0)}(x,y,t) + \varepsilon \sigma_{ij}^{(1)}(x,y,t) + \cdots \tag{7.9}$$

for the solutions of the problem, where

$$e_{ij}^{(0)}(x,y,t) = \frac{1}{2}\left(\frac{\partial u_i^{(0)}}{\partial x_j} + \frac{\partial u_j^{(0)}}{\partial x_i} \right) + \frac{1}{2}\left(\frac{\partial u_i^{(1)}}{\partial y_j} + \frac{\partial u_j^{(1)}}{\partial y_i} \right)$$

$$e_{ij}^{(1)}(x,y,t) = \frac{1}{2}\left(\frac{\partial u_i^{(1)}}{\partial x_j} + \frac{\partial u_j^{(1)}}{\partial x_i} \right) + \frac{1}{2}\left(\frac{\partial u_i^{(2)}}{\partial y_j} + \frac{\partial u_j^{(2)}}{\partial y_i} \right) \tag{7.10}$$

and

$$\sigma_{ij}^{(0)}(x, y, t) = c_{ijkh}(y)\frac{\partial u_k^{(0)}}{\partial x_h} + c_{ijkh}(y)\frac{\partial u_k^{(1)}}{\partial y_h}$$

(7.11)

$$\sigma_{ij}^{(1)}(x, y, t) = c_{ijkh}(y)\frac{\partial u_k^{(1)}}{\partial x_h} + c_{ijkh}(y)\frac{\partial u_k^{(2)}}{\partial y_h}$$

Using Equations 7.7 and 7.9 in Equation 7.1 and retaining the terms $O(\varepsilon^{-1})$ and $O(\varepsilon^0)$, we find

$$\frac{\partial \sigma_{ij}^{(0)}}{\partial y_j} = 0$$

(7.12)

$$\frac{\partial \sigma_{ij}^{(0)}}{\partial x_j} + \frac{\partial \sigma_{ij}^{(1)}}{\partial y_j} + P_i = \rho^{(\varepsilon)}\frac{\partial^2 u_i^{(0)}}{\partial t^2}$$

(7.13)

Equation 7.11 is now substituted into Equation 7.12, giving

$$\frac{\partial}{\partial y_j}\left(c_{ijkh}(y)\frac{\partial u_k^{(1)}(x, y, t)}{\partial y_h}\right) = -\frac{\partial u_k^{(0)}(x, t)}{\partial x_l}\frac{\partial c_{ijkl}(y)}{\partial y_j}$$

(7.14)

Now if we consider x and t as parameters in this last equation, then, writing

$$u_n^{(1)}(x, y, t) = N_n^{kl}(y)\frac{\partial u_k^{(0)}(x, t)}{\partial x_l}$$

(7.15)

we obtain the following local problems for determining the Y-periodic functions $N_n^{kl}(y)$ $(n, k, l = 1, 2, 3)$:

$$\frac{\partial}{\partial y_j}\left(c_{ijnh}(y)\frac{\partial N_n^{kl}(y)}{\partial y_h}\right) = -\frac{\partial c_{ijkl}(y)}{\partial y_j}$$

(7.16)

Introducing the notation

$$\tau_{ij}^{kl}(y) = c_{ijnh}(y)\frac{\partial N_n^{kl}(y)}{\partial y_h}$$

(7.17)

Equation 7.16 can be written in the form

$$\frac{\partial \tau_{ij}^{kl}(y)}{\partial y_j} = - \frac{\partial c_{ijkl}(y)}{\partial y_j} \tag{7.18}$$

which is familiar from the ordinary theory of elasticity.
  Upon substituting Equation 7.15 we find

$$\sigma_{ij}^{(0)}(x, y, t) = \left(c_{ijkl}(y) + c_{ijnh} \frac{\partial N_n^{kl}(y)}{\partial y_n} \right) \frac{\partial u_k^{(0)}(x, t)}{\partial x_l}$$

$$= \left(c_{ijkl}(y) + \tau_{ij}^{kl}(y) \right) \frac{\partial u_k^{(0)}(x, t)}{\partial x_l} \tag{7.19}$$

$$\sigma_{ij}^{(1)}(x, y, t) = c_{ijnh}(y) N_n^{kl}(y) \frac{\partial^2 u_k^{(0)}(x, t)}{\partial x_l \, \partial x_h} + c_{ijkh}(y) \frac{\partial u_k^{(2)}}{\partial y_h} \tag{7.20}$$

  The equations of motion (equilibrium) of the homogenized body can now be derived from Equation 7.13 which, taken together with Equation 7.20, may be considered as a system of equations for determining the functions $u_k^{(2)}(x, y, t)$ in Y. Because the functions $\sigma_{ij}^{(1)}$ are Y-periodic in y, it follows that the equations

$$\frac{\partial \sigma_{ij}^{(1)}}{\partial y_j} = \rho^{(\varepsilon)} \frac{\partial^2 u_i^{(0)}}{\partial t^2} - P_i - \frac{\partial \sigma_{ij}^{(0)}}{\partial x_j} \tag{7.21}$$

will have a unique solution if

$$\frac{1}{|Y|} \int_Y \left[\rho^{(\varepsilon)} \frac{\partial^2 u_i^{(0)}}{\partial t^2} - P_i - \frac{\partial \sigma_{ij}^{(0)}}{\partial x_j} \right] dy = 0 \tag{7.22}$$

where the integral is performed over the volume $|Y|$ of the unit cell of the problem. Note that Equation 7.22 is a consequence of the relation

$$\int_Y \frac{\partial \sigma_{ij}^{(1)}}{\partial y_j} \, dy = \int_{\partial Y} \sigma_{ij}^{(1)} n_j \, dS_J = 0 \tag{7.23}$$

where n_j is the unit vector in the outward normal direction to the surface ∂Y, and the integral has opposite signs at the opposite sides of this surface.

The volume-averaging procedure applied to the right-hand side of Equation 7.21 now leads to the following macroscopic equations of motion:

$$\frac{\partial \langle \sigma_{ij}^{(0)} \rangle}{\partial x_j} + P_i = \langle \rho \rangle \frac{\partial^2 u_i^{(0)}(x,t)}{\partial t^2} \qquad (7.24)$$

where

$$\langle \sigma_{ij}^{(0)} \rangle = \frac{1}{|Y|} \int_Y \sigma_{ij}^{(0)} \, dy = \langle c_{ijkl} \rangle \frac{\partial u_k^{(0)}(x,t)}{\partial x_l} \qquad (7.25)$$

and

$$\langle c_{ijkl} \rangle = \frac{1}{|Y|} \int_Y \left(c_{ijkl}(y) + \tau_{ij}^{kl}(y) \right) dy \qquad (7.26)$$

are the effective elastic constants of the composite.

This method for treating elasticity theory problems corresponds to the so-called zeroth-order approximation of Pobedrya (1983a, 1983b, 1984, 1985) in which one starts by solving the problem given by Equation 7.16 for the local functions $N_n^{kl}(y)$ and then proceeds to calculate the effective moduli given by Equation 7.26. This is followed by the application of the effective moduli method to the homogenized problem, that is, the solution of Equations 7.24 through 7.26 satisfying the conditions

$$u_i^{(0)} = 0 \quad \text{on } \partial_i \Omega$$

$$\qquad (7.27)$$

$$\sigma_{ij}^{(0)} n_j = 0 \quad \text{on } \partial_2 \Omega$$

is found, and then the stress tensor components are calculated using Formula 7.19. We thus see that we have only to solve the homogenized problem given by Equations 7.24 through 7.27 for approximately predicting the microstresses and microdisplacements in the unit cell. If the number of the cells is sufficiently large in the system, Formula 7.19 estimates quite closely the state of stress in an individual cell of the composite; for laminated elastic composites, this fact has been verified by comparison with exact results due to Gorbachev (1979).

In his 1985 review Pobedrya discusses in considerable detail the application of his theory to the mechanics of a physically nonlinear, nonhomoge-

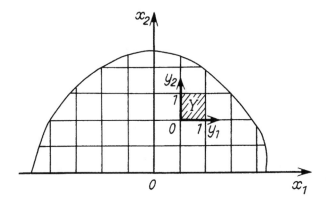

FIGURE 7.2. Half-plane with a periodic inhomogeneity and its unit cell.

neous deformable solid with a regular structure. The mechanical behavior of the solid is examined in terms of displacement and stress variables, and the relationship between the stress and (infinitesimal) strain tensors is taken in the very general operator form

$$\sigma_{ij} = F_{ij}(x, \text{grad } \mathbf{u}) \tag{7.28}$$

in which $\mathbf{u}$ is the displacement vector, and the presence of the coordinate x reflects the nonhomogeneity of the medium.

We now propose to discuss what modifications must be introduced into the general homogenization scheme when boundary effects are or may be important (Bakhvalov and Panasenko 1984, Sanchez-Palencia 1987). Consider a two-dimensional elasticity problem for a half-plane with a periodic inhomogeneity and with specified stresses at the boundary (cf. Figure 7.2). The equations of equilibrium are taken in the form

$$\frac{\partial \sigma_{i\alpha}^{(\varepsilon)}}{\partial x_\alpha} = 0 \qquad (|x_1| < \infty, \, x_2 > 0) \tag{7.29}$$

and the appropriate elasticity relations are written as

$$\sigma_{i\alpha}^{(\varepsilon)}(x) = c_{i\alpha k\beta}(y) \frac{\partial u_k^{(\varepsilon)}(x)}{\partial x_\beta} \tag{7.30}$$

with functions $c_{i\alpha k\beta}(y)$ being 1-periodic in the variables

$$y = x/\varepsilon \qquad x = (x_1, x_2) \qquad (\alpha, \beta = 1, 2, i, k = 1, 2, 3)$$

The stress boundary conditions are stated by

$$\sigma_{i2}(x_1, 0) = p_i(x_1) \tag{7.31}$$

To get started we represent the displacements and stress fields by expansions of the form

$$u_k^{(\varepsilon)} = u_k^{(0)}(x) + \varepsilon u_k^{(1)}(x, y) + \cdots \tag{7.32}$$

$$\sigma_{i\alpha}^{(\varepsilon)} = \sigma_{i\alpha}^{(0)}(x, y) + \varepsilon \sigma_{i\alpha}^{(1)}(x, y) + \cdots \tag{7.33}$$

where

$$\sigma_{i\alpha}^{(0)}(x, y) = c_{i\alpha k\beta}(y) \left(\frac{\partial u_k^{(0)}(x)}{\partial x_\beta} + \frac{\partial u_k^{(1)}(x, y)}{\partial y_\beta} \right) \tag{7.34}$$

$$\sigma_{i\alpha}^{(1)}(x, y) = c_{i\alpha k\beta}(y) \left(\frac{\partial u_k^{(1)}(x, y)}{\partial x_\beta} + \frac{\partial u_k^{(2)}(x, y)}{\partial y_\beta} \right) \tag{7.35}$$

Using Equation 7.33 in Equations 7.29 and 7.31 yields the following set of problems:

$$\frac{\partial \sigma_{i\alpha}^{(0)}(x, y)}{\partial y_\alpha} = 0 \tag{7.36}$$

$$\sigma_{i2}^{(0)}\big|_{x_2=0} = p_i(x_1) \tag{7.37}$$

and

$$\frac{\partial \sigma_{i\alpha}^{(0)}(x, y)}{\partial x_\alpha} + \frac{\partial \sigma_{i\alpha}^{(1)}(x, y)}{\partial y_\alpha} = 0 \tag{7.38}$$

$$\sigma_{i2}^{(1)}\big|_{x_2=0} = 0 \tag{7.39}$$

Combining Equations 7.34 and 7.35 gives

$$\frac{\partial}{\partial y_\alpha}\left[c_{i\alpha k\beta}(y)\frac{\partial u_k^{(1)}(x,y)}{\partial y_\beta}\right] = -\frac{\partial c_{i\alpha k\beta}(y)}{\partial y_\alpha}\frac{\partial u_k^{(0)}(x)}{\partial x_\beta} \tag{7.40}$$

$$c_{i2k\beta}(y)\Big|_{y_2=0}\frac{\partial u_k^{(1)}(x,y)}{\partial y_\beta}\Big|_{y_2=0} = -c_{i2k\beta}(y)\Big|_{y_2=0}\frac{\partial u_k^{(0)}(x)}{\partial x_\beta}\Big|_{x_2=0} + p_i(x_1) \tag{7.41}$$

and

$$\frac{\partial}{\partial y_\alpha}\left[c_{i\alpha k\beta}(y)\frac{\partial u_k^{(2)}(x,y)}{\partial y_\beta}\right]$$

$$= -\frac{\partial c_{i\alpha k\beta}(y)}{\partial y_\alpha}\frac{\partial u_k^{(1)}(x,y)}{\partial x_\beta} - c_{i\alpha k\beta}(y)$$

$$\times\left[\frac{\partial^2 u_k^{(1)}(x,y)}{\partial y_\beta\,\partial x_\alpha} + \frac{\partial^2 u_k^{(1)}(x,y)}{\partial y_\alpha\,\partial x_\beta} + \frac{\partial^2 u_k^{(0)}(x)}{\partial x_\alpha\,\partial x_\beta}\right] \tag{7.42}$$

$$c_{i2k\beta}(y)\Big|_{y_2=0}\frac{\partial u_k^{(2)}(x,y)}{\partial y_\beta}\Big|_{y_2=0} = -c_{i2k\beta}(y)\Big|_{y_2=0}\frac{\partial u_k^{(1)}(x,y)}{\partial x_\beta} \tag{7.43}$$

A solution for the problem given by Equations 7.40 and 7.41 is assumed here in the form

$$u_n^{(1)}(x,y) = N_n^{k\beta}(y)\frac{\partial u_k^{(0)}(x)}{\partial x_\beta} + u_n^{(1,2)}(x,y) \tag{7.44}$$

where the function $N_n^{k\beta}(y)$ is 1-periodic in both y_1 and y_2 and is to be determined from the local problem

$$\frac{\partial}{\partial y_\alpha}\left[c_{i\alpha n\gamma}(y)\frac{\partial N_n^{k\beta}(y)}{\partial y_\gamma}\right] = -\frac{\partial c_{i\alpha k\beta}(y)}{\partial y_\alpha} \tag{7.45}$$

on the periodicity cell, whereas $u_n^{(1,2)}(x, y)$ is 1-periodic only in y_1 and is solved from the problem

$$\frac{\partial}{\partial y_\alpha}\left[c_{i\alpha k\beta}(y)\frac{\partial u_k^{(1,2)}(x, y)}{\partial y_\beta}\right] = 0 \qquad (0 < y_1 < 1, y_2 > 0) \quad (7.46)$$

$$c_{i2k\beta}(y)\frac{\partial u_k^{(1,2)}(x, y)}{\partial y_\beta}\bigg|_{y_2=0}$$

$$= -\left[c_{i2k\beta}(y) + c_{i2n\gamma}(y)\frac{\partial N_n^{k\beta}(y)}{\partial y_\gamma}\right]\bigg|_{y_2=0}\frac{\partial u_k^{(0)}}{\partial x_\beta}\bigg|_{x_2=0} + p_i(x_1) \quad (7.47)$$

$$u_k^{(1,2)}(x, y) \to 0 \quad \text{as} \quad y_2 \to +\infty \qquad (7.48)$$

Integrating Equation 7.46 over the strip $0 < y_1 < 1$, $y_2 > 0$, using Equations 7.47 and 7.48, and recalling the periodicity of $u_k^{(1,2)}(x, y)$ in y_1, the following solvability condition for the problem given by Equations 7.46 through 7.48 is obtained:

$$C_{i2k\beta}^*\frac{\partial u_k^{(0)}(x)}{\partial x_\beta}\bigg|_{x_2=0} = p_i(x_1) \qquad (7.49)$$

$$C_{i2k\beta}^* = \int_0^1 C_{i2k\beta}(y_1, 0)\, dy_1 \qquad (7.50)$$

$$C_{i2k\beta}(y_1, 0) = \left[c_{i2k\beta}(y) + c_{i2n\gamma}(y)\frac{\partial N_n^{k\beta}(y)}{\partial y_\gamma}\right]\bigg|_{y_2=0} \qquad (7.51)$$

We next employ Equation 7.49 to eliminate $p_i(x)$ from Equation 7.47 and represent $u_k^{(1,2)}(x, y)$ in the form

$$u_n^{(1,2)}(x, y) = N_n^{(1)k\beta}(y)\frac{\partial u_k^{(0)}(x)}{\partial x_\beta} \qquad (7.52)$$

where the function $N_n^{(1)k\beta}(y)$ is 1-periodic in y_1 and satisfies the equation

$$\frac{\partial}{\partial y_\alpha}\left[c_{i\alpha n\gamma}(y)\frac{\partial N_n^{(1)k\beta}(y)}{\partial y_\gamma}\right] = 0 \qquad (0 < y_1 < 1, y_2 > 0) \quad (7.53)$$

and the conditions

$$c_{i2n\gamma}(y)\Big|_{y_2=0} \frac{\partial N_n^{(1)k\beta}(y)}{\partial y_\gamma}\Bigg|_{y_2=0} = \left[C_{i2k\beta}^* - C_{i2k\beta}(y_1, 0)\right] \quad (7.54)$$

$$N_n^{(1)k\beta}(y) \to 0 \quad \text{as } y_2 \to \infty \quad (7.55)$$

Substituting Equations 7.44 and 7.52 into Equations 7.38 and 7.43 results in the following problem for determining $u_k^{(2)}(x, y)$:

$$\frac{\partial}{\partial y_\alpha}\left[c_{i\alpha k\beta}(y)\frac{\partial u_k^{(2)}(x, y)}{\partial y_\beta}\right]$$

$$= -\frac{\partial}{\partial y_\alpha}\left[c_{i\alpha n\gamma}(y)N_n^{k\beta}(y)\right]\frac{\partial^2 u_k^{(0)}(x)}{\partial x_\beta\, \partial x_\gamma}$$

$$-\left[c_{i\alpha k\beta}(y) + c_{i\alpha n\gamma}(y)\frac{\partial N_n^{k\beta}(y)}{\partial y_\gamma}\right]\frac{\partial^2 u_k^{(0)}(x)}{\partial x_\alpha\, \partial x_\beta}$$

$$-\frac{\partial}{\partial y_\alpha}\left[c_{i\alpha n\gamma}(y)N_n^{(1)k\beta}(y)\right]\frac{\partial^2 u_k^{(0)}(x)}{\partial x_\beta\, \partial x_\gamma}$$

$$- c_{i\gamma n\alpha}(y)\frac{\partial N_n^{(1)k\beta}(y)}{\partial y_\gamma}\frac{\partial^2 u_k^{(0)}(x)}{\partial x_\alpha\, \partial x_\beta} \quad (7.56)$$

$$c_{i2k\beta}(y)\Big|_{y_2=0} \frac{\partial u_k^{(2)}(x, y)}{\partial y_\beta}\Bigg|_{y_2=0}$$

$$= -c_{i2n\gamma}(y)\Big|_{y_2=0}\left[N_n^{k\beta}(y) + N_n^{(1)k\beta}(y)\right]_{y_2=0}\frac{\partial^2 u_k^{(0)}(x)}{\partial x_\beta\, \partial x_\gamma}\Bigg|_{x_2=0} \quad (7.57)$$

By analogy with Equation 7.44, the solution of Equation 7.56 is taken in the form

$$u_k^{(2)}(x, y) = u_k^{(2, 1)}(x, y) + u_k^{(2, 2)}(x, y) \quad (7.58)$$

where the function $u_k^{(2,1)}(x, y)$ is 1-periodic in y_1 and y_2 and solves the local problem

$$\frac{\partial}{\partial y_\alpha}\left[c_{i\alpha k\beta}(y)\frac{\partial u_k^{(2,1)}(x,y)}{\partial y_\beta}\right] = -\frac{\partial}{\partial y_\alpha}\left[c_{i\alpha n\gamma}(y)N_n^{k\beta}(y)\right]\frac{\partial^2 u_k^{(0)}(x)}{\partial x_\beta \partial x_\gamma}$$

$$-\left[c_{i\alpha n\gamma}(y)\frac{\partial N_n^{k\beta}(y)}{\partial y_\gamma}\right.$$

$$\left.+c_{i\alpha k\beta}(y)\right]\frac{\partial^2 u_k^{(0)}(x)}{\partial x_\alpha \partial x_\beta}\qquad(7.59)$$

whereas the function $u_k^{(2,2)}(x, y)$ is periodic only in y_1 and satisfies (1) Equation 7.59 in the region $0 < y_1 < 1$, $y_2 > 0$ (with $N_n^{k\beta}$ replaced by $N_n^{(1)k\beta}$) and (2) the conditions

$$c_{i2k\beta}(y)\frac{\partial u_k^{(2,2)}(x,y)}{\partial y_\beta}\bigg|_{y_2=0} = -c_{i2k\beta}(y)\frac{\partial u_k^{(2,1)}(x,y)}{\partial y_\beta}\bigg|_{y_2=0}$$

$$-c_{i2n\gamma}(y)\big|_{y_2=0}(N_n^{k\beta}(y)$$

$$+N_n^{(1)k\beta}(y))\big|_{y_2=0}\frac{\partial^2 u_k^{(0)}}{\partial x_\beta \partial x_\gamma}\bigg|_{x_2=0}\qquad(7.60)$$

$$u_k^{(2,2)}(x, y) \to 0 \quad \text{as } y_2 \to \infty \qquad(7.61)$$

The solvability condition for the local problem given by Equation 7.59 now gives the following averaged equation for $u_k^{(0)}(x)$ in the half-plane $x_2 > 0$:

$$\langle c_{i\alpha k\beta}\rangle \frac{\partial^2 u_k^{(0)}(x)}{\partial x_\alpha \partial x_\beta} = 0 \qquad(7.62)$$

where the quantities

$$\langle c_{i\alpha k\beta}\rangle = \int_Y\left[c_{i\alpha k\beta}(y) + c_{i\alpha n\gamma}(y)\frac{\partial N_n^{k\beta}(y)}{\partial y_\gamma}\right]dy_1\,dy_2 \qquad(7.63)$$

are the effective elastic moduli.

We thus conclude that, in the framework of the effective moduli method, the foregoing half-plane elasticity problem reduces to that of solving Equations 7.62 subjected to the initial conditions given by Equation 7.49. To perform this solution, one only needs to calculate the local functions $N_n^{k\beta}(y)$ from Equation 7.45, but to determine the zeroth-order displacement and stress fields of the problem, the local functions $N_n^{(1)k\beta}(y)$ must be found from the system given by Equations 7.53 through 7.55. The displacements and stresses are then calculated from

$$u_n = u_n^{(0)}(x) + \varepsilon(N_n^{k\beta}(y) + N_n^{(1)k\beta}(y))\frac{\partial u_k^{(0)}(x)}{\partial x_\beta} + \cdots \quad (7.64)$$

$$\sigma_{i\alpha} = \left[c_{i\alpha k\beta}(y) + c_{i\alpha n\gamma}(y)\left(\frac{\partial N_n^{k\beta}(y)}{\partial y_\gamma} + \frac{\partial N_n^{(1)k\beta}(y)}{\partial y_\gamma} \right) \right] \frac{\partial u_k^{(0)}(x)}{\partial x_\beta} + \cdots$$

$$(7.65)$$

The foregoing discussion goes through equally well if, instead of stresses, one considers material displacements as being specified at the boundary $x_2 = 0$. So if we write

$$u_i^{(\varepsilon)}\big|_{x_2=0} = v_i(x_1) \quad (7.66)$$

the homogenized problem reduces to the solution in the plane $x_2 > 0$ of Equation 7.62 subjected to the condition

$$u_i^{(0)}\big|_{x_2=0} = v_i(x_1) \quad (7.67)$$

The stress and displacement fields will again be determined by Equations 7.64 and 7.65, but the boundary layer type functions $N_n^{(2)k\beta}(y)$ will have to be found from the problem

$$\frac{\partial}{\partial y_\alpha}\left[c_{i\alpha n\gamma}(y)\frac{\partial N_n^{(2)k\beta}(y)}{\partial y_\gamma} \right] = 0 \quad (0 < y_1 < 1, y_2 > 0) \quad (7.68)$$

$$N_n^{(2)k\beta}(y)\big|_{y_2=0} = -N_n^{k\beta}(y)\big|_{y_2=0} + h_n^{(2)k\beta} \quad (7.69)$$

$$N_n^{(2)k\beta} \to 0 \quad \text{as } y_2 \to \infty \quad (7.70)$$

and must be 1-periodic in y_1.

As shown by Panasenko (1979) (see also Bakhvalov and Panasenko 1984) the constants $h_n^{(2)k\beta}$ in Equation 7.69 can be determined in a unique manner from the solvability condition for the problem given by Equations 7.68 through 7.70. To this aim, one first solves the problem given by Equation 7.68 under the condition

$$\tilde{N}_n^{(2)k\beta}(y)\big|_{y_2=0} = -N_n^{k\beta}(y)\big|_{y_2=0} \tag{7.71}$$

in the class of functions that are bounded and 1-periodic in y_1. This problem is solved in a unique manner (see Bakhvalov and Panasenko 1984), and the constants $\tilde{h}_n^{(2)k\beta}$ have the property

$$\lim_{y_2 \to \infty} \tilde{N}_n^{(2)k\beta}(y) = \tilde{h}_n^{(2)k\beta} \tag{7.72}$$

so that, setting

$$h_n^{(2)k\beta} = -\tilde{h}_n^{(2)k\beta} \tag{7.73}$$

it can be seen that

$$N_n^{(2)k\beta} = \tilde{N}_n^{(2)k\beta} - \tilde{h}_n^{(2)k\beta} \tag{7.74}$$

is a unique solution to the problem given by Equations 7.68 through 7.70. We note also that because of the differentiation involved the constants $h_n^{(2)k\beta}$ will be absent from the resulting local stress formula, Equation 7.65.

If the boundary conditions are of mixed type—for example, if

$$\sigma_{12}\big|_{x_2=0} = p_l(x_1) \qquad (l = 1, 3) \tag{7.75}$$

$$u_2\big|_{x_2=0} = v_2(x_1)$$

—then the homogenized problem reduces to that of solving Equation 7.62 in the half-plane $x_2 > 0$ under the conditions

$$C_{l2k\beta}^* \frac{\partial u_k^{(0)}}{\partial x_\beta}\bigg|_{x_2=0} = p_l(x_1) \qquad (l = 1, 3) \tag{7.76}$$

$$u_2^{(0)}\big|_{x_2=0} = v_2(x_1)$$

and the local functions $N_n^{(3)k\beta}(y)$ entering the displacement and stress expressions (Equations 7.64 and 7.65, respectively) are 1-periodic in y_1 and must be found from the problem

$$\frac{\partial}{\partial y_\alpha}\left[c_{i\alpha n\gamma}(y)\frac{\partial N_n^{(3)k\beta}(y)}{\partial y_\gamma}\right] = 0 \qquad (0 < y_1 < 1, \, y_2 = 0) \quad (7.77)$$

$$c_{l2n\gamma}\frac{\partial N_n^{(3)k\beta}(y)}{\partial y_\gamma}\Bigg|_{y_2=0} = C_{l2k\beta}^* - C_{l2k\beta}(y_1,0) \qquad (l = 1,3) \quad (7.78)$$

$$N_2^{(3)k\beta}(y)\big|_{y_2=0} = -N_2^{k\beta}(y)\big|_{y_2=0} + h_2^{(3)k\beta} \qquad (7.79)$$

$$N_h^{(3)k\beta}(y) \to 0 \qquad \text{as } y_2 \to \infty \qquad (7.80)$$

The constants $h_2^{(3)k\beta}$ in Equation 7.79 are determined using the same procedure followed in the problem given by Equations 7.68 through 7.70 and are absent again in the final expressions for the local stress distribution.

A consistent homogenization analysis of the static, half-plane, boundary value elasticity problem was given by Iosifyan et al. (1982). The authors obtained a complete asymptotic expansion in powers of ε and were able to show its validity by properly estimating the remainder of the expansion.

8. LAMINATED COMPOSITES: EFFECTIVE PROPERTIES AND FRACTURE CRITERIA

We have seen in Section 7 that to calculate the effective moduli $\langle c_{ijkl}\rangle$ within the framework of the zeroth-order approximation requires that the local functions $N_n^{kl}(y)$ be determined from the local problem given by Equations 7.16 on the unit cell. This problem is solved without difficulty for the case of a laminated composite made up by nonhomogeneous layers (or laminae) periodically repeating themselves along a certain direction. If we take this direction to be along the x_3 axis as in Figure 8.1, the (true) elastic moduli c_{ijkl} will depend on $y_3 = x/\varepsilon$ alone, and Equations 7.16 take the form

$$\frac{d}{dy_3}\left(c_{i3n3}(y_3)\frac{dN_n^{kl}}{dy_3}\right) = -\frac{dc_{i3kl}(y_3)}{dy_3} \qquad (0 \le y_3 \le 1) \quad (8.1)$$

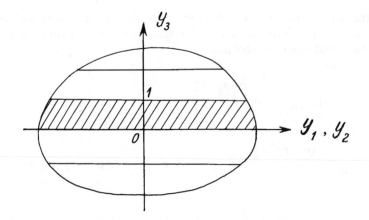

FIGURE 8.1. Layered composite material.

which by integration gives

$$c_{i3n3}(y_3)\frac{dN_n^{kl}}{dy_3} = -c_{i3kl}(y_3) + A_{ikl} \tag{8.2}$$

where the A_{ikl} are constants of integration. From Equation 8.2,

$$\frac{dN_n^{kl}}{dy_3} = -c_{n3i3}^{-1}(y_3)c_{i3kl}(y_3) + c_{n3i3}^{-1}(y_3)A_{ikl} \tag{8.3}$$

where the 3×3 matrix c_{n3i3}^{-1} is inverse to the matrix c_{i3n3}, that is, c_{k3i3}^{-1} $c_{i3n3} = \delta_{kn}$. Equation 8.3 is again integrated to give

$$\langle c_{n3i3}^{-1}\rangle A_{ikl} = \langle c_{n3q3}^{-1}c_{q3kl}\rangle \tag{8.4}$$

where

$$\langle \cdots \rangle = \int_0^1 (\cdots)\, dy_3$$

and where we have used the fact that the function $N_n^{kl}(y_3)$ is 1-periodic in y_3, that is, $N_n^{kl}(0) = N_n^{kl}(1)$.

From Equation 8.4,

$$A_{ikl} = \langle c_{i3n3}^{-1}\rangle^{-1}\langle c_{n3q3}^{-1}c_{q3kl}\rangle \tag{8.5}$$

giving

$$\tau_{ij}^{kl}(y_3) = c_{ijn3}(y_3)\frac{dN_n^{kl}}{dy_3}$$

$$= -c_{ijn3}(y_3)c_{n3p3}^{-1}(y_3)c_{p3kl}(y_3)$$

$$+ c_{ijn3}(y_3)c_{n3s3}^{-1}(y_3)\langle c_{s3p3}^{-1}\rangle^{-1}\langle c_{p3q3}^{-1}c_{q3kl}\rangle \qquad (8.6)$$

which when substituted into Equation 7.26 yields

$$\langle C_{ijkl}\rangle = \langle c_{ijkl}\rangle - \langle c_{ijn3}c_{n3p3}^{-1}c_{p3kl}\rangle$$

$$+ \langle c_{ijn3}c_{n3s3}^{-1}\rangle\langle c_{s3p3}^{-1}\rangle^{-1}\langle c_{p3q3}^{-1}c_{q3kl}\rangle \qquad (8.7)$$

for the effective moduli of the composite (Pobedrya 1984).

The use of these results may be illustrated by specifying the symmetry of the anisotropic medium within the unit cell of the composite. In the important special case of the hexagonal symmetry 6*mm*, taking axis x_3 in the direction of the symmetry axis, the medium is characterized by five moduli, $c_{11}(y_3)$, $c_{12}(y_3)$, $c_{13}(x_3)$, $c_{33}(y_3)$, and $c_{44}(y_3)$, and the set of nonzero local functions, $N_1^{13} = N_2^{23}$, $N_3^{11} = N_3^{22}$, and N_3^{33}, are found from the equations

$$\frac{dN_1^{13}}{dy_3} = -1 + \frac{A_{113}}{c_{44}(y_3)}$$

$$\frac{dN_3^{11}}{dy_3} = -\frac{c_{13}(y_3)}{c_{33}(y_3)} + \frac{A_{311}}{c_{33}(y_3)} \qquad (8.8)$$

$$\frac{dN_3^{33}}{dy_3} = -1 + \frac{A_{333}}{c_{33}(y_3)}$$

which are readily obtained from Equation 8.3. It follows by integrating Equation 8.8 that

$$A_{113} = \frac{1}{\langle c_{44}^{-1}\rangle} \qquad A_{311} = \frac{\langle c_{13}c_{33}^{-1}\rangle}{\langle c_{33}^{-1}\rangle} \qquad A_{333} = \frac{1}{\langle c_{33}^{-1}\rangle} \qquad (8.9)$$

and the effective moduli are found to be, in matrix notation,

$$\tilde{c}_{11} = \tilde{c}_{22} = \langle c_{11} \rangle - \langle c_{13}^2 c_{33}^{-1} \rangle + \frac{\langle c_{13} c_{33}^{-1} \rangle^2}{\langle c_{33}^{-1} \rangle}$$

$$\tilde{c}_{33} = \frac{1}{\langle c_{33}^{-1} \rangle} \qquad \tilde{c}_{23} = \tilde{c}_{13} = \frac{\langle c_{13} c_{33}^{-1} \rangle}{\langle c_{33}^{-1} \rangle}$$

$$\tilde{c}_{12} = \langle c_{12} \rangle - \langle c_{13}^2 c_{33}^{-1} \rangle + \frac{\langle c_{13} c_{33}^{-1} \rangle^2}{\langle c_{33}^{-1} \rangle} \qquad (8.10)$$

$$\tilde{c}_{44} = \frac{1}{\langle c_{44}^{-1} \rangle} \qquad \tilde{c}_{55} = \tilde{c}_{44}$$

$$\tilde{c}_{66} = \langle c_{66} \rangle = \frac{1}{2}[\langle c_{11} \rangle - \langle c_{12} \rangle]$$

Setting $c_{11} = c_{33}$, $c_{13} = c_{12}$, and $c_{66} = c_{44}$ in Equations 8.10 yields the effective moduli for the case when the layer material has cubic symmetry (with three independent elastic moduli). If, further, $c_{11} = c = \lambda + 2G$, $c_{13} = c_{12} = \lambda$, and $c_{66} = c_{44} = G$, the effective moduli of a composite with isotropic laminae are obtained,

$$\tilde{c}_{11} = \tilde{c}_{22} = \langle \lambda + 2G \rangle - \left\langle \frac{\lambda^2}{\lambda + 2G} \right\rangle + \left\langle \frac{\lambda}{\lambda + 2G} \right\rangle^2 \frac{1}{\langle (\lambda + 2G)^{-1} \rangle}$$

$$\tilde{c}_{33} = \frac{1}{\langle (\lambda + 2G)^{-1} \rangle} \qquad \tilde{c}_{23} = \tilde{c}_{13} = \left\langle \frac{\lambda}{\lambda + 2G} \right\rangle \frac{1}{\langle (\lambda + 2G)^{-1} \rangle}$$

$$\tilde{c}_{12} = \langle \lambda \rangle - \left\langle \frac{\lambda^2}{\lambda + 2G} \right\rangle + \left\langle \frac{\lambda}{\lambda + 2G} \right\rangle^2 \frac{1}{\langle (\lambda + 2G)^{-1} \rangle} \qquad (8.11)$$

$$\tilde{c}_{44} = \frac{1}{\langle G^{-1} \rangle} \qquad \tilde{c}_{55} = \frac{1}{\langle G^{-1} \rangle} \qquad \tilde{c}_{66} = \langle G \rangle$$

where λ and G are the Lamé constants. It is seen that a homogenized laminated composite comprising isotropic layers acts like a transversely isotropic material possessing five independent elastic moduli: c_{11}, c_{33}, c_{12}, $c_{23} = c_{13}$, $c_{44} = c_{55}$.

In the important practical case of a two-phase composite material (see Figure 8.2), for homogeneous and isotropic components, the effective moduli are given by

$$\tilde{c}_{11} = \tilde{c}_{22} = (\lambda_1 + 2G_1)\gamma + (\lambda_2 + 2G_2)(1 - \gamma)$$

$$- \left[\frac{\lambda_1^2}{\lambda_1 + 2G_1}\gamma + \frac{\lambda_2^2}{\lambda_2 + 2G_2}(1 - \gamma) \right]$$

$$+ \left[\frac{\lambda_1\gamma}{\lambda_1 + 2G_1} + \frac{\lambda_2(1 - \gamma)}{\lambda_2 + 2G_2} \right]^2 \left[(\lambda_1 + 2G_1)^{-1}\gamma \right.$$

$$\left. + (\lambda_2 + 2G_2)^{-1}(1 - \gamma) \right]^{-1}$$

$$\tilde{c}_{33} = \left[(\lambda_1 + 2G_1)^{-1}\gamma + (\lambda_2 + 2G_2)(1 - \gamma) \right]^{-1}$$

$$\tilde{c}_{13} = \tilde{c}_{23} = \left[\frac{\lambda_1}{\lambda_1 + 2G_1}\gamma + \frac{\lambda_2}{\lambda_2 + 2G_2}(1 - \gamma) \right]$$

$$\times \left[(\lambda_1 + 2G_1)^{-1}\gamma + (\lambda_2 + 2G_2)^{-1}(1 - \gamma) \right]^{-1}$$

$$\tilde{c}_{12} = [\lambda_1\gamma + \lambda_2(1 - \gamma)] - \left[\frac{\lambda_1^2}{\lambda_1 + 2G_1}\gamma + \frac{\lambda_2^2}{\lambda_2 + 2G_2}(1 - \gamma) \right]$$

$$+ \left[\frac{\lambda_1\gamma}{\lambda_1 + 2G_1} + \frac{\lambda_2(1 - \gamma)}{\lambda_2 + 2G_2} \right]^2 \left[(\lambda_1 + 2G_1)^{-1}\gamma \right.$$

$$\left. + (\lambda_2 + 2G_2)^{-1}(1 - \gamma) \right]^{-1}$$

$$\tilde{c}_{44} = \left[G_1^{-1}\gamma + G_2^{-1}(1 - \gamma) \right]^{-1} \qquad \tilde{c}_{55} = \tilde{c}_{44}$$

$$\tilde{c}_{66} = [G_1\gamma + G_2(1 - \gamma)] \tag{8.12}$$

The simple formulas we have derived for the effective moduli of composite materials make it possible to gain a useful insight into the manner in which the geometrical and mechanical properties of the individual components affect the strength properties of the composite as a whole; also, and no less important, the overall strength criterion can be deduced

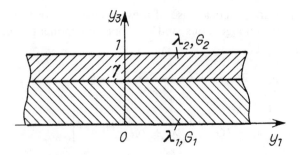

FIGURE 8.2. Unit cell of a two-component laminate composite with homogeneous isotropic components.

from the knowledge of the individual macroscopic strength criteria (Gorbachev and Pobedrya 1985).

Following Gorbachev and Pobedrya (1985), we begin by introducing the stress concentration tensor $A_{ijpq}(y)$, which relates $\langle \sigma_{ij}^{(0)} \rangle$ to the layer stresses $\sigma_{ij}^{(0)}$. Making use of Equation 7.25 we determine the averaged strain components

$$e_{kl}^0 = \frac{1}{2}\left(\frac{\partial u_k^{(0)}(x)}{\partial x_l} + \frac{\partial u_l^{(0)}(x)}{\partial x_k} \right)$$

and substitute them into Equation 7.19 to obtain

$$\sigma_{ij}(x, y) \cong \sigma_{ij}^{(0)}(x, y) = A_{ijpq}\langle \sigma_{pq}^{(0)} \rangle \tag{8.13}$$

where

$$A_{ijpq}(y) = \left(c_{ijkl}(y) + \tau_{ij}^{kl}(y) \right)\langle C_{klpq} \rangle^{-1} \tag{8.14}$$

with tensors $\langle C_{ijkl} \rangle$ and $\langle C_{klpq} \rangle^{-1}$ mutually inverse in the sense

$$\langle C_{ijkl} \rangle \langle C_{klpq} \rangle^{-1} = \tfrac{1}{2}(\delta_{ip}\delta_{jk} + \delta_{iq}\delta_{jp})$$

The stress concentration tensor components now can be rewritten with the use of Equation 8.6 giving

$$A_{ijpq}(y_3) = \left[c_{ijkl}(y_3) - c_{ijn3}(y_3)c_{n3p3}^{-1}(y_3)c_{p3kl}(y_3) \right.$$

$$\left. + c_{ijn3}(y_3)c_{n3s3}^{-1}(y_3)\langle c_{s3p3}^{-1} \rangle^{-1}\langle c_{p3q3}^{-1}c_{q3kl} \rangle \right]\langle C_{klpq} \rangle^{-1} \tag{8.15}$$

If the layer material is isotropic,

$$c_{ijkl}(y_3) = \frac{E(y_3)}{1 + \nu(y_3)} \left[\frac{1}{2}(\delta_{ik}\delta_{jl} + \delta_{il}\delta_{jk}) + \frac{\nu(y_3)}{1 - 2\nu(y_3)} \delta_{ij}\delta_{kl} \right]$$

$$(8.16)$$

where $E(y_3)$ denotes Young's modulus and $\nu(y_3)$ denotes Poisson's ratio, it is easily shown that

$$A_{i3pq} = A_{3ipq} = \tfrac{1}{2}(\delta_{ip}\delta_{3q} + \delta_{iq}\delta_{3p}) \qquad (8.17)$$

and that the remaining components are given by (Gorbachev and Pobedrya 1985)

$$A_{1111} = A_{2222} = \frac{E(y_3)}{1 - \nu^2(y_3)} \frac{\langle E/(1 - \nu^2) \rangle - \nu(y_3)\langle E\nu/(1 - \nu^2) \rangle}{\langle E/(1 - \nu) \rangle\langle E/(1 + \nu) \rangle}$$

$$A_{1122} = A_{2211} = \frac{E(y_3)}{1 - \nu^2(y_3)} \frac{\nu(y_3)\langle E/(1 - \nu^2) \rangle - \langle E\nu/(1 - \nu^2) \rangle}{\langle E/(i - \nu) \rangle\langle E/(1 + \nu) \rangle}$$

$$A_{1133} = A_{2233} = \frac{\nu(y_3)}{1 - \nu(y_3)} - \frac{E(y_3)}{1 - \nu(y_3)} \frac{\langle \nu/(1 - \nu) \rangle}{\langle E/(1 - \nu) \rangle} \qquad (8.18)$$

$$A_{1212} = \frac{1}{2} \frac{E(y_3)/(1 + \nu(y_3))}{\langle E/(1 - \nu) \rangle} = \tfrac{1}{2}(A_{1111} - A_{1122})$$

An integral fracture criterium for piezoelectric and laminated composites was received in Parton (1976), Parton and Kudryavtsev (1986) (so-called PK-criterium).

As a fracture criterion for an individual composite component, a very general form due to Goldenblat and Kopnov (1968) (see also Malmeister et al. 1980) will be used, which is

$$P_{ij}(y)\sigma_{ij} + P_{ijkl}(y)\sigma_{ij}\sigma_{kl} + \cdots = 1 \qquad (8.19)$$

where P_{ij}, $P_{ijkl} \ldots$ are the stress strength tensors of the material. Substituting Equation 8.13 and applying the mean operator, this becomes

$$P_{ij}^* \langle \sigma_{ij}^{(0)} \rangle + P_{ijkl}^* \langle \sigma_{ij}^{(0)} \rangle\langle \sigma_{kl}^{(0)} \rangle + \cdots = 1 \qquad (8.20)$$

where P_{ij}^*, P_{ijkl}^*, ... are the effective stress strength tensors defined by

$$P_{ij}^* = \langle P_{mn} A_{mnij} \rangle$$

$$P_{ijkl}^* = \langle P_{mnpq} A_{mnij} A_{pqkl} \rangle \qquad (8.21)$$

$$\vdots$$

As a special case of the individual fracture criterion we take the form

$$P_{ijkl}(y)\sigma_{ij}\sigma_{kl} = 1 \qquad (8.22)$$

which expresses the well-known shape variation energy criterion and in which

$$P_{ijkl}(y) = \frac{1}{2\tau_b^2(y_3)}\left[\frac{1}{2}(\delta_{ik}\delta_{jl} + \delta_{il}\delta_{jk}) - \frac{1}{3}\delta_{ij}\delta_{kl}\right] \qquad (8.23)$$

with τ_b denoting the shear strength of the material. The nonzero strength tensor components P_{ijkl}^* as given by Gorbachev and Pobedrya (1985) are

$$P_{1111}^* = P_{2222}^* = \frac{1}{6}\left\langle \frac{1}{\tau_b^2}(A_{1111} - A_{1122})^2 + \frac{1}{\tau_b^2}A_{1111}^2 + \frac{1}{\tau_b^2}A_{1122}^2 \right\rangle$$

$$P_{1133}^* = P_{2233}^* = -\frac{1}{6}\left\langle \frac{1}{\tau_b^2}(1 - A_{1133})(A_{1111} + A_{1122}) \right\rangle$$

$$P_{1122}^* = -\frac{1}{6}\left\langle \frac{1}{\tau_b^2}(A_{1111} - A_{1122})^2 - \frac{2}{\tau_b^2}A_{1111}A_{1122} \right\rangle$$

$$\qquad\qquad\qquad\qquad\qquad\qquad\qquad\qquad (8.24)$$

$$P_{3333}^* = \frac{1}{3}\left\langle \frac{1}{\tau_b^2}(1 - A_{1133})^2 \right\rangle$$

$$P_{1313}^* = P_{2323}^* = \frac{1}{4}\left\langle \frac{1}{\tau_b^2} \right\rangle$$

$$P_{1212}^* = \left\langle \frac{1}{\tau_b^2}A_{1212}^2 \right\rangle$$

Following Gorbachev and Pobedrya (1985), we introduce at this point the so-called engineering strength properties, σ_L, σ_F, τ_L, τ_F, and σ_p, related to the tensor components P^*_{ijkl} by the equations

$$\sigma_F = 1/\sqrt{P^*_{3333}} \qquad \sigma_L = 1/\sqrt{P^*_{1111}}$$

$$\tau_F = 1/2\sqrt{P^*_{1313}} \qquad \tau_L = 1/2\sqrt{P^*_{1212}} \tag{8.25}$$

$$\sigma_p = 1/\sqrt{P^*_{3333} + 4P^*_{1111} + 4P^*_{1133} - 4P^*_{1212}}$$

where σ_L (respectively σ_F) is the strength for stretching or compressing along (respectively, perpendicular to) the layer, τ_L (respectively, τ_F) is the strength for shear along (respectively, perpendicular to) the layer, and σ_p is the uniform compression strength. With the quantities so defined, the fracture criterion

$$P^*_{ijkl}\langle \sigma^{(0)}_{ij}\rangle\langle \sigma^{(0)}_{kl}\rangle = 1 \tag{8.26}$$

for a laminated composite can be rewritten as

$$\frac{1}{\sigma_L^2}\left(\langle \sigma^{(0)}_{11}\rangle^2 + \langle \sigma^{(0)}_{22}\rangle^2\right) + \frac{1}{\sigma_F^2}\langle \sigma^{(0)}_{33}\rangle^2 + \frac{1}{\tau_L^2}\langle \sigma^{(0)}_{12}\rangle^2$$

$$+ \frac{1}{\tau_F^2}\left(\langle \sigma^{(0)}_{13}\rangle^2 + \langle \sigma^{(0)}_{23}\rangle^2\right) + \left(\frac{2}{\sigma_L^2} - \frac{1}{\tau_L^2}\right)\langle \sigma^{(0)}_{11}\rangle\langle \sigma^{(0)}_{22}\rangle$$

$$+ \frac{1}{2}\left(\frac{1}{\sigma_p^2} + \frac{1}{\tau_L^2} - \frac{1}{\sigma_F^2} - \frac{4}{\sigma_L^2}\right)(\langle \sigma^{(0)}_{11}\rangle + \langle \sigma^{(0)}_{22}\rangle)\langle \sigma^{(0)}_{33}\rangle = 1 \tag{8.27}$$

It will be understood that the average stress components $\langle \sigma^{(0)}_{ij}\rangle$ in this last equation are determined by the effective modulus method, that is, by solving Equations 7.24 through 7.27 under appropriate boundary conditions. We note also that Equation 8.27 may as well be used as a composite material plasticity criterion if we replace τ_b by the simple shear yield strength and consider σ_L, σ_F, τ_L, τ_F, and σ_p as the engineering yield strength properties.

The effective elastic moduli of a composite have also been derived by Duvaut (1976) and Pobedrya (1984). In their 1977 paper, Pobedrya and Gorbachev apply the homogenization method to the static elasticity problem (formulated in terms of stress) and calculate the components of the effective compliance tensor for a composite system; in later papers (Pobedrya 1983a, 1983b, 1984), expressions for the effective thermal conductivity tensor may be found. As regards the dynamical elasticity problem for composite materials, an important point to be made is that in the asymptotic expansions involved, higher-order terms in ε must be retained. As discussed by Karimov (1986) in connection with the free vibrations of a finite-thickness composite layer, the fourth-order terms retained in the asymptotic expansions given by Equations 7.7 through 7.9 reveal the so-called wave filter effect, in which free vibrations prove to be nonexistent in some frequency range(s).

9. EFFECTIVE CHARACTERISTICS OF UNIDIRECTIONAL FIBER COMPOSITES

The two-dimensional local problems arising on the unit cell of such composites are readily amenable to a treatment by the analytical methods of the theory of (doubly periodic) functions of a complex variable. The unit cell in this case is obtained by cutting the material by a plane normal to the fiber direction and may have, for example, the form of a square with a circular inclusion corresponding to the circular cross section of the fiber (see Figure 9.1). If both the matrix and fiber materials are homogeneous and the fibers are aligned along the axis of x_3, the elastic tensor of the composite is a piecewise-constant, doubly periodic function of y_1 and y_2, and the local problem at hand is stated as

$$\frac{\partial \tau_{\alpha j}^{kl}(F)}{\partial y_\alpha} = 0 \quad \text{in region } Y_F \text{ occupied by fibers}$$

$$\frac{\partial \tau_{\alpha j}^{kl}(M)}{\partial y_\alpha} = 0 \quad \text{in region } Y_M \text{ occupied by the matrix}$$

$$(9.1)$$

$$\tau_{\alpha j}^{kl}(F) = c_{\alpha j n \beta}^{(F)} \frac{\partial N_n^{kl}(F)}{\partial y_\beta}$$

$$\tau_{\alpha j}^{kl}(M) = c_{\alpha j n \beta}^{(M)} \frac{\partial N_n^{kl}(M)}{\partial y_\beta}$$

$$(9.2)$$

where $\alpha, \beta = 1, 2$ and $j, k, l, n = 1, 2, 3$.

We specify, then, that the solution be doubly periodic in y_1 and y_2 and we also require that the following perfect contact conditions be satisfied on the boundary Γ:

$$N_n^{kl}(F)\big|_\Gamma = N_n^{kl}(M)\big|_\Gamma \tag{9.3}$$

$$\left(\tau_{\alpha j}^{kl}(F) + c_{\alpha jkl}^{(F)}\right) n_\alpha \big|_\Gamma = \left(\tau_{\alpha j}^{kl}(M) + c_{\alpha jkl}^{(M)}\right) n_\alpha \big|_\Gamma \tag{9.4}$$

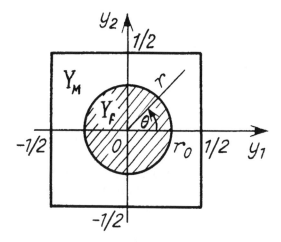

FIGURE 9.1. Unit cell of a unidirectional fiber composite.

where $c_{\alpha jkl}^{(F)}$ and $c_{\alpha jkl}^{(M)}$ are the elastic modulus tensors of the fiber and matrix materials, respectively, and n_α are the components of the unit vector normal to Γ ($\alpha = 1, 2$).

Exactly how many problems of the type given by Equations 9.1 and 9.2 must be solved in each particular case depends, of course, on the symmetry properties of the matrix and fiber materials (note that the double-index superscript kl takes the values 11, 22, 33, 23, 31, 12). If both are isotropic, the problem decomposes into four similar plane strain problems for τ^{11}, τ^{22}, τ^{33}, and τ^{12} and two antiplane problems for τ^{23} and τ^{13}. In the case of homogeneous isotropic materials, the plane strain problems are of the form

$$\frac{\partial \tau_{11}^{kl}}{\partial y_1} + \frac{\partial \tau_{12}^{kl}}{\partial y_2} = 0 \qquad \frac{\partial \tau_{12}^{kl}}{\partial y_1} + \frac{\partial \tau_{22}^{kl}}{\partial y_2} = 0 \qquad (9.5)$$

$$\tau_{11}^{kl} = c_{11}(y)\frac{\partial N_1^{kl}}{\partial y_1} + c_{12}(y)\frac{\partial N_2^{kl}}{\partial y_2}$$

$$\tau_{22}^{kl} = c_{12}(y)\frac{\partial N_1^{kl}}{\partial y_1} + c_{11}(y)\frac{\partial N_2^{kl}}{\partial y_2} \qquad (9.6)$$

$$\tau_{12}^{kl} = \frac{1}{2}(c_{11}(y) - c_{12}(y))\left(\frac{\partial N_2^{kl}}{\partial y_1} + \frac{\partial N_1^{kl}}{\partial y_2}\right)$$

where

$$c_{11}(y) = \begin{cases} c_{11}^{(F)} = \lambda_F + 2G_F & (y \in Y_F) \\ c_{11}^{(M)} = \lambda_M + 2G_M & (y \in Y_M) \end{cases}$$

$$c_{12}(y) = \begin{cases} c_{12}^{(F)} = \lambda_F & (y \in Y_F) \\ c_{12}^{(M)} = \lambda_M & (y \in Y_M) \end{cases}$$

λ_M, G_M, λ_F, and G_F are the Lamé constants of the matrix and fiber materials, and $kl = 11, 22, 33, 12$. The problems for determining the local functions $\tau_{\alpha 3}^{23}$ and $\tau_{\alpha 3}^{13}$ are in this case

$$\frac{\partial \tau_{13}^{kl}}{\partial y_1} + \frac{\partial \tau_{23}^{kl}}{\partial y_2} = 0 \qquad\qquad (9.7)$$

$$\tau_{13}^{kl} = c_{44}(y)\frac{\partial N_3^{kl}}{\partial y_1} \qquad \tau_{23}^{kl} = c_{44}(y)\frac{\partial N_3^{kl}}{\partial y_2} \qquad (9.8)$$

where

$$(kl = 23, 13) \qquad c_{44}(y) = \begin{cases} c_{44}^{(F)} = G_F & (y \in Y_F) \\ c_{44}^{(M)} = G_M & (y \in Y_M) \end{cases}$$

The solution of the systems given by Equations 9.5 and 9.6 and by Equations 9.7 and 9.8 must be doubly periodic in y_1 and y_2 and must satisfy Conditions 9.3 and 9.4 on the fiber–matrix interface. If both the matrix and fiber materials are homogeneous and isotropic, the boundary conditions take the form

$$[N_n^{kl}]_\Gamma = 0 \qquad\qquad (9.9)$$

$$[(\tau_{11}^{11} + c_{11})n_1 + \tau_{12}^{11}n_2]_\Gamma = 0 \qquad [\tau_{12}^{11}n_1 + (\tau_{22}^{11} + c_{12})n_2]_\Gamma = 0 \quad (9.10)$$

$$[(\tau_{11}^{22} + c_{12})n_1 + \tau_{12}^{22}n_2]_\Gamma = 0 \qquad [\tau_{12}^{11}n_1 + (\tau_{22}^{22} + c_{11})n_2]_\Gamma = 0 \quad (9.11)$$

$$[(\tau_{11}^{33} + c_{12})n_1 + \tau_{12}^{33}n_2]_\Gamma = 0 \qquad [\tau_{12}^{33}n_1 + (\tau_{22}^{33} + c_{12})n_2]_\Gamma = 0 \quad (9.12)$$

$$[\tau_{11}^{12}n_1 + (\tau_{12}^{12} + c_{66})n_2]_\Gamma = 0 \qquad [(\tau_{12}^{12} + c_{66})n_1 + \tau_{22}^{12}n_2]_\Gamma = 0 \quad (9.13)$$

where $c_{66} = \frac{1}{2}(c_{11} - c_{12})$ and $[A]$ means the jump in the value of the quantity A at the boundary Γ.

It should be noted here that, with one possible exception of circular fiber cross section, no analytical methods may be expected to be applicable to the local problems given by Equations 9.5 and 9.6 or by Equations 9.7 and 9.8, and in most cases more or less elaborate numerical techniques, such as the final-element method, must be employed.

This numerical work is, of course, simplified if some symmetry elements occur either in the geometry of the unit cell or in the constituent properties, or both, because a local problem for the entire unit cell may be reduced in this case to a boundary value problem for only a part of the cell. As an example of such a reduction, consider the system given by Equations 9.5 and 9.6 for $kl = 11$ for a composite with isotropic components and with a unit cell shown in Figure 9.1. We assume $c_{11}(y)$ and $c_{12}(y)$ to be even functions of y_1 and y_2 and consider the system of equations

$$\frac{\partial \tau_{11}^{11}}{\partial y_1} + \frac{\partial \tau_{12}^{11}}{\partial y_2} = -\frac{\partial c_{11}}{\partial y_1} \qquad \frac{\partial \tau_{12}^{11}}{\partial y_1} + \frac{\partial \tau_{22}^{11}}{\partial y_2} = -\frac{\partial c_{12}}{\partial y_2} \qquad (9.14)$$

and

$$\tau_{11}^{11} = c_{11}(y)\frac{\partial N_1^{11}}{\partial y_1} + c_{12}(y)\frac{\partial N_2^{11}}{\partial y_2}$$

$$\tau_{22}^{11} = c_{12}(y)\frac{\partial N_1^{11}}{\partial y_1} + c_{11}(y)\frac{\partial N_2^{11}}{\partial y_2} \qquad (9.15)$$

$$\tau_{12}^{11} = \frac{1}{2}(c_{11}(y) - c_{12}(y))\left(\frac{\partial N_2^{11}}{\partial y_1} + \frac{\partial N_1^{11}}{\partial y_2}\right)$$

From Equations 9.14 we observe that the local functions τ_{11}^{11} and τ_{22}^{11} are even, and τ_{12}^{11} is odd, in y_1 and y_2, and from Equations 9.15 we see that N_1^{11} (N_2^{11}) is even in y_2 (y_1) and odd in y_1 (y_2). Because the functions N_1^{11}, N_2^{11}, and τ_{12}^{11} must be continuous at the boundary of the unit cell and across the lines $y_1 = 0$, $y_2 = 0$, it then follows that the problem of interest in this case is the plane boundary value problem given by Equations 9.14 and 9.15 in the region $0 < y_1 < \frac{1}{2}$, $0 < y_2 < \frac{1}{2}$, and the boundary conditions to be satisfied are

$$N_1^{11} = 0 \quad \text{and} \quad \tau_{12}^{11} = 0 \quad \text{if } y_1 = 0, \, y_1 = \tfrac{1}{2}, 0 < y_2 < \tfrac{1}{2}$$

$$\qquad\qquad\qquad\qquad\qquad\qquad\qquad\qquad\qquad\qquad (9.16)$$

$$N_2^{11} = 0 \quad \text{and} \quad \tau_{12}^{11} = 0 \quad \text{if } y_2 = 0, \, y_2 = \tfrac{1}{2}, 0 < y_1 < \tfrac{1}{2}$$

Note that the derivatives $\partial c_{11}/\partial y_1$ and $\partial c_{12}/\partial y_2$ on the right-hand side of Equations 9.14 exhibit a delta function behavior as a result of the discontinuities in the coefficients c_{11} and c_{12}.

The solutions of the local problems given by Equations 9.5 through 9.8 for a composite with homogeneous isotropic components have been obtained by Mol'kov and Pobedrya (1985, 1988), who employ the Muskhelishvili complex potentials to estimate the components of the effective elastic modulus tensor. Numerical methods for treating unit cell local problems are discussed by Pobedrya and Sheshenin (1979), Sheshenin (1980), and Leontyev (1984), among others.

We now wish to proceed to the analytical treatment of the two-dimensional local problems set on the unit cell shown in Figure 9.1. Two of these problems, Equations 9.7 and 9.8, are relatively simple and reduce to the determination of doubly periodic functions $N_3^{kl}(y)$ ($kl = 23, 13$) satisfying Laplace's equation in the regions Y_F and Y_M and satisfying Conditions 9.13 at the boundary Γ. This is easily achieved by representing the harmonic functions $N_3^{kl}(y)$ by the series expansion given by Equation 6.40 in the region Y_M and by a Taylor series expansion in powers of $z = y_1 + iy_2$ in the region Y_F; the coefficients in these expansions will be found from the infinite systems of simultaneous algebraic equations resulting from Conditions 9.13.

Grigolyuk and Fil'shtinskii (1970) have found that the theory of doubly periodic functions of a complex variable provides an effective means for treatment of the local problems given by Equations 9.5 and 9.6, to which we turn next. The case when $kl = 11$ will be used to illustrate this approach. We begin by representing the biharmonic stress functions in the usual manner in terms of the analytical functions $\varphi(z)$ and $\psi(z)$ for $z = y_1 + iy_2$ and express the local functions $\tau_{11}^{11}, \tau_{22}^{11}, \tau_{12}^{11}, N_1^{11}$, and N_2^{11} in the form (Grigolyuk and Fil'shtinskii 1970)

$$\tau_{11}^{11} + \tau_{22}^{11} = 2\left[\varphi'(\mathscr{Z}) + \overline{\varphi'(\mathscr{Z})}\right] \tag{9.17}$$

$$\tau_{22}^{11} - \tau_{11}^{11} + 2i\tau_{12}^{11} = 2[\overline{\mathscr{Z}}\varphi''(\mathscr{Z}) + \psi'(\mathscr{Z})]$$

$$2G(N_1^{11} + iN_2^{11}) = \kappa\varphi(\mathscr{Z}) - \mathscr{Z}\overline{\varphi'(z)} - \overline{\psi(\mathscr{Z})} \tag{9.18}$$

where we have written

$$G = \begin{cases} G_F & (z \in Y_F) \\ G_M & (z \in Y_M) \end{cases} \quad \text{and} \quad \kappa = \frac{\lambda + 3G}{\lambda + G} = \begin{cases} \kappa_F & (z \in Y_F) \\ \kappa_M & (z \in Y_M) \end{cases}$$

If we introduce polar coordinates $(r = |z|, \theta)$ as shown in Figure 9.1, the boundary conditions given by Equations 9.10 at $\Gamma(r = r_0)$ take the form

$$\left[\tfrac{1}{2}(\tau_{11}^{11} + \tau_{22}^{11}) - \tfrac{1}{2}e^{2i\theta}(\tau_{22}^{11} - \tau_{11}^{11} + 2i\tau_{12}^{11})\right]_{r=r_0-0}$$

$$= \left[\tfrac{1}{2}(\tau_{11}^{11} + \tau_{22}^{11}) - \tfrac{1}{2}e^{2i\theta}(\tau_{22}^{11} - \tau_{11}^{11} + 2i\tau_{12}^{11})\right]_{r=r_0+0}$$

$$+ \tfrac{1}{2}(c_{11}^{(M)} + c_{12}^{(M)} - c_{11}^{(F)} - c_{12}^{(F)})$$

$$+ \tfrac{1}{2}(c_{11}^{(M)} - c_{12}^{(M)} - c_{11}^{(F)} + c_{12}^{(F)})e^{2i\theta} \tag{9.19}$$

Making use of Equations 9.17 and introducing the notation

$$\Phi_F(z) = \varphi_F'(z) \qquad \Psi_F(z) = \psi_F'(z)$$

$$\Phi_M(z) = \varphi_M'(z) \qquad \Psi_M(z) = \psi_M'(z)$$

for the analytical functions in the fiber and matrix regions (Y_F and Y_M, respectively), Equation 9.19 becomes

$$\Phi_F(t) + \overline{\Phi_F(t)} - e^{2i\theta}(\overline{t}\Phi_F'(t) + \Psi_F(t))$$

$$= \Phi_M(t) + \overline{\Phi_M(t)} - e^{2i\theta}(\overline{t}\Phi_M'(t) + \Psi_M(t))$$

$$+ \tfrac{1}{2}(c_{11}^{(M)} + c_{12}^{(M)} - c_{11}^{(F)} - c_{12}^{(F)})$$

$$+ \tfrac{1}{2}(c_{11}^{(M)} + c_{12}^{(M)} - c_{11}^{(F)} + c_{12}^{(F)})e^{2i\theta} \tag{9.20}$$

where $t = r_0 \exp(i\theta)$ is an arbitrary point on Γ.

To proceed further it is necessary to specialize Condition 9.9 by expressing the continuity of the functions N_1^{11} and N_2^{11} across the contour $\Gamma(r = r_0)$. Using Equation 9.18 we obtain

$$\left[\kappa_F \varphi_F(t) - \overline{t\varphi_F'(t)} - \overline{\psi_F(t)}\right] = \frac{G_M}{G_F}\left[\kappa_M \varphi_M(t) - \overline{t\varphi_M'(t)} - \overline{\psi_M(t)}\right]$$

$$\tag{9.21}$$

This is now differentiated along the direction s tangent to Γ, giving

$$\left[-\kappa_F\overline{\Phi_F(t)} + \Phi_F(t) - e^{2i\theta}(t\Phi'_F(t) + \Psi_F(t))\right]$$

$$= \frac{G_M}{G_F}\left[-\kappa_M\overline{\Phi_M(t)} + \Phi_M(t) - e^{2i\theta}(t\Phi'_M(t) + \Psi_M(t))\right] \quad (9.22)$$

where it is recalled that

$$\frac{d}{ds} = -\frac{\partial}{\partial y_1}\sin\theta + \frac{\partial}{\partial y_2}\cos\theta$$

$$\frac{\partial}{\partial y_1} = \frac{\partial}{\partial \mathscr{Z}} + \frac{\partial}{\partial \overline{\mathscr{Z}}} \quad \text{and} \quad \frac{\partial}{\partial y_2} = i\left(\frac{\partial}{\partial \mathscr{Z}} - \frac{\partial}{\partial \overline{\mathscr{Z}}}\right)$$

In view of the symmetry of the unit cell with respect to the y_1 and y_2 axes, the functions $\Phi_M(z)$ and $\Psi_M(z)$ can be represented in the respective forms (Grigolyuk and Fil'shtinskii 1970)

$$\Phi_M(\mathscr{Z}) = \alpha_0 + \sum_{k=0}^{\infty} \alpha_{2k+2}r_0^{2k+2}\frac{\wp^{(2k)}(\mathscr{Z})}{(2k+1)!} \quad (9.23)$$

$$\Psi_M(\mathscr{Z}) = \beta_0 + \sum_{k=0}^{\infty} \beta_{2k+2}\frac{r_0^{2k+2}\wp^{(2k)}(\mathscr{Z})}{(2k+1)!} - \sum_{k=0}^{\infty} \alpha_{2k+2}\frac{r_0^{2k+2}Q^{(2k+1)}(\mathscr{Z})}{(2k+1)!}$$

$$(9.24)$$

where α_{2k} and β_{2k} ($k = 0, 1, 2, \ldots$) are real constants; $\wp(z)$ is the doubly periodic Weierstrass elliptic function with periods $\omega_1 = 1$ and $\omega_2 = i$; the special meromorphic function $Q(z)$ is defined by

$$Q(\mathscr{Z}) = \sum_{m,n}'\left\{\frac{\overline{P}}{(\mathscr{Z} - P)} - 2\mathscr{Z}\frac{\overline{P}}{P^3} - \frac{\overline{P}}{P^2}\right\} \quad (9.25)$$

for $P = m + in$, $\overline{P} = m - in$ and satisfies the relations

$$Q^{(k)}(\mathscr{Z} + 1) - Q^{(k)}(\mathscr{Z}) = \wp^{(k)}(\mathscr{Z})$$

$$Q^{(k)}(\mathscr{Z} + i) - Q^{(k)}(\mathscr{Z}) = -i\wp^{(k)}(\mathscr{Z}) \quad (k = 1, 2, \ldots)$$

The functions $\Phi_M(z)$ and $\Psi_M(z)$ as given by Equations 9.23 and 9.24 ensure a symmetrical doubly periodic distribution of the local stresses τ_{11}^{11}

and τ_{12}^{11}, and it can be shown (Grigolyuk and Fil'shtinskii 1970) that the constants α_0 and β_0 relate to α_2 and β_2 by

$$\alpha_0 = \frac{\pi}{2}\beta_2 r_0^2 \qquad \beta_0 = (\gamma_1 + \pi)\alpha_2 r_0^2 \qquad \gamma_1 = 2Q\left(\frac{1}{2}\right) - \wp\left(\frac{1}{2}\right) \quad (9.26)$$

To obtain the derivatives $\wp^{(2k)}(z)$ and $Q^{(2k+1)}(z)$ involved in Equations 9.23 and 9.24 we use the well-known Laurent expansions of these functions about the point $z = 0$,

$$\frac{\wp^{(2k)}(\mathscr{Z})}{(2k+1)!} = \frac{1}{\mathscr{Z}^{2k+2}} + \sum_{j=0}^{\infty} r_{j,k}\mathscr{Z}^{2j} \qquad (k = 0,1,2,\dots) \quad (9.27)$$

$$\frac{Q^{(2k+1)}(\mathscr{Z})}{(2k+1)!} = \sum_{j=0}^{\infty} s_{j,k} z^{2j} \qquad (k = 0,1,2,\dots) \quad (9.28)$$

where

$$r_{j,k} = \frac{(2j+2k+1)!}{(2j)!(2k+1)!}\frac{g_{j+k+1}}{2^{2j+2k+2}} \qquad (r_{0,0} = 0)$$

$$g_{j+k+1} = {\sum_{m,n}}'\left(\frac{p}{2}\right)^{-2j-2k-2}$$

$$s_{j,k} = \frac{(2j+2k+2)!}{(2j)!(2k+2)!}\frac{p_{j+k+1}}{2^{2j+2k+2}} \qquad (s_{0,0} = 0)$$

$$p_{j+k+1} = {\sum_{m,n}}'\left(\frac{\bar{p}}{2}\right)\left(\frac{p}{2}\right)^{-2j-2k-3}$$

Coming back to Equations 9.23 and 9.24, the Laurent expansions for Φ_M and Ψ_M are then given by

$$\Phi_M(\mathscr{Z}) = \alpha_0 + \sum_{k=0}^{\infty}\alpha_{2k+2}\frac{r_0^{2k+2}}{\mathscr{Z}^{2k+2}} + \sum_{k=0}^{\infty}\sum_{j=0}^{\infty}\alpha_{2k+2}r_0^{2k+2}r_{j,k}\mathscr{Z}^{2j} \quad (9.29)$$

$$\Psi_M(\mathscr{Z}) = \beta_0 + \sum_{k=0}^{\infty}\beta_{2k+2}\frac{r_0^{2k+2}}{\mathscr{Z}^{2k+2}} + \sum_{k=0}^{\infty}\sum_{j=0}^{\infty}\beta_{2k+2}r_0^{2k+2}r_{j,k}\mathscr{Z}^{2j}$$

$$- \sum_{k=0}^{\infty}\sum_{j=0}^{\infty}(2k+2)\alpha_{2k+2}r_0^{2k+2}s_{j,k}\mathscr{Z}^{2j} \quad (9.30)$$

Because the functions $\Phi_F(z)$ and $\Psi_F(z)$ are both regular in Y_F, we are justified in representing them by the Taylor series

$$\Phi_F(z) = \sum_{k=0}^{\infty} a_{2k} z^{2k} \qquad \Psi_F(z) = \sum_{k=0}^{\infty} b_{2k} z^{2k} \qquad (9.31)$$

with a_{2k} and b_{2k} real constants.

Substituting the expansions given by Equations 9.29 through 9.31 into Conditions 9.20 and 9.22 on Γ and equating coefficients of equal powers of $\exp(i\theta)$ yields the following infinite system of algebraic equations for the constants α_{2k} and β_{2k+2} ($k = 2, 3, \dots$):

$$(2k+1)\alpha_{2k} - \beta_{2k+2} + \sum_{j=1}^{\infty} \alpha_{2j+2} A_{k,j}^{(1)} + \sum_{j=2}^{\infty} \beta_{2j+2} B_{k,j}^{(1)}$$

$$= \frac{(\kappa_F - \kappa_M G_M/G_F)}{2\Delta_\alpha \Delta_\alpha^*} (c_{11}^{(M)} - c_{12}^{(M)} - c_{11}^{(F)} + c_{12}^{(F)}) \qquad (9.32)$$

$$-\frac{(\kappa_M G_M/G_F + 1)}{(G_M/G_F - 1)} \alpha_{2k} + \sum_{j=1}^{\infty} \alpha_{2j+2} A_{k,j}^{(2)} + \sum_{j=2}^{\infty} \beta_{2j+2} B_{k,j}^{(2)}$$

$$= \frac{\gamma_{k,0} r_0^{2k}}{2\Delta_\alpha \Delta_\alpha^*} (c_{11}^{(M)} - c_{12}^{(M)} - c_{11}^{(F)} + c_{12}^{(F)})$$

$$+ \frac{r_{k-1,0} r_0^{2k}(1 - \kappa_F)}{4\Delta_\beta} (c_{11}^{(M)} + c_{12}^{(M)} - c_{11}^{(F)} - c_{12}^{(F)})$$

$$(k = 2, 3, \dots) \quad (9.33)$$

where

$$A_{k,j}^{(1)} = \frac{(\kappa_F - \kappa_M G_M/G_F)}{(\kappa_F + G_M/G_F)} \left[r_{k,j} + r_0^2 r_{k,0} \frac{(G_M/G_F - 1)}{\Delta_\alpha \Delta_\alpha^*} \gamma_{1,j} \right] r_0^{2j+2B+2}$$

$$B_{k,j}^{(1)} = \left(\kappa_F - \kappa_M \frac{G_M}{G_F} \right) \frac{(G_M/G_F - 1)}{\Delta_\alpha \Delta_\alpha^*} r_{k,0} r_{0,j} r_0^{2k+2j+6}$$

$$A_{k,j}^{(2)} = \left[\gamma_{k,j} - \frac{(G_M/G_F - 1)}{\Delta_\alpha \Delta_\alpha^*} r_0^2 \gamma_{k,0} \gamma_{1,j} \right.$$

$$\left. + \frac{1}{\Delta_\beta} \left((1 - \kappa_M) \frac{G_M}{G_F} - 1 + \kappa_F \right) r_0^2 r_{k-1,0} r_{0,j} \right] r_0^{2j+2k}$$

$$B_{k,j}^{(2)} = \left[-r_{k-1,j} + \frac{(G_M/G_F - 1)}{\Delta_\alpha \Delta_\alpha^*} r_0^2 \gamma_{k,0} r_{0,j} \right] r_0^{2k+2j}$$

$$\gamma_{k,0} = r_0^2 r_{k,0} - (2k - 2)r_{k-1,0} + 2s_{k-1,0}$$

$$- \frac{(\kappa_F - \kappa_M G_M/G_F)}{(\kappa_F + G_M/G_F)} r_0^6 r_{1,0} r_{k-1,1} - 3r_0^2 r_{k-1,1}$$

$$\gamma_{k,j} = r_0^2 r_{k,j} - (2k - 2)r_{k-1,j} + (2j + 2)s_{k-1,j}$$

$$- \frac{(\kappa_F - \kappa_M G_M/G_F)}{(\kappa_F + G_M/G_F)} r_0^6 r_{1,j} r_{k-1,1}$$

$$\Delta_\alpha = -1 - \kappa_M \frac{G_M}{G_F} - \left(\frac{G_M}{G_F} - 1 \right)(r_0^4 r_{1,0} - (\gamma_1 + \pi)r_0^2)$$

$$\Delta_\beta = \frac{G_M}{G_F} \left((1 - \kappa_M) \frac{\pi r_0^2}{2} - 1 \right) - \frac{1}{2}(1 - \kappa_F)(\pi r_0^2 - 1)$$

$$\Delta_\alpha^* = 1 - \frac{3}{\Delta_\alpha} \left(\frac{G_M}{G_F} - 1 \right) r_0^4 r_{0,1} - \frac{1}{\Delta_\alpha} \left(\frac{G_M}{G_F} - 1 \right)$$

$$\frac{(\kappa_F - \kappa_M G_M/G_F)}{(\kappa_F + G_M/G_F)} r_0^8 r_{1,0} r_{0,1}$$

The remaining coefficients in the expansions given by Equations 9.29 through 9.31 can be calculated from

$$\alpha_2 = -\frac{(G_M/G_F - 1)}{\Delta_\alpha \Delta_\alpha^*} \sum_{j=1}^{\infty} \gamma_{1,j} \alpha_{2j+2} r_0^{2j+2}$$

$$+ \frac{(G_M/G_F - 1)}{\Delta_\alpha \Delta_\alpha^*} \sum_{j=2}^{\infty} \beta_{2j+2} r_0^{2j+2} r_{0,j}$$

$$- \frac{1}{2\Delta_\alpha \Delta_\alpha^*} (c_{11}^{(M)} - c_{12}^{(M)} - c_{11}^{(F)} + c_{12}^{(F)}) \qquad (9.34)$$

$$\beta_2 = \frac{(1 - \kappa_B)}{4\Delta_\beta}(c_{11}^{(M)} + c_{12}^{(M)} - c_{11}^{(F)} - c_{12}^{(F)})$$

$$-\frac{1}{\Delta_\beta}\left((1 - \kappa_M)\frac{G_M}{G_F} - (1 - \kappa_F)\right)\sum_{j=1}^{\infty}\alpha_{2j+2}r_0^{2j+2}r_{0,j} \quad (9.35)$$

$$\beta_4 = \left[3 + \frac{(\kappa_F - \kappa_M G_M/G_F)}{(\kappa_F + G_M/G_F)}r_0^4 r_{1,0}\right]\alpha_2$$

$$+ \frac{(\kappa_F - \kappa_M G_M/G_F)}{(\kappa_F + G_M/G_F)}\sum_{j=1}^{\infty}\alpha_{2j+2}r_0^{2j+4}r_{1,j} \quad (9.36)$$

$$a_0 = \frac{1}{2}(\pi r_0^2 - 1) + \sum_{j=1}^{\infty}\alpha_{2j+2}r_0^{2j+2}r_{0,j}$$

$$+ \frac{1}{4}(c_{11}^{(M)} + c_{12}^{(M)} - c_{11}^{(F)} - c_{12}^{(F)}) \quad (9.37)$$

$$a_2 r_0^2 = \frac{(1 + \kappa_M)G_M/G_F}{(\kappa_F + G_M/G_F)}\left[r_0^4 r_{1,0}\alpha_2 + \sum_{j=1}^{\infty}\alpha_{2j+2}r_0^{2j+4}r_{1,j}\right] \quad (9.38)$$

$$a_{2k}r_0^{2k} = r_0^{2k+2}r_{k,0}\alpha_2 + (2k + 1)\alpha_{2k} - \beta_{2k+2}$$

$$+ \sum_{j=1}^{\infty}\alpha_{2j+2}r_0^{2j+2k+2}r_{k,j} \quad (k = 2, 3, \ldots) \quad (9.39)$$

$$b_0 = -a_2 r_0^2 - (1 - (\gamma_1 + \pi)r_0^2 + r_0^4 r_{1,0})\alpha_2 - \sum_{j=1}^{\infty}\left(r_0^2 r_{1,j}\right.$$

$$+ (2j + 2)s_{0,j})\alpha_{2j+2}r_0^{2j+2} + r_0^4 r_{0,1}\beta_4$$

$$+ \sum_{j=2}^{\infty}\beta_{2j+2}r_0^{2j+2}r_{0,j}$$

$$- \frac{1}{2}(c_{11}^{(M)} - c_{12}^{(M)} - c_{11}^{(F)} + c_{12}^{(F)}) \quad (9.40)$$

$$b_{2k-2}r_0^{2k-2} = -(2k-1)a_{2k}r_0^{2k} - (r_0^{2k+2}r_{k,0} - (2k-2)r_0^{2k}r_{k-1,0}$$

$$+ 2s_{k-1,0}r_0^{2k})\alpha_2 + r_0^{2k}r_{k-1,0}\beta_2 - \alpha_{2k} - \sum_{j=1}^{\infty}(r_0^2 r_{k,j}$$

$$- (2k-2)z_{k-1,j} + (2j+2)s_{k-1,j})\alpha_{2j+2}r_0^{2j+2k}$$

$$+ \sum_{j=1}^{\infty} \beta_{2j+2}r_0^{2j+2k}r_{k-1,j} \qquad (k = 2,3,\ldots) \qquad (9.41)$$

All the coefficients in Equations 9.29 through 9.31 having been determined, Formula 9.17 now can be employed for calculating the local stresses τ_{11}^{11}, τ_{22}^{11}, and τ_{12}^{11}, and the solution of the local problem given by Equations 9.5 and 9.6 for $kl = 11$ is thus completed. The cases $kl = 22, 33, 12$, which we consider next, differ little if at all from the case $kl = 1$ and are treated in very much the same way, thus providing all the information necessary for determining the effective elastic moduli

$$\tilde{c}_{ijkl} = \int_Y \left(c_{ijkl}(y) + \tau_{ij}^{kl}(y) \right) dy_1\, dy_2 \qquad (9.42)$$

$Y = Y_F \cup Y_M$ denoting the region occupied by the unit cell of the (unidirectional fiber) composite. It is useful to point out that, in view of the symmetry of the preceding local problems,

$$\int_Y \tau_{12}^{kl}\, dy_1\, dy_2 = 0 \qquad (kl = 11, 22, 33)$$

$$\int_Y \tau_{11}^{12}\, dy_1\, dy_2 = \int_Y \tau_{22}^{12}\, dy_1\, dy_2 = 0$$

$$\int_Y \tau_{13}^{23}\, dy_1\, dy_2 = \int_Y \tau_{23}^{13}\, dy_1\, dy_2 = 0$$

and it is therefore readily shown that, changing to the two-index notation, the effective elastic matrix of such a composite is of the form

$$\tilde{c}_{\alpha\beta} = \begin{pmatrix} \tilde{c}_{11} & \tilde{c}_{12} & \tilde{c}_{13} & 0 & 0 & 0 \\ \tilde{c}_{12} & \tilde{c}_{11} & \tilde{c}_{13} & 0 & 0 & 0 \\ \tilde{c}_{13} & \tilde{c}_{13} & \tilde{c}_{33} & 0 & 0 & 0 \\ 0 & 0 & 0 & \tilde{c}_{44} & 0 & 0 \\ 0 & 0 & 0 & 0 & \tilde{c}_{44} & 0 \\ 0 & 0 & 0 & 0 & 0 & \tilde{c}_{66} \end{pmatrix} \qquad (9.43)$$

where

$$\tilde{c}_{11} = \int_Y (c_{11}(y) + \tau_{11}^{11}(y)) \, dy_1 \, dy_2$$

$$\tilde{c}_{12} = \int_Y (c_{12}(y) + \tau_{11}^{22}(y)) \, dy_1 \, dy_2$$

$$\tilde{c}_{13} = \int_Y (c_{12}(y) + \tau_{11}^{33}(y)) \, dy_1 \, dy_2$$

$$\tilde{c}_{33} = \int_Y (c_{11}(y) + \tau_{33}^{33}(y)) \, dy_1 \, dy_2 \qquad (9.44)$$

$$= \int_Y \left[c_{11}(y) + \frac{c_{12}(y)}{c_{11}(y) + c_{12}(y)} (\tau_{11}^{33} + \tau_{22}^{33}) \right] dy_1 \, dy_2$$

$$\tilde{c}_{44} = \int_Y (c_{44}(y) + \tau_{23}^{23}(y)) \, dy_1 \, dy_2$$

$$\tilde{c}_{66} = \int_Y (c_{66}(y) + \tau_{12}^{12}(y)) \, dy_1 \, dy_2$$

Note that the local function τ_{33}^{33} entering the $\tilde{c}_{33}$ expression is defined by

$$\tau_{33}^{33} = c_{12} \left(\frac{\partial N_1^{33}}{\partial y_1} + \frac{\partial N_2^{33}}{\partial y_2} \right) \qquad (9.45)$$

or, using Equations 9.6,

$$\tau_{33}^{33} = \frac{c_{12}}{c_{12} + c_{11}} (\tau_{11}^{33} + \tau_{22}^{33}) \qquad (9.46)$$

10. PLANE ELASTICITY PROBLEM FOR A PERIODIC COMPOSITE WITH A CRACK

It is a common practice in performing the stress analysis of a composite with an ideally smooth macroscopic crack (macrocrack) that the homogeneous composite medium being studied is replaced with a homogeneous anisotropic medium whose response is supposed to be equivalent to that of the actual composite in a certain averaged sense (see, e.g., Parton and

Morozov 1978, 1989, Cherepanov 1983, Parton 1992). The main attraction of the method is that the calculation of the averaged stress field in such a composite reduces to an elasticity theory problem for a homogeneous anisotropic material with a mathematical cut within it.

If the material has a periodic structure, the averaged (or effective) properties of the equivalent medium can be estimated by means of the homogenization method, which also gives asymptotically correct results for the local structure of the stress field in the composite. Parton and Kudryavtsev (1986) adopted this approach in their analysis of the stress field arising in the neighborhood of a macrocrack in a laminated composite with a periodic structure; the stress intensity factors of the composite are expressed by the authors in terms of the constituent properties and the parameters determining the position of the crack in the material. In this section, we discuss a refined version of this approach (developed by Parton et al. 1988a, 1990a), in which the boundary effect occurring in the vicinity of the crack is taken into account by introducing additional solutions of the boundary layer type discussed in Section 7. For the sake of simplicity, only the plane formulation will be considered.

We are thus interested in the state of stress of a periodically nonhomogeneous (composite) medium containing a rectilinear macrocrack whose length is much greater than the unit cell dimensions. It is assumed that the elastic medium possesses a doubly periodic nonhomogeneity in the (x_1, x_2) plane and that the rims of the (tunnel) crack are parallel to the boundary of the unit cell, as illustrated in Figure 10.1. The problem so stated is relevant, for example, to the study of a fiber composite with a tunnel crack lying in the plane normal to the fiber direction or of a laminated composite with a plane crack normal or parallel to the laminae.

Let the rectilinear crack lie in the interface between two unit cells of the (unbounded) composite, and let its rims be subjected to a prescribed set of self-balanced normal and tangential loads. To determine the state of stress and strain in the neighborhood of such a crack, an asymptotic method will be applied to the equations of elasticity in the periodically nonhomogeneous half-plane $x_2 > 0$ under mixed boundary conditions on $x_2 = 0$. These latter specify the stresses σ_{i2} ($i = 1, 2, 3$) on the segment $|x_1| < a$ and the stresses σ_{12} and σ_{32} and the displacements u_2 at $|x_1| > a$ (see Figure 10.2). To satisfy these types of conditions within the framework of the asymptotic method, two auxiliary plane problems will be considered in the region $x_2 > 0$, and for their solution the results derived in Section 7 will be employed.

The first of these problems is stated as

$$\sigma_{i\alpha, \alpha}(x_1, x_2) = 0 \tag{10.1}$$

$$\sigma_{i\alpha} = c_{i\alpha k\beta}(y_1, y_2)u_{k, \beta} \tag{10.2}$$

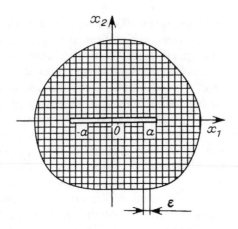

FIGURE 10.1. Macroscopic crack in a composite material.

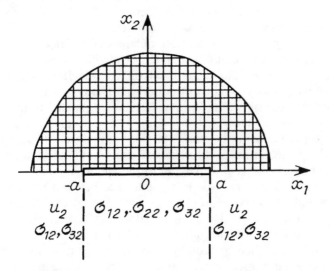

FIGURE 10.2. Boundary conditions on the line $x_2 = 0$.

where the elastic coefficients $c_{i\alpha k\beta}(y_1, y_2)$ are 1-periodic functions of the variables $y = x_\alpha/\varepsilon$; a comma (respectively, a vertical line) with the indices denotes partial differentiation with respect to x_1, x_2 (respectively, y_1, y_2); Latin indices range from 1 to 3, and Greek indices from 1 to 2.

The conditions to be satisfied at the boundary of the half-plane are

$$\sigma_{i2}(x_1, 0) = p_i(x_1) \qquad (i = 1, 2, 3) \tag{10.3}$$

In Section 7, the asymptotic analysis of the problem given by Equations 10.1 through 10.3 yielded Formulas 7.64 and 7.65, from which the displacement and stress fields can be estimated to leading terms in ε. The local functions $N_n^{k\beta}(y)$ and boundary layer functions $N_n^{(1)k\beta}(y)$ involved in these formulas are determined from the problems given by Equation 7.45 and by Equations 7.53 through 7.55, respectively.

In the second auxiliary problem for the half-plane $x_2 > 0$, the (mixed) boundary conditions at $x_2 = 0$ are of the form

$$\sigma_{l2}(x_1, 0) = p_l(x_1) \qquad (l = 1, 3)$$

$$u_2(x_1, 0) = v_2(x_1)$$

(10.4)

In this case, it can be shown by the same procedure used in Section 7 that the displacements and stresses are given (to the same accuracy) by formulas very similar to Equations 7.64 and 7.65 in which the functions $N_n^{(1)k\beta}(y)$ are replaced by boundary layer type functions $N_n^{(3)k\beta}(y)$ obtained from the problem given by Equations 7.77 through 7.80.

With these results at our disposal we can now proceed to the stress analysis of the rectilinear crack of normal rupture in the interface between two rectangular unit cells in a composite material. If we assume that there are no tangential stresses on the rims of the crack ($\sigma_{12} = \sigma_{13} = 0$) and that $\sigma_{22} = p_2(x_1)$ at $x_2 = \pm 0$, $|x_1| < a$, then, because of the symmetry of the stress state with respect to the x_1 axis, we may reduce the problem to one concerning the upper half-plane $x_2 > 0$ with the boundary conditions of the form (cf. Figure 10.2)

$$\sigma_{l2} = 0 \qquad \text{for } x_2 = 0, |x_1| < \infty, l = 1, 3$$

$$\sigma_{22} = p_2(x_1) \qquad \text{for } x_2 = 0, |x_1| < a \qquad (10.5)$$

$$u_2 = 0 \qquad \text{for } x_2 = 0, |x_1| > a$$

Referring to Figure 10.3, several characteristic regions should be distinguished in the nonhomogeneous half-plane subjected to these conditions. To begin with, there is region IV in the vicinity of the points $x_1 = \pm a$, $x_2 = 0$, which has the form of a rectangle with dimensions $\tilde{\varepsilon}_1 + \tilde{\varepsilon}_2$ and $\tilde{\varepsilon}$ and which contain a finite (and sufficiently small) number of unit cells within itself. The state of stress (strain) in this region must be determined directly from the solution of the elasticity problem rather than by applying the asymptotic process of the homogenization method. In the unbounded strip $0 < x_2 < \tilde{\varepsilon}$, $|x_1| < \infty$ (of which region IV is a part), region I ($0 < x_2 <$

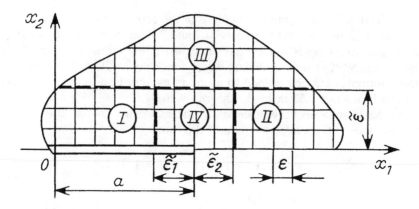

FIGURE 10.3. Auxiliary regions I–IV in the vicinity of a macrocrack in a composite material of periodic structure.

$\tilde{\varepsilon}$, $|x_1| < a$) and region II ($0 < x_2 < \tilde{\varepsilon}$, $|x_1| > a$) may be recognized, in which the asymptotic solutions of, respectively, the first and the second of the auxiliary problems just discussed may be utilized. We should remark here that the functions $N_n^{(1)k\beta}(y)$ and $N_n^{(3)k\beta}(y)$ determined from the problems given by Equations 7.53 through 7.55 and by Equations 7.77 through 7.80 are boundary layer type solutions and it is therefore possible to set

$$N_n^{(1)k\beta}(y) = 0 \quad \text{and} \quad N_n^{(3)k\beta}(y) = 0$$

in region III ($x_2 > \tilde{\varepsilon}$). The magnitude of $\tilde{\varepsilon}$ is thus found from the conditions $N_n^{(1)k\beta}(y_1, \tilde{\varepsilon}) \approx 0$ and $N_n^{(3)k\beta}(y_1, \tilde{\varepsilon}) \approx 0$.

The magnitudes of $\tilde{\varepsilon}_1$ and $\tilde{\varepsilon}_2$ may be obtained by noting that the stress perturbation resulting from the change of boundary conditions in the vicinity of the tip of the crack is localized within region IV and does not spread outside it. Therefore, the characteristic dimension of the perturbed area can be estimated from the knowledge of the asymptotic expressions for the stress field in the vicinity of the crack. Note also that, for the sake of convenience, the values of $\tilde{\varepsilon}$, $\tilde{\varepsilon}_1$, and $\tilde{\varepsilon}_2$ should be selected to make the boundaries of region IV coincide with those of the unit cells of the composite (see Figure 10.3).

With the boundary layer solutions available, stresses on the two vertical sides ($x_1 = a - \tilde{\varepsilon}_1$, $0 < x_2 < \tilde{\varepsilon}$ and $x_1 = a + \tilde{\varepsilon}_2$, $0 < x_2 < \tilde{\varepsilon}$) and the upper horizontal side ($a - \tilde{\varepsilon}_1 < x_1 < a + \varepsilon_2$, $x_2 = \tilde{\varepsilon}$) of region IV can be specified. Because the boundary conditions of the segment $a - \tilde{\varepsilon}_1 < x_1 < a + \tilde{\varepsilon}_2$, $x_2 = 0$ are known (see Equations 10.5), it is possible in principle to solve the plane elasticity problem in region IV—for example, by numerical methods. The problem is given by

$$\sigma_{\alpha\beta,\beta} = 0$$

$$\sigma_{\alpha\beta} = c_{\alpha\beta\lambda\mu}(y)u_{\lambda,\mu}$$

$$\sigma_{12} = 0 \qquad \text{for } x_2 = 0,\ a - \tilde{\varepsilon}_1 < x_1 < a + \tilde{\varepsilon}_2$$

$$\sigma_{22} = p_2(x_1) \quad \text{for } x_2 = 0,\ a - \tilde{\varepsilon}_1 < x_1 < a$$

$$u_2 = 0 \qquad \text{for } x_2 = 0,\ a < x_1 < a + \tilde{\varepsilon}_2 \qquad (10.6)$$

and

$$\sigma_{\mu 1} = \left[c_{\mu 1 \lambda \beta}(y) + c_{\mu 1 n \gamma}(y)\left(N_{n|\gamma}^{\lambda\beta} + N_{n|\gamma}^{(1)\lambda\beta} \right) \right] u_{\lambda,\beta}^{(0)}$$

$$\text{for } x_1 = a - \tilde{\varepsilon}_1,\ 0 < x_2 < \tilde{\varepsilon}$$

$$\sigma_{\mu 1} = \left[c_{\mu 1 \lambda \beta}(y) + c_{\mu 1 n \gamma}(y)\left(N_{n|\gamma}^{\lambda\beta} + N_{n|\gamma}^{(3)\lambda\beta} \right) \right] u_{\lambda,\beta}^{(0)}$$

$$\text{for } x_1 = a + \tilde{\varepsilon}_2,\ 0 < x_2 < \tilde{\varepsilon}$$

$$\sigma_{\mu 2} = \left[c_{\mu 2 \lambda \beta}(y) + c_{\mu 2 n \gamma}(y) N_{n|\gamma}^{\lambda\beta} \right] u_{\lambda,\beta}^{(0)}$$

$$\text{for } x_2 = \tilde{\varepsilon},\ a - \tilde{\varepsilon}_1 < x_1 < a + \tilde{\varepsilon}_2 \quad (10.7)$$

It should be noted in connection with Equations 10.7 that they do not contain the constants involved in Conditions 7.79 and that the functions $u_\alpha^{(0)}(x)$ are solutions of the following homogenized problem

$$\langle C_{\lambda\alpha\mu\beta} \rangle u_{\mu,\alpha\beta}^{(0)} = 0 \qquad \text{for } x_2 > 0$$

$$C_{12\,\alpha\beta}^{*} u_{\alpha,\beta}^{(0)} = 0 \qquad \text{for } x_2 = 0,\ |x_1| < \infty$$

$$C_{22\,\alpha\beta}^{*} u_{\alpha,\beta}^{(0)} = p_2(x_1) \quad \text{for } x_2 = 0,\ |x_1| < a \qquad (10.8)$$

$$u_2^{(0)} = 0 \qquad \text{for } x_2 = 0,\ |x_1| > a$$

with coefficients defined by Equations 7.50 and 7.63.

The special significance of the separation in the composite of a local region IV with known boundary conditions on its perimeter is that this

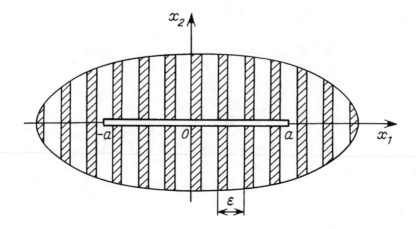

FIGURE 10.4. Macrocrack in a laminated two-component composite.

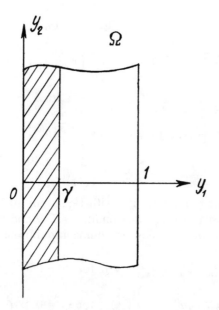

FIGURE 10.5. Unit cell of a laminated two-component composite.

region is amenable to a reasonably rigorous stress analysis for various positions of the crack tip in the unit cell of the composite.

The foregoing approach will now be applied to the problem of a macrocrack in a laminated periodic composite, shown in Figure 10.4, the unit cell of which consists of two (isotropic) layers (see Figure 10.5) characterized by the parameters E_1, ν_1 and E_2, ν_2.

The local problems given by Equation 7.45 are solved in an elementary fashion in this case and, according to Pobedrya (1984), the nonzero 1-periodic functions are given by

$$
N_1^{11} =
\begin{cases}
\dfrac{(c_{11}^{(2)} - c_{11}^{(1)})(1 - \gamma)}{(1 - \gamma)c_{11}^{(1)} + \gamma c_{11}^{(2)}} \left(y_1 - \dfrac{\gamma}{2} \right) & \text{for } 0 < y_1 < \gamma \\[4mm]
\dfrac{(c_{11}^{(2)} - c_{11}^{(1)})\gamma}{(1 - \gamma)c_{11}^{(1)} + \gamma c_{11}^{(2)}} \left(\dfrac{1 + \gamma}{2} - y_1 \right) & \text{for } \gamma < y_1 < 1
\end{cases}
$$

$$
N_1^{22} =
\begin{cases}
\dfrac{(c_{12}^{(2)} - c_{12}^{(1)})(1 - \gamma)}{(1 - \gamma)c_{11}^{(1)} + \gamma c_{11}^{(2)}} \left(y_1 - \dfrac{\gamma}{2} \right) & \text{for } 0 < y_1 < \gamma \\[4mm]
\dfrac{(c_{12}^{(2)} - c_{12}^{(1)})\gamma}{(1 - \gamma)c_{11}^{(1)} + \gamma c_{11}^{(2)}} \left(\dfrac{1 + \gamma}{2} - y_1 \right) & \text{for } \gamma < y_1 < 1 \quad (10.9)
\end{cases}
$$

$$
N_2^{12} = N_2^{21} = N_3^{31} =
\begin{cases}
\dfrac{(c_{44}^{(2)} - c_{44}^{(1)})(1 - \gamma)}{(1 - \gamma)c_{44}^{(1)} + \gamma c_{44}^{(2)}} \left(y_1 - \dfrac{\gamma}{2} \right) \\[2mm]
\qquad \text{for } 0 < y_1 < \gamma \\[4mm]
\dfrac{(c_{44}^{(2)} - c_{44}^{(1)})\gamma}{(1 - \gamma)c_{44}^{(1)} + \gamma c_{44}^{(2)}} \left(\dfrac{1 + \gamma}{2} - y_1 \right) \\[2mm]
\qquad \text{for } \gamma < y_1 < 1
\end{cases}
$$

where the superscripts (1) and (2) on the elastic properties c_{ik} refer to the layers of the composite. Note that

$$
C_{i2k\beta}^{*} = \langle C_{i2k\beta} \rangle \tag{10.10}
$$

because the relevant functions $C_{i2k\beta}$ are independent of y_2.

The solution of the relevant boundary layer problem given by Equations 7.53 through 7.55 is nontrivial only for the functions

$$
N_1^{(1)11}, N_2^{(1)11} \quad \text{and} \quad N_1^{(1)22}, N_2^{(1)22} \tag{10.11}
$$

whereas nonzero solutions of the problem given by Equations 7.77 through 7.80 only exist for the boundary layer functions

$$
N_1^{(3)21} = N_1^{(3)12}(y_1, y_2) \quad \text{and} \quad N_2^{(3)21} = N_2^{(3)12}(y_1, y_2) \tag{10.12}
$$

a result that is easily deduced by the use of Expressions 10.9 and 10.10 for the local functions $N_n^{k\beta}$. Functions 10.11 and 10.12 can be obtained numerically from the appropriate boundary layer problems.

In the special case we consider, the homogenized medium is isotropic, with x_1 as the axis of symmetry, and its effective characteristics are as follows:

$$\tilde{c}_{11} = \left[\gamma \frac{(1 + \nu_1)(1 - 2\nu_1)}{E_1(1 - \nu_1)} + (1 - \gamma) \frac{(1 + \nu_2)(1 - 2\nu_2)}{E_2(1 - \nu_2)} \right]^{-1}$$

$$\tilde{c}_{12} = \tilde{c}_{11} \left[\gamma \frac{\nu_1}{1 - \nu_1} + (1 - \gamma) \frac{\nu_2}{1 - \nu_2} \right]$$

$$\tilde{c}_{22} = \gamma \frac{E_1}{1 - \nu_1^2} + (1 - \gamma) \frac{E_2}{1 - \nu_2^2} + \frac{(\tilde{c}_{12})^2}{\tilde{c}_{11}}$$ (10.13)

$$\tilde{c}_{55} = \left[\gamma \frac{2(1 + \nu_1)}{E_1} + (1 - \gamma) \frac{2(1 + \nu_2)}{E_2} \right]^{-1}$$

Using Equation 10.10, it turns out (Liebowitz 1968) that the homogenized problem given by Equations 10.8 is solved analytically to give

$$u_1^{(0)} = 2 \, \mathrm{Re}[r_1 \varphi(\mathcal{Z}_1) + r_2 \psi(\mathcal{Z}_2)]$$

$$u_2^{(0)} = 2 \, \mathrm{Re}[q_1 \varphi(\mathcal{Z}_1) + q_2 \psi(\mathcal{Z}_2)]$$

$$\sigma_{11} = 2 \, \mathrm{Re}\left[\mu_1^2 \varphi'(\mathcal{Z}_1) + \mu_2^2 \psi'(\mathcal{Z}_2) \right]$$ (10.14)

$$\sigma_{22} = 2 \, \mathrm{Re}[\varphi'(\mathcal{Z}_1) + \psi'(\mathcal{Z}_2)]$$

$$\sigma_{12} = -2 \, \mathrm{Re}[\mu_1 \varphi'(\mathcal{Z}_1) + \mu_2 \psi'(\mathcal{Z}_2)]$$

where

$$\mathcal{Z}_1 = x_1 + \mu_1 x_2 \qquad\qquad \mathcal{Z}_2 = x_1 + \mu_2 x_2$$

$$r_1 = \tilde{a}_{11} \mu_1^2 + \tilde{a}_{12} \qquad\qquad r_2 = \tilde{a}_{11} \mu_2^2 + \tilde{a}_{12}$$

$$q_1 = \frac{\tilde{a}_{12} \mu_1^2 + \tilde{a}_{22}}{\mu_1} \qquad\qquad q_2 = \frac{\tilde{a}_{12} \mu_2^2 + \tilde{a}_{22}}{\mu_2}$$

$$\tilde{a}_{11} = \tilde{c}_{22}\left(\tilde{c}_{11}\tilde{c}_{22} - \tilde{c}_{12}^2\right)^{-1} \qquad \tilde{a}_{12} = -\tilde{c}_{12}\left(\tilde{c}_{11}\tilde{c}_{22} - \tilde{c}_{12}^2\right)^{-1}$$

$$\tilde{a}_{22} = \tilde{c}_{11}\left(\tilde{c}_{11}\tilde{c}_{22} - \tilde{c}_{12}^2\right)^{-1} \qquad \tilde{a}_{55} = \left(2\tilde{c}_{55}\right)^{-1}$$

$$\varphi(\mathscr{Z}_1) = -\frac{\mu_2}{\mu_2 - \mu_1}\frac{p}{2}\left[\sqrt{\mathscr{Z}_1^2 - a^2} - \mathscr{Z}_1\right]$$

$$\psi(\mathscr{Z}_2) = \frac{\mu_1}{\mu_2 - \mu_1}\frac{p}{2}\left[\sqrt{\mathscr{Z}_2^2 - a^2} - \mathscr{Z}_2\right]$$

and where

$$\mu_1 = \alpha_1 + i\beta_1 \quad \text{and} \quad \mu_2 = \alpha_2 + i\beta_2 \qquad (\beta_1 > 0, \ \beta_2 > 0, \ \beta_1 \neq \beta_2)$$

are the roots of the characteristic equation

$$\tilde{a}_{11}\mu^4 + 2(\tilde{a}_{12} + \tilde{a}_{55})\mu^2 + \tilde{a}_{22} = 0$$

The final formulation of the elastic problem for region IV of Figure 10.3 is obtained from Equations 10.7 by substituting the local functions $N_n^{k\beta}$ given by Equations 10.9 and the numerically calculated Functions 10.11 and 10.12.

Although straightforward in principle, the final element of the solution of the boundary layer problems given by Equations 7.53 through 7.55 has been given a very close examination by the authors of this book, in view of the importance of the final result for the construction of the solution for region IV. Numerical computations performed on a 494-element (540-site) grid with aspect ratio of $1:10$ showed that at a distance as short as only four length units from the lower edge, the functions sought for differed very little from zero, a fact that enabled us to reduce the aspect ratio to $1:4$ (i.e., to set $\tilde{\varepsilon} = 4\varepsilon$) in the subsequent work. A high degree of confidence in the numerical results so obtained was provided by the observation that they changed only slightly when a similar calculation using square eight-site elements (that is, with nearly twice as many sites) was carried out. The results were then substituted into Equation 10.7, and the computation procedure was limited to region IV, this being represented by 376 elements and 419 sites. The tip of the crack was enclosed by a singular finite element, and the displacement field was approximated with the use of the known solution for the crack-containing region. It turned out that even without reducing grid spacing, the singular finite element scheme provided high accuracy in estimating stress intensity factors and in modeling the singular stress field in the vicinity of the crack tip. For the details of the procedure, the reader is referred to Parton and Boriskovsky (1989, 1990).

Numerical results were obtained for $\gamma = 0.667$, equivalent to the $2:1$ ratio of the layer thicknesses in the unit cell. The length of the crack was taken to be $2a = 20.333\varepsilon$, the tip was assumed to lie in the interior of the wider layer, and the values of the parameters $\tilde{\varepsilon}_1$ and $\tilde{\varepsilon}_2$ (see Figure 10.3) were chosen to be 2.333ε and 2.663ε, respectively. Of the two cases considered in the study, one was obtained from the other by interchanging the layer materials, so that in Case I the tip of the crack lay in the stiffer material and in Case II it lay in the softer material. The elastic parameters of the layers were $E = 20$ and 1.5 GPa and, respectively, $\nu_1 = 0.3$ and 0.446. In either case, the values of the stress intensity factors and the local stress fields in the vicinity of the crack tip were calculated.

The stress intensity factor

$$\frac{K_I}{p\sqrt{\pi a}} = \lim_{x_1 \to a+0} \left[\sqrt{2\pi(x_1 - a)}\, \sigma_{22}(x_1, 0) \right]$$

was found to be about 3.21 in Case I and about 1.18 in Case II. The local stress distributions σ_{11}/p and σ_{22}/p shown in Figures 10.6 and 10.7 (Case I) and in Figures 10.8 and 10.9 (Case II) are seen to exhibit strong oscillations in the vicinity of the crack tip. The stresses σ_{22}, for example, undergo a marked jump at the interface between the layers, the values in the stiffer material exceeding by a factor of 10 or more those in the softer one (see Figures 10.7 and 10.9).

The reliability of the results obtained was estimated by fine-grid test calculations of the stress intensity factors for the entire plane composite with a central crack and the same geometric and material parameters. The accuracy, in this sense, was found to be within 12%, or probably even better if the error of the test calculations themselves (unavoidable for highly nonhomogeneous media) were to be taken into account. It was also of interest to compare the preceding approach with the approximating scheme developed by Parton and Kudryavtsev (1986). This latter was found to give strongly underestimated values for the stress intensity factor (1.15 and 0.73 in the first and second cases, respectively), indicative of the considerable role of the near-crack local boundary effects.

11. HOMOGENIZATION OF THE GEOMETRICALLY NONLINEAR ELASTICITY PROBLEM FOR A PERIODIC COMPOSITE. ELASTIC STABILITY EQUATIONS

The problem of predicting the stability of composite material structures is as important as it is complex. Because of the inherent high inhomogeneity of such materials, the equations of nonlinear elasticity (of the geometri-

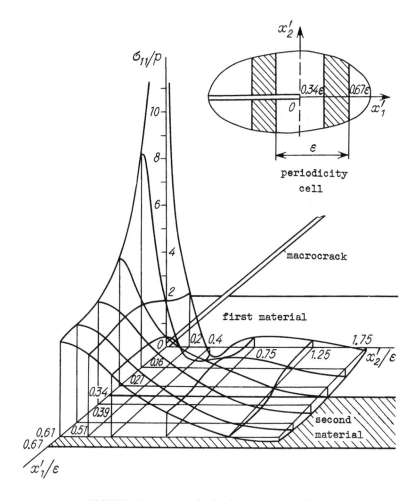

FIGURE 10.6. Local distribution of σ_{11}/p in Case I.

cally nonlinear elasticity in the case of small undercritical deformations) are virtually intractable unless simplifying assumptions are made or the homogenization principles adopted.

In this section we follow the presentation of Parton and Kalamkarov (1988a, 1988b), who employ the asymptotic method of homogenization in constructing the equations of the geometrically nonlinear theory of elasticity for composites with a regular structure. Based on these equations, the elastic stability of three-dimensional bodies made of such materials is examined, under the assumption that functions determining the stress conditions at the moment of buckling vary on a scale much larger than the unit cell dimensions.

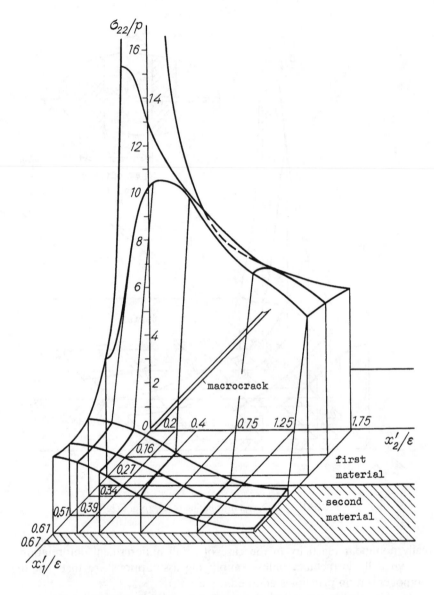

FIGURE 10.7. Local distribution of σ_{22}/p in Case I.

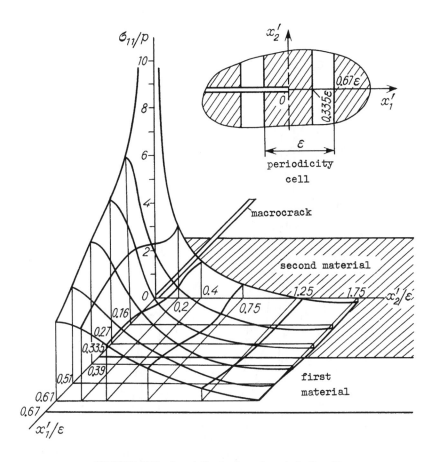

FIGURE 10.8. Local distribution of σ_{11}/p in Case II.

For small tensile and shear deformations, referring the body to a curvilinear coordinate system $(\theta_1, \theta_2, \theta_3)$, the nonlinear equations of motion may be represented in the form (cf. Novozhilov 1948, Guz' 1986)

$$\nabla_j(\sigma^{ij} + \sigma^{lj} \nabla_l u^i) + P^{*i} = \rho \ddot{u}^i \qquad (11.1)$$

where ∇_j denotes a covariant derivative, P^{*i} is the body force after the deformation, indices range from 1 through 3, and generalized stresses are considered to be indistinguishable from the true stresses, in view of the small deformation assumption.

The mechanical constitutive equations are taken to be Hooke's law (see, e.g., Sedov 1972).

$$\sigma^{ij} = c^{ijkl} e_{kl} \qquad (11.2)$$

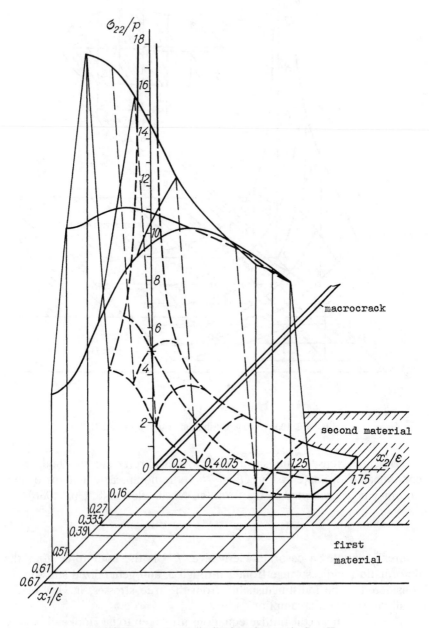

FIGURE 10.9.   Local distribution of σ_{22}/p in Case II.

in which the strain tensor is defined by

$$2e_{kl} = \nabla_l u_k + \nabla_k u_l + \nabla_k u_m \nabla_l u^m \tag{11.3}$$

The boundary conditions on stress and displacements on the surface $S = S_1 \cup S_2$ of the body are

$$(\sigma^{ij} + \sigma^{lj} \nabla_l u^i)n_j \overset{S_1}{=} p^{*i} \qquad u_i \overset{S_2}{=} u_i^* \tag{11.4}$$

where n_j is the unit vector normal to the surface of the nondeformed body and p^{*i} are the external surface tractions acting in the deformed body. Note that both Equations 11.1 and 11.4 take a more compact form if we introduce

$$t^{ij} = \sigma^{ij} + \sigma^{li} \nabla_l u^j \tag{11.5}$$

the (nonsymmetric) Kirchhoff stress tensor (Guz' 1986).

Consider the problem given by Equations 11.1 through 11.4 for a body made from a periodic composite with the unit cell $Y : \{-h_1/2 < y_1 < h_i/2\}$, $y_1 = \theta_i/\varepsilon$, $\varepsilon \ll 1$, $h_i \sim 1$ $(i = 1, 2, 3)$. The mass density and elastic coefficients are represented in this case by piecewise-smooth periodic functions of the "rapid" variables $y = (y_1, y_2, y_3)$, with Y as the unit cell.

We solve the problem by asymptotically expanding the solution in powers of the small parameter ε in the following manner:

$$u_i = u_i^{(0)}(\theta, t) + \varepsilon u_i^{(1)}(\theta, t, y) + \varepsilon^2 u_i^{(2)}(\theta, t, y) + \cdots \tag{11.6}$$

where $u_i^{(m)}(\theta, t, y)$ $(m = 1, 2, \dots)$ are periodic functions of y with a unit cell Y, and $\theta = (\theta_1, \theta_2, \theta_3)$. In accordance with the two-scale expansion method, and remembering that the metric tensor g_{ij} is independent of the "rapid" coordinates y, the operator ∇_i will be replaced everywhere by $\nabla_i + (1/\varepsilon) \partial/\partial y_i$.

From Equations 11.2, 11.3, 11.5, and 11.6, introducing the notation $\partial u_j/\partial y_i = u_{j|i}$, the leading terms in the stress and strain expressions are found to be given by

$$2e_{kl}^{(0)} = \nabla_l u_k^{(0)} + \nabla_k u_l^{(0)} + \nabla_k u_m^{(0)} \nabla_l u^{(0)m} + u_{l|k}^{(1)} + u_{k|l}^{(1)}$$

$$+ u_{m|k}^{(1)} \nabla_l u^{(0)m} + u_{|l}^{(1)m} \nabla_k u_m^{(0)} + u_{m|k}^{(1)} u_{|l}^{(1)m}$$

$$\sigma^{(0)ij} = c^{ijkl} e_{kl}^{(0)} \tag{11.7}$$

$$t^{(0)ij} = \sigma^{(0)ij} + \sigma^{(0)li} \left(\nabla_l u^{(0)j} + u_{|l}^{(1)j} \right)$$

Setting to zero the terms $O(\varepsilon^{-1})$ and $O(\varepsilon)$ in Equations 11.1 and 11.4 results in

$$t^{(0)ji}_{|j} = 0 \qquad \nabla_j t^{(0)ji} + t^{(1)ji}_{|j} + P^{*i} = \rho \ddot{u}^{(0)i} \tag{11.8}$$

$$t^{(0)ji} n_j \overset{S_1}{=} p^{*i} \qquad u^{(0)}_i \overset{S_2}{=} u^*_i \tag{11.9}$$

Expressions 11.7 for $t^{(0)ij}$ are now substituted into the first of Equations 11.8, and products of three and more rapid-coordinate derivatives $\nabla_i u^{(0)}_j$ are discarded because of their smallness. The solution of the resulting equation for $u^{(1)}_k$ may be taken in the form

$$u^{(1)}_k = N^{mn}_k \nabla_n u^{(0)}_m + W^{mnpq}_k \nabla_n u^{(0)}_m \nabla_q u^{(0)}_p \tag{11.10}$$

if the $N^{mn}_k(y)$ and $W^{mnpq}_k(y)$ are periodic functions of period Y solving the local problems

$$C^{ijmn}_{|i} = 0 \qquad C^{ijmn} = c^{ijmn} + c^{ijkl} N^{mn}_{k|l} \tag{11.11}$$

$$B^{ijmnpq}_{|i} + C^{ilmn} g^{kj} N^{pq}_{k|li} = 0$$

and

$$B^{ijmnpq} = c^{ijkl} W^{mnpq}_{k|l} + \tfrac{1}{2} c^{ijnq} g^{mp} + c^{ijkn} g^{ms} N^{pq}_{s|k}$$

$$+ \tfrac{1}{2} c^{ijkl} g^{sr} N^{mn}_{s|k} N^{pq}_{r|l} \tag{11.12}$$

with the perfect contact continuity conditions

$$[N^{mn}_k] = 0 \qquad \left[C^{ijmn} n^{(k)}_i\right] = 0 \tag{11.13}$$

$$[W^{mnpq}_k] = 0 \qquad \left[\left(B^{ijmnpq} + C^{ilmn} g^{kj} N^{pq}_{k|l}\right) n^{(k)}_i\right] = 0 \tag{11.14}$$

on surfaces where discontinuities in material characteristics occur; $n^{(k)}_i$ represents the normal to the discontinuity surface.

The local problem given by Equations 11.11 and 11.13 coincides with the corresponding local problem obtained in Section 7 in the process of homogenization of linear elasticity equations. The second local problem, Equations 11.12 and 11.14, is identical in structure, is linear in the unknown functions, and incorporates the solution of the first problem as its input data. Either problem has a unique solution, to within a constant term (see Bakhvalov and Panasenko 1984).

Combining Equations 11.7 and 11.10 results in the forms

$$\sigma^{(0)ij} = C^{ijmn}\,\nabla_n u_m^{(0)} + B^{ijmnpq}\,\nabla_n u_m^{(0)}\,\nabla_q u_p^{(0)}$$

$$t^{(0)ij} = \sigma^{(0)ij} + C^{ilmn}\,\nabla_n u_m^{(0)}\,\nabla_l u^{(0)j} + g^{kj}C^{ilmn}N_{k|l}^{pq}\,\nabla_n u_m^{(0)}\,\nabla_q u_p^{(0)} \tag{11.15}$$

which when subjected to the unit cell average operation

$$\langle\varphi\rangle = \frac{1}{|Y|}\int_Y \varphi\,dy$$

yield the pair of equations

$$\langle\sigma^{(0)ij}\rangle = \langle C^{ijmn}\rangle\nabla_n u_m^{(0)} + \langle B^{ijmnpq}\rangle\nabla_n u_m^{(0)}\,\nabla_q u_p^{(0)} \tag{11.16}$$

$$\langle t^{(0)ij}\rangle = \langle\sigma^{(0)ij}\rangle + \langle\sigma^{(0)li}\rangle\nabla_l u^{(0)j} \tag{11.17}$$

where use has been made of the fact that $\langle C^{ilmn}N_{k|l}^{pq}\rangle = 0$ because of Equations 11.11 and the periodicity in y.

Application of the same average operator to the second of Equations 11.8 gives

$$\nabla_j\!\left[\langle\sigma^{(0)lj}\rangle(g_l^i + \nabla_l u^{(0)i})\right] + P^{*i} = \langle\rho\rangle\ddot{u}^{(0)i}$$

$$\langle\sigma^{(0)lj}\rangle(g_l^i + \nabla_l u^{(0)i})n_j \overset{S_1}{=} p^{*i} \tag{11.18}$$

$$u_i^{(0)} \overset{S_l}{=} u_i^*$$

after use of Equations 11.17 and the periodicity in y.

Equations 11.16 and 11.18 comprise the homogenized problem of small strain nonlinear elasticity and, together with Relations 11.15, provide the zeroth approximation to the local structure of the stress field in the material under study. In the limiting case $c^{ijkl} = $ constant (homogeneous material), we find that

$$C^{ijmn} = c^{ijmn} \quad\text{and}\quad B^{ijmnpq} = \tfrac{1}{2}g^{mp}c^{ijnq}$$

from Equations 11.11 and 11.12, and the problem given by Equations 11.16 and 11.18 reduces to a known one.

We can now proceed to the problem of the elastic stability of three-dimensional bodies made from composite materials with a regular structure. Because our discussion will be limited to the leading term in Expansion 11.6, the superscript (0) will be omitted in the homogenized problem given by Equations 11.16 and 11.18.

Instead, we append it, unbracketed, on functions related to the (unperturbed) state of equilibrium, and, as is customary, we describe the perturbed state by the functions $u_i^0 + v_i$, $\sigma^{0ij} + \sigma^{ij}$, and $e_{ij}^0 + e_{ij}$, in which the second terms denote perturbations from the corresponding equilibrium values. We substitute these expressions into Equations 11.16 and 11.18 and make use of the fact that the unperturbed quantities satisfy the equations of equilibrium.

Linearizing with respect to the displacement perturbations v_i then yields

$$\nabla_j\left[\langle\sigma^{lj}\rangle(g_i^i + \nabla_l u^{0i}) + \langle\sigma^{0lj}\rangle\nabla_l v^i\right] = \langle\rho^0\rangle\ddot{v}^i$$

$$\left[\langle\sigma^{lj}\rangle(g_i^i + \nabla_l u^{0i}) + \langle\sigma^{0lj}\rangle\nabla_l v^i\right]n_j \overset{S_1}{=} f^i \qquad (11.19)$$

$$v_i \overset{S_2}{=} 0$$

and from Equation 11.16 it follows that

$$\langle\sigma^{ij}\rangle = \langle C^{ijmn}\rangle\nabla_n v_m + 2\langle B^{ijmnpq}\rangle\nabla_n u_m^0 \nabla_q v_p \qquad (11.20)$$

Equations 11.19 were derived under the assumption that the external body forces are independent of displacements. For surface tractions, we have

$$f^i = p^{*i}(u_k^0 + v_k) - p^{*i}(u_k^0) \qquad (11.21)$$

For the special case of the follow-up loading, a formula for $\mathbf{f}^i$ may be found in the books by Novozhilov (1948) and Guz' (1986).

The equilibrium displacements u_i^0 are determined from the problem given by Equations 11.16 and 11.18, in which all nonlinear terms may be omitted because the turning angles are negligibly small in the state of equilibrium. This results in the following linear problem:

$$\nabla_j\langle\sigma^{0ij}\rangle + P^i = 0$$

$$\langle\sigma^{0ij}\rangle = \langle C^{ijmn}\rangle\nabla_n u_m^0$$

$$\langle\sigma^{0ij}\rangle n_j \overset{S_1}{=} p^i \qquad\qquad (11.22)$$

$$u_i^0 \overset{S_2}{=} u_i^*$$

In view of the smallness of $\nabla_l u^{0i}$, Relations 11.19 reduce to the problem

$$\nabla_j \left[\langle \sigma^{ij} \rangle + \langle \sigma^{0lj} \rangle \nabla_l v^i \right] = \langle \rho^0 \rangle \ddot{v}^i$$

$$(\langle \sigma^{ij} \rangle + \langle \sigma^{0lj} \rangle \nabla_l v^i) n_j \overset{S_1}{=} f^i \qquad (11.23)$$

$$v_i \overset{S_2}{=} 0$$

which can be simplified further by assuming that the turning angles in the perturbed state are much greater than the strain components (Novozhilov 1948, Guz' 1986).

It will be understood that, strictly speaking, the problems given by Equations 11.19 through 11.23 do not describe the stability of the original composite medium, but rather relate to the homogenized medium whose effective properties are determined from the local problems given by Equations 11.11 through 11.14. Therefore, for the critical loads so obtained to the realistic, it is a necessary condition that the unit cell be small on the length scale for the variation of the functions determining the stress (strain) state of the composite at the moment of the loss of stability.

4

Coupled Fields and Mechanics of Periodic Composites

One of the most important applications of homogenization ideas is for analyzing coupled fields in deformable composite media with a regular structure. This section treats problems concerned with the coupling of elastic variables to thermal and/or electrical variables. In particular, the effective properties of piezoelectric laminar and unidirectional fiber composites are derived, and the behavior of electrically conducting, periodically nonhomogeneous piezoelectric media is considered. For a homogenized model of a porous elastic body containing a viscous fluid, it is shown that the homogenized equations of motion are in fact those of linear viscoelasticity. The use of the effective property results is illustrated by considering the problem of designing a laminar composite with prescribed stiffness and thermal properties.

12. THERMOELASTICITY OF COMPOSITES WITH REGULAR STRUCTURE

Thermoelasticity may perhaps be considered as a classical example of a theory involving the interaction of physical fields of different natures. Ene (1983) and Parton et al. (1986b, 1987) have applied the method of homogenization to the coupled thermoelasticity analysis of a regularly nonhomogeneous composite. Paša (1983) has proved that the solution of a nonhomogeneous problem converges to that of the homogenized coupled thermoelasticity problem in the limit as the period of the structure tends to zero.

Following Parton et al. (1987), we consider a non-steady-state problem of coupled thermoelasticity for a body made of an anisotropic nonhomogeneous material (a composite). Let the position of a typical point of the body be denoted by three coordinates (η_1, η_2, η_3) of a curvilinear system of axes, and let the unit cell of the structure, Y, be defined by the inequalities

$$-h_i/2 < y_i < h_i/2$$

where $y_i = \eta_i/\varepsilon$, $\varepsilon \ll 1$ and $h_i \sim 1$ $(i = 1, 2, 3)$.

The linear version of the equation of motion given by Equation 11.1 is

$$\nabla_j \sigma^{ij} + P^i = \rho \ddot{u}^i \tag{12.1}$$

where, once again, ∇_j denotes a covariant derivative.

The stress and strain tensors are related to the temperature change by the equation

$$\sigma^{ij} = c^{ijkl} e_{kl} - \beta^{ij}\theta$$

$$\beta^{ij} = c^{ijkl}\alpha_{kl}^T \tag{12.2}$$

known as the Duhamel–Neumann law (Kovalenko 1970, Pobedrya and Sheshenin 1981), the strain tensor being defined by

$$2e_{kl} = \nabla_l u_k + \nabla_k u_l \tag{12.3}$$

The heat flow vector is expressed in terms of temperature by the Fourier law

$$q^i = -\lambda^{ij} \nabla_j \theta \tag{12.4}$$

where λ^{ij} denotes the thermal conductivity components, and the heat balance equation is taken in the form (Pobedrya and Sheshenin 1981)

$$-\nabla_i q^i - T_0 \beta^{ij} \dot{e}_{ij} = c_v \dot{\theta} - f \tag{12.5}$$

where c_v is the volumetric specific heat and f is the density of internal heat sources. As in Section 11 (Chapter 3), all coefficients in Relations 12.1 through 12.5 are considered to be piecewise-smooth periodic functions of the coordinates y_1, y_2, y_3 with unit cell Y.

Equations 12.1 and 12.5 together with Relations 12.2 through 12.4 form a closed system of equations of linear coupled thermoelasticity for an anisotropic nonhomogeneous solid. To this system we must adjoin appropriate boundary and initial conditions. By analogy with Equation 11.4, the placement and traction conditions on the surface $\partial\Omega = \partial_1\Omega \cup \partial^2\Omega$ of the body are taken to be

$$\sigma^{ij}\mathbf{n}_j =_{\partial_1 \Omega} \mathbf{p}^i \qquad u_i =_{\partial_2 \Omega} u_i^* \tag{12.6}$$

and the heat exchange conditions are written as

$$q^j \mathbf{n}_j =_{\partial \Omega} \alpha_S T - q_S^* \tag{12.7}$$

where α_S represents the heat transfer coefficient and q_S^* is the external heat flow. In the special case of the heat exchange of the third kind we have

$$q_S^* = \alpha_S T_S^* \tag{12.8}$$

where T_S is the ambient temperature.

Finally, the initial conditions (at $t = 0$, t being time) are specified in the form

$$u_i = u_i^0 \qquad \dot{u}_i = v_i^0 \qquad T = T^0 \tag{12.9}$$

Because the degree of strain–temperature coupling is usually insignificant in practical applications, it is a common practice in problems of the kind we are discussing to neglect effects of deformation on the temperature distribution in the material. It is known, however, that the inclusion of the coupling may even change the qualitative nature of the solution of a dynamic problem, as exemplified by the work of Podstrigach and Shvets (1978) on the thermoelasticity of thin shells. On the other hand, there is evidence that in some polymeric materials thermoelastic coupling may be strong enough to be considered, especially when impact-type loads are applied (Kovalenko 1970). As shown by Lukovkin et al. (1983), a jump in temperature due to an impact load may play a crucial role in the fracture of glassy polymers, the reason lying in the specifics of deformation and the low thermal conductivity of such materials rather than in a large thermoelastic coupling. Because composite materials very often have polymers as their matrices, clearly the preceding facts are of special significance in the present context.

The solution of the problem will be sought by assuming the asymptotic small-parameter expansions of the form

$$u_i = u_i^{(0)}(\eta, t) + \varepsilon u_i^{(1)}(\eta, t, y) + \varepsilon^2 u_i^{(2)}(\eta, t, y) + \cdots$$

$$\theta = \theta^{(0)}(\eta, t) + \varepsilon \theta^{(1)}(\eta, t, y) + \varepsilon^2 \theta^{(2)}(\eta, t, y) + \cdots \tag{12.10}$$

where $u_i^{(m)}(\eta, t, y)$ and $\theta^{(m)}(\eta, t, y)$ $(m = 1, 2, \dots)$ are periodic functions of y with unit cell Y.

Without going through the details of the homogenization procedure here, the leading terms in Expansions 12.10 are found to be given by

$$u_k = u_k^{(0)}(\eta, t) + \varepsilon \left[N_k^{mn} \nabla_n u_m^{(0)} + M_k \theta^{(0)} \right] + \cdots$$

$$\theta = \theta^{(0)}(\eta, t) + \varepsilon W^n \nabla_n \theta^{(0)} + \cdots$$

(12.11)

where $N_k^{mn}(y)$, $M_k(y)$, and $W^n(y)$ are periodic functions with unit cell Y, of which N_k^{mn} solves the local problem given by Equations 11.11 and 11.13, resulting from the homogenization of the elasticity theory problem, and M_k and W^n are solutions of the following local problems:

$$S_{|i}^{ij} = 0 \qquad S^{ij} = \beta^{ij} - c^{ijkl} M_{k|l}$$

(12.12)

$$\Lambda_{|i}^{ij} = 0 \qquad \Lambda^{ij} = \lambda_{ij} + \lambda^{ik} W_{|k}^j$$

(12.13)

The continuity conditions to be satisfied across the surfaces of discontinuity in material properties are similar to those given by Equation 11.13, that is,

$$[M_k] = 0 \qquad \left[S^{ij} n_i^{(k)} \right] = 0$$

(12.14)

$$[W^k] = 0 \qquad \left[\Lambda^{ij} n_i^{(k)} \right] = 0$$

(12.15)

In accordance with Equations 12.11, the leading-order terms in the stress tensor and heat flow vector expansions are given by

$$\sigma^{(0)ij} = C^{ijmn} \nabla_n u_m^{(0)} - S^{ij} \theta^{(0)}$$

$$q^{(0)i} = -\Lambda^{ij} \nabla_j \theta^{(0)}$$

(12.16)

In a manner parallel to that discussed in Section 11, we set to zero the terms $O(1)$ in Equations 12.1 and 12.5 and take a volume average over the unit cell Y to obtain

$$\nabla_j \langle \sigma^{(0)ij} \rangle + P^i = \langle \rho \rangle \ddot{u}^{(0)i}$$

$$- \nabla_i \langle q^{(0)i} \rangle - T_0 \langle B^{ij} \rangle \nabla_j \dot{u}_i^{(0)} = \langle C_v \rangle \dot{\theta}^{(0)} - f$$

(12.17)

where we have introduced

$$B^{ij} = C^{klij} \alpha_{kl}^T$$

$$C_v = c_v + T_0 \alpha_{kl}^T (\beta^{kl} - S^{kl})$$

(12.18)

Averaging Equations 12.16 results in the equations

$$\langle \sigma^{(0)ij} \rangle = \langle C^{ijmn} \rangle \nabla_n u_m^{(0)} - \langle S^{ij} \rangle \theta^{(0)}$$

$$\langle q^{(0)i} \rangle = -\langle \Lambda^{ij} \rangle \nabla_j \theta^{(0)}$$

(12.19)

which are the constitutive relations of the homogenized medium, and whose coefficients represent the (effective) elastic and thermal properties of the medium. It is important to note that

$$\langle B^{ij} \rangle = \langle S^{ij} \rangle$$

(12.20)

To prove this, combine Equations 11.11, 12.2, and 12.18 to give

$$B^{ij} = \beta^{ij} + \beta^{mn} N_{m|n}^{ij}$$

(12.21)

Using the periodicity in y and the local problem given by Equations 12.12, observe that

$$\langle \beta^{mn} N_{m|n}^{ij} \rangle = -\langle N_m^{ij} \beta_{|n}^{mn} \rangle$$

$$= -\langle N_m^{ij} \left(c^{mnkl} M_{k|l} \right)_{|n} \rangle$$

$$= c^{mnkl} M_{k|l} N_{m|n}^{ij} \rangle$$

(12.22)

Similarly, from the local problem given by Equations 11.11, obtain

$$-\langle c^{ijkl} M_{k|l} \rangle = \langle M_k c_{|l}^{ijkl} \rangle$$

$$= -\langle M_k \left(c^{klmn} N_{m|n}^{ij} \right)_{|l} \rangle$$

$$= \langle c^{klmn} N_{m|n}^{ij} M_{k|l} \rangle$$

(12.23)

Comparing Equations 12.22 and 12.23 and using the symmetry properties of the elastic modulus tensor, we now find that

$$\langle \beta^{mn} N_{m|n}^{ij} \rangle = -\langle c^{ijkl} M_{k|l} \rangle$$

(12.24)

which proves Equation 12.20 when combined with Equations 12.12 and 12.21.

To obtain the boundary and initial conditions of the homogenized problem, take a volume average of Equations 12.6 and 12.7 and retain only the leading-order terms in Expansions 12.11 to arrive at

$$\langle \sigma^{(0)ij} \rangle n_j =_{\partial_1 \Omega} p^i \qquad u_i =_{\partial_2 \Omega} u_i^*$$

$$\langle q^{(0)j} \rangle n_j =_{\partial \Omega} \langle \alpha_S \rangle T^{(0)} - q_s^* \tag{12.25}$$

$$u_i^{(0)} = u_i^0 \qquad \dot{u}_i^{(0)} = v_i^0 \qquad T^{(0)} = T^0 \quad \text{for } t = 0$$

Relations 12.17 through 12.20 and Equations 12.25 describe the homogenized problem of thermoelasticity for the medium being examined. The effective properties of the medium are calculated from the solutions of the local problems given by Equations 11.11, 11.13, 12.12, 12.14, and 12.13 and 12.15, and Relations 12.16 provide the zeroth-order approximation to the local structure of the stress and heat flow fields. Note that the solutions of the local problems are only unique to within constant terms, an ambiguity that is easily removed by introducing the conditions

$$\langle N_k^{mn} \rangle = 0 \qquad (N_k^{mn} \leftrightarrow M_k \leftrightarrow W^n) \tag{12.26}$$

Averaging Equations 12.11 then yields the formulas

$$\langle u_k \rangle = u_k^{(0)}(\eta, t) \quad \text{and} \quad \langle \theta \rangle = \theta^{(0)}(\eta, t) \tag{12.27}$$

clarifying the meaning of the leading-order terms in Expansions 12.11.

If coupling effects may be neglected, the term in $\nabla_j \dot{u}_i^{(0)}$ in the heat balance equation (Equations 12.17) should be omitted, and the effective specific heat of the system should be written as $\langle C_v \rangle = \langle c_v \rangle$, equivalent to the use of the rule of mixtures.

13. FIBER COMPOSITES: LOCAL STRESSES AND EFFECTIVE PROPERTIES

We shall be concerned in this section with predicting the mechanical behavior of unidirectional fiber-reinforced composites with a regular structure. The unit cell in this case is obtained by cutting the material by a plane orthogonal to the direction of fibers and may be represented, in particular, by a square containing an inclusion in the form of the fiber cross section. If the fibers are aligned along the y_3 axis, then, for homogeneous matrix and fiber materials, all the characteristics of the composite will be described by piecewise-smooth, doubly periodic functions of y_1 and

y_2. The local problem given by Equations 11.11 reduces in this case to the following system:

$$\tau_{|\alpha}^{\alpha jmn}(F) = 0 \quad \text{in the region occupied by the fibers}$$

$$\tau_{|\alpha}^{\alpha jmn}(M) = 0 \quad \text{in the region occupied by the matrix}$$

$$(13.1)$$

$$\tau^{\alpha jmn}(F) = c^{\alpha jk\lambda}(F)n_{k|\lambda}^{mn}(F)$$

$$\tau^{\alpha jmn}(M) = c^{\alpha jkl}(M)N_{k|\lambda}^{mn}(M)$$

(where $\alpha, \lambda = 1, 2$ *and* $j, k, m, n = 1, 2, 3$) whose solutions, apart from double periodicity property, must also satisfy Conditions 11.13 of the perfect bond on the interface Γ, that is,

$$N_k^{mn}(F)|_\Gamma = N_k^{mn}(M)|_\Gamma$$

$$(13.2)$$

$$(\tau^{\alpha jmn}(F) + c^{\alpha jmn}(F))n_\alpha^{(k)}|_\Gamma = (\tau^{\alpha jmn}(M) + c^{\alpha jmn}(M))n_\alpha^{(k)}|_\Gamma$$

where $c^{\alpha jmn}(F)$ and $c^{\alpha jmn}(M)$ are elasticity tensors of, respectively, the fiber and matrix materials, and $n_\alpha^{(k)}$ is the unit vector in the outward normal direction to the surface Γ ($\alpha = 1, 2$).

As already discussed in Section 9 (Chapter 3), the specific structure of the local problems given by Equations 13.1 and 13.2 depends on the symmetry properties of the matrix and fiber materials. If both materials are isotropic, for example, we have to solve four similar plane strain problems for $\tau^{\alpha\beta 11}$, $\tau^{\alpha\beta 22}$, $\tau^{\alpha\beta 33}$, and $\tau^{\alpha\beta 12}$, and two antiplane problems for $\tau^{\alpha 323}$ and $\tau^{\alpha 313}$.

It was noted earlier that if there are some elements of symmetry in the geometry of the unit cell or in the properties of the constituents of the composite, a unit cell local problem reduces to a boundary value problem for only a part of the cell, with a resultant reduction in the amount of numerical work.

A similar analysis can be carried out in the case of unidirectional fiber composites for the local problems given by Equations 12.12 through 12.14 and by Equations 12.13 and 12.15. Because

$$\langle S^{ij} \rangle = \langle B^{ij} \rangle = \langle \beta^{ij} + \beta^{mn} N_{m|n}^{ij} \rangle \tag{13.3}$$

from Equations 12.20 and 12.21, it follows that the effective thermoelastic coefficients $\langle S^{ij} \rangle$ can be calculated using the solution of the local elasticity problem given by Equations 13.1 and 13.2.

The method used by Parton et al. (1987) in their finite-element analysis of the local problems given by Equations 13.1 and 13.2 consists of dividing a quarter of the unit cell of the composite into a set of triangular elements, numbering from 350 to 400 for different composite combinations. The good agreement between the effective moduli predicted and the relevant results known from earlier work (Pobedrya 1984, Mol'kov and Pobedrya 1985) encouraged the application of the solutions so obtained to the calculation of the effective properties of various types of fiber composites.

The authors considered, in particular, a system of transversely isotropic carbon and organic fibers of circular cross section embedded in an isotropic plastic (EDT-10) matrix. The properties of the fiber materials are listed in Table 13.1, reproduced from Maksimov et al. (1983); the characteristics of the matrix material are (from the same source) $E_M = 3.236$ GPa and $\nu_M = 0.42$. The engineering properties of these systems were computed by the use of the local problem solutions and are represented in Table 13.2 together with the corresponding experimental results taken, again, from Maksimov et al. (1983). The agreement between the two sets of values may be considered as quite satisfactory, especially if one notes that, of the experimental values shown, only E_3 had been measured directly. The other four had been computed from approximate formulas whose inaccuracy accounts at least in part for the discrepancies in the values of E_1 and G_{13}.

The effects of unit cell geometry and fiber volume fraction on the effective properties of a composite were examined by Parton et al. (1987) by applying their method to a composite of Al–B–Si fibers in an epoxy matrix, the epoxy being either pure or reinforced with a fine elastomer dispersion (which is an effective means for enhancing the impact strength of a material (see, e.g., Lukovkin et al. 1983). The elastic and thermal properties of the fiber and matrix materials are shown in Table 13.3.

Of the four major geometries considered in the study (see Figure 13.1), the first three differ in their matrix materials as follows:

I   Pure epoxy
II A macroscopically isotropic polydisperse medium (see Christensen 1979) formed by spherical rubber inclusions of volume fraction 0.2 embedded in epoxy
III A transversely isotropic material formed by rubber fibers of volume fraction 0.2 embedded in epoxy (note the elongated shape of the inclusions)

In composition IV, the matrix material is again pure epoxy, but this time there is a thin rubber coating around each fiber; the volume fraction of the rubber in the region outside the fibers is again 0.2.

TABLE 13.1
Fiber Elastic Properties

	Organic fiber	Carbon fiber
E_3 (GPa)	127.500	225.600
E_1 (GPa)	3.236	7.751
G_{13} (GPa)	2.354	59.720
ν_{13}	0.29	0.30
ν_{12}	0.17	0.20

TABLE 13.2
Elastic Properties of Unidirectional Fiber-Reinforced
Composite Materials

	Organoplastic (fiber volume fraction 0.48)		Carboplastic (fiber volume fraction 0.40)	
	Predicted value	Experiment	Predicted value	Experiment
E_3 (GPa)	62.610	62.763 ± 4.630	92.150	87.279 ± 6.982
E_1 (GPa)	3.761	3.727 ± 0.333	7.035	5.884 ± 0.628
G_{13} (GPa)	1.814	1.569 ± 0.177	4.341	3.825 ± 0.451
ν_{13}	0.365	0.37 ± 0.06	0.374	0.38 ± 0.05
ν_{12}	0.489	0.45 ± 0.06	0.506	0.46 ± 0.06

The results of the numerical calculations are summarized in Tables 13.4 through 13.6, in which all the effective properties of the composites are related to the corresponding fiber material properties (see Table 13.3). The notation adopted in the tables is as follows: γ denotes the fiber volume fraction; K_{12} is the bulk elastic modulus for plane deformation normal to the fiber direction,

$$K_{12} = \tfrac{1}{2}(\langle C^{2222} \rangle + \langle C^{1122} \rangle)$$

K_F is the bulk modulus for the plane deformation of an isotropic fiber material. Based on the data given in Table 13.3, it is found that $K_F = 54.953$. The compositions are labeled in accordance with Figure 13.1.

Referring to Equations 12.18, the effective heat capacity is given by the expression

$$\langle C_v \rangle = \langle c_v \rangle + \left\langle T_0 \alpha_{kl}^T (\beta^{kl} - S^{kl}) \right\rangle \tag{13.4}$$

in which the first term corresponds to the rule of mixtures, whereas the second corrects for thermoelastic coupling effects and turns out to be negligibly small. For example, for $\gamma = 0.6$, with constituent properties

TABLE 13.3
Elastic and Thermal Properties of Some Matrix and Fiber Materials ($T = 293$ K)

Material				Properties			
	E (GPa)	ν	G (GPa)	α^T (K)$^{-1}$	β (N$/m^2 \cdot K$)	λ (Wt$/$m $\cdot$ K)	C_v (J$/m^3 \cdot$ K)
Al–B–Si fiber	73.0	0.23	29.68	5×10^{-6}	6.765×10^5	0.9	1.826×10^6
Epoxy	3.43	0.35	1.27	6×10^{-5}	6.852×10^5	0.198	1.533×10^6
Rubber	2.94×10^{-3}	0.4998	9.8×10^{-4}	2×10^{-4}	1.47×10^6	0.22	1.623×10^6
Epoxy with fine rubber dispersion (volume fraction 0.2)	2.22	0.393	7.974×10^{-1}	8.632×10^{-5}	8.938×10^5	0.202	1.551×10^6

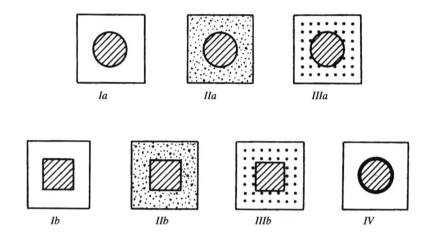

FIGURE 13.1.   Unit cell of the unidirectional fiber composite under study.

TABLE 13.4
Predicted Elastic and Thermal Effective Properties of Unidirectional
Glass-Reinforced Plastics ($T = 293$ K)

	Compositions					
	Ia	**IIa**	**IIIa**	**Ia**	**IIa**	**IIIa**
Properties	$\gamma = 0.2$	$\gamma = 0.2$	$\gamma = 0.2$	$\gamma = 0.4$	$\gamma = 0.4$	$\gamma = 0.4$
E_3/E_F	0.238	0.225	0.230	0.429	0.420	0.424
K_{12}/K_F	0.157	0.146	0.148	0.245	0.232	0.234
ν_{31}/ν_F	1.234	1.400	1.268	1.134	1.170	1.152
G_{32}/G_F	0.123	0.104	0.109	0.212	0.188	0.194
G_{12}/G_F	0.120	0.102	0.102	0.200	0.180	0.180
$\langle \Lambda^{11} \rangle / \lambda_F$	0.316	0.321	0.321	0.427	0.431	0.431
$\langle \Lambda^{33} \rangle / \lambda_F$	0.376	0.380	0.380	0.532	0.535	0.535
$\langle S^{11} \rangle / \beta_F$	1.012	1.294	1.272	1.010	1.265	1.244
$\langle S^{33} \rangle / \beta_F$	1.011	1.274	1.223	1.009	1.224	1.184

taken from Table 13.3, the rule of mixtures gives $\langle c_v \rangle = 1.709$ for the combinations Ia and Ib of Figure 13.1, whereas the (room-temperature) values of the correction term are 0.108 and 0.101 J/m^{3}K (that is, seven orders of magnitude less) for, respectively, the circular (composition Ia) and square (composition Ib) cross sections.

The agreement (or otherwise) between the predicted effective properties (shown in Tables 13.4 through 13.6) and their Hashin–Shtrikman variational estimates is illustrated in Figures 13.2 through 13.6 which show the effective engineering constants of the Ia composition as functions of the fiber volume fraction γ and in which the curves indicating the

TABLE 13.5
Predicted Elastic and Thermal Effective Properties of Unidirectional Glass-Reinforced Plastics ($T = 293$ K)

	Compositions						
Properties	**Ia** $\gamma = 0.6$	**Ib** $\gamma = 0.6$	**IIa** $\gamma = 0.6$	**IIb** $\gamma = 0.6$	**IIIa** $\gamma = 0.6$	**IIIb** $\gamma = 0.6$	**IV** $\gamma = 0.6$
E_3/E_F	0.619	0.620	0.613	0.614	0.616	0.617	0.616
K_{12}/K_F	0.347	0.422	0.331	0.405	0.333	0.408	0.333
ν_{31}/ν_F	1.081	1.060	1.104	1.077	1.092	1.068	1.090
G_{32}/G_F	0.316	0.391	0.282	0.361	0.291	0.368	0.223
G_{12}/G_F	0.285	0.366	0.261	0.342	0.261	0.342	0.275
$\langle \Lambda^{11} \rangle / \lambda_F$	0.556	0.603	0.560	0.607	0.560	0.607	0.560
$\langle \Lambda^{33} \rangle / \lambda_F$	0.688	0.688	0.690	0.690	0.690	0.690	0.690
$\langle S^{11} \rangle / \beta_F$	1.009	1.008	1.231	1.205	1.213	1.189	1.119
$\langle S^{33} \rangle / \beta_F$	1.007	1.006	1.173	1.162	1.145	1.134	1.097

TABLE 13.6
Predicted Elastic and Thermal Effective Properties of Unidirectional Glass-Reinforced Plastics ($T = 293$ K)

	Compositions					
Properties	**Ia** $\gamma = 0.7$	**Ib** $\gamma = 0.7$	**IIa** $\gamma = 0.7$	**IIb** $\gamma = 0.7$	**IIIa** $\gamma = 0.7$	**IIIb** $\gamma = 0.7$
E_3/E_F	0.715	0.715	0.710	0.710	0.712	0.712
K_{12}/K_F	0.415	0.496	0.395	0.476	0.398	0.480
ν_{31}/ν_F	1.061	1.044	1.078	1.057	1.070	1.050
G_{32}/G_F	0.389	0.463	0.347	0.427	0.358	0.436
G_{12}/G_F	0.338	0.428	0.309	0.401	0.309	0.401
$\langle \Lambda^{11} \rangle / \lambda_F$	0.635	0.682	0.639	0.685	0.639	0.685
$\langle \Lambda^{33} \rangle / \lambda_F$	0.766	0.766	0.767	0.767	0.767	0.767
$\langle S^{11} \rangle / \beta_F$	1.008	1.007	1.209	1.181	1.192	1.166
$\langle S_{33} \rangle / \beta_F$	1.006	1.005	1.145	1.133	1.122	1.111

Hashin–Shtrikman upper and lower bounds of the corresponding properties are drawn. Also shown are the values computed for composition Ib (closed squares) and composition IV (closed circles). Although for the E_3 property the agreement is all but perfect, it is seen from the figures that in other properties discrepancies up to a factor of 2 occur. In Figure 13.7, effective thermal conductivities are plotted as functions of γ and, for the purpose of comparison, the lower-bound curve is shown (Christensen 1979). For the effective property $\langle \Lambda^{33} \rangle$, the rule of mixtures is seen to be valid. Note that, at a qualitative level, very similar behavior is exhibited by compositions II and III.

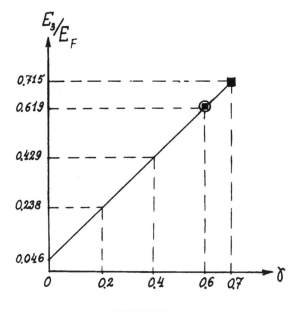

FIGURE 13.2.

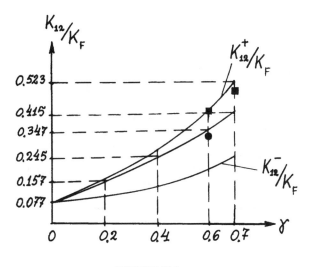

FIGURE 13.3.

Based on the analysis of Tables 13.4 through 13.6, the following general conclusions are made:

1. With constituent volume fractions respectively equal in any two composites, geometric structure generally influences the effective

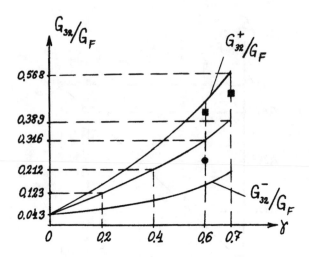

FIGURE 13.4.

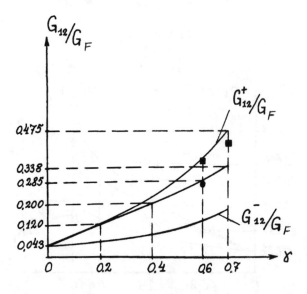

FIGURE 13.5.

properties. For example, the properties E_3, K_{12}, and G_{32} increase and ν_{31}, $\langle S^{11} \rangle$, and $\langle S^{33} \rangle$ decrease in passing from composition II to composition III in Figure 13.1. The properties G_{12} and $\langle \Lambda^{11} \rangle$ remain the same throughout.

2.   At $\gamma = 0.6$, components II, III, and IV are identical in terms of the volume fractions of their three components. The properties E_3, K_{12},

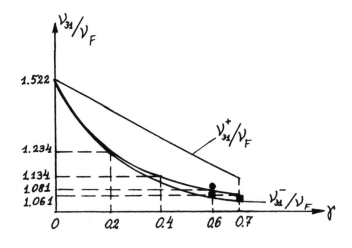

FIGURE 13.6.

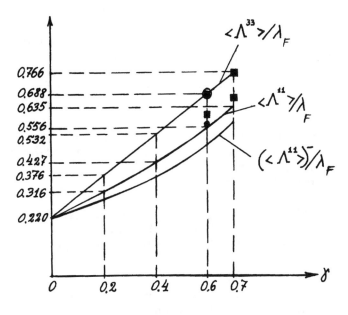

FIGURE 13.7.

and $\langle \Lambda^{11} \rangle$ coincide for compositions IIIa and IV, but in the latter case a higher value of G_{12} and lower values of ν_{31}, G_{32}, $\langle S^{11} \rangle$, and $\langle S^{33} \rangle$ are observed. Also, G_{12} in IV is greater than in IIa.

3.   For equal constituent volume contents, the shape of fiber cross section is a factor controlling the effective properties of the compos-

ite. The values of K_{12}, G_{32}, G_{12}, and $\langle \Lambda^{11} \rangle$, for example, increase from the circular to the square cross section (that is, from compositions I–IIIa to I–IIIb in Figure 13.1). For ν_{31}, $\langle S^{11} \rangle$, and $\langle S^{33} \rangle$, the reverse is the case.

To summarize, the results obtained from the solutions of the local problems make it possible to elucidate the effects of both component content and unit cell geometry on the mechanical behavior of a composite, thereby enabling the prediction and optimization of the properties of the composite.

Note also that, based on the numerical solutions of the local problems, the distribution of stress over the unit cell of the composite can be determined.

With knowledge of the distributions of the functions C^{ijmn} and S^{ij} and of the solution of the homogenized problem, Formula 12.6 will give, with a high degree of accuracy, the distribution of local stresses for various loading conditions and given temperature gradients. In particular, if the temperature is constant and macroscopic deformation uniform ($e_{11}^{(0)} = 0$), the functions C^{1111}, C^{2211}, and C^{3311} coincide with the respective stress components $\sigma_{11}^{(0)}$, $\sigma_{22}^{(0)}$, $\sigma_{33}^{(0)}$, and $\sigma_{12}^{(0)}$. This means that, apart from the first three components, this type of deformation also gives rise to the fourth, $\sigma_{12}^{(0)}$, which, as seen from Figure 13.8, is less than but of the same order of magnitude as the others. We note also, again with reference to Figure 13.8, that the maximum stress occurs in the vicinity of the fiber–matrix interface (see also Parton et al. 1989c).

Once maximum stress regions have been located and the values of the maxima calculated, the strength criteria known for the homogeneous component materials can be employed to determine failure conditions for the entire composite system.

14. LAMINATED COMPOSITE WITH PRESCRIBED THERMOELASTIC PROPERTIES. COMPOSITE MATERIAL DESIGN

The problem we consider here is concerned with the design of laminated composites and may be formulated as follows: Given prescribed overall properties and a set of local component properties of the composite, secure the former by varying the latter. The solution of the problem has been given by Kolpakov and Rakin (1986) and Alekhin et al. (1988) based on the earlier work of Pontryagin et al. (1976) and Alekseev et al. (1979). In this section we reproduce the basic results of Kolpakov and Rakin (1986), derived for a laminated composite with a periodic structure under the assumption that the components of the composite are isotropic and have equal Poisson ratios. The latter condition is approximately

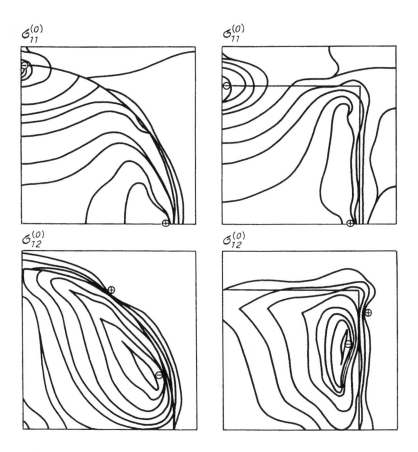

FIGURE 13.8. Local stress distributions for the case of macrodeformation, $e_{11}^{(0)} = 1$ (compositions Ia and Ib). Extremum stress values (in gigapascals): (Ia) $\sigma_{11}^+ = 78.675$, $\sigma_{11}^- = -93.078$, $\sigma_{12}^+ = 0.865$, $\sigma_{12}^- = -25.436$; (Ib) $\sigma_{11}^+ = 90.501$, $\sigma_{11}^- = -116.177$, $\sigma_{12}^+ = 5.144$, $\sigma_{12}^- = -61.232$.

fulfilled in most practical laminated composites with metal ($\nu \approx \frac{1}{3}$) or polymeric ($\nu \approx 0.4$) components.

For a laminated periodic composite with a characteristic lamina thickness $\varepsilon \ll 1$, the thermal and mechanical properties of interest are of course functions of only one variable, say, x_1. In the case we consider these are $c_v(x_1/\varepsilon)$ (the local bulk heat capacity), $\lambda(x_1/\varepsilon)$ (the local thermal conductivity), $E(x_1/\varepsilon)$ (Young's modulus), and $\alpha(x_1/\varepsilon)$ (the coefficient of linear expansion), all of which are 1-periodic in their argument, x_1 varying from zero to ε. If ε is allowed to approach zero, the solutions of the corresponding heat conduction and elasticity problems tend to those of the respective problems for a homogeneous anisotropic material described by the effective properties. Under the isotropic component assumption, the effective thermal and mechanical properties of the homogenized medium are given by (cf. Pobedrya 1984)

$$\tilde{c}_v = \langle c_v \rangle \tag{14.1}$$

$$\tilde{\lambda}_{11} = \left\langle \frac{1}{\lambda} \right\rangle^{-1} \qquad \tilde{\lambda}_{22} = \tilde{\lambda}_{33} = \langle \lambda \rangle \tag{14.2}$$

$$\tilde{h}_{11} = \frac{(1 + v)(1 - 2v)}{1 - v} \left\langle \frac{1}{E} \right\rangle + \frac{2v^2}{1 - v} \frac{1}{\langle E \rangle} \tag{14.3}$$

$$\tilde{h}_{22} = \tilde{h}_{33} = \frac{1}{\langle E \rangle}$$

$$\tilde{h}_{12} = \tilde{h}_{13} = \tilde{h}_{23} = -\frac{v}{\langle E \rangle}$$

$$\tilde{h}_{55} = \tilde{h}_{66} = \frac{1 + v}{2} \left\langle \frac{1}{E} \right\rangle \qquad \tilde{h}_{44} = \frac{1 + v}{2} \frac{1}{\langle E \rangle}$$

$$\tilde{\alpha}_{11} = \frac{1 + v}{1 - v} \langle \alpha \rangle - \frac{2v}{1 - v} \frac{\langle E\alpha \rangle}{\langle E \rangle} \tag{14.4}$$

$$\tilde{\alpha}_{22} = \tilde{\alpha}_{33} = \frac{\langle E\alpha \rangle}{\langle E \rangle}$$

In Equations 14.1 through 14.4,

$$\langle f \rangle = \int_0^\varepsilon f(x_1/\varepsilon)\, dx_1 = \int_0^1 f(t)\, dt$$

$\tilde{\lambda}_{11}$, $\tilde{\lambda}_{22}$, and $\tilde{\lambda}_{33}$ are the (only) nonzero elements of the heat conductivity tensor; $\tilde{h}_{\alpha\beta}$ is the elastic compliance tensor (in matrix notation, where α, $\beta = 1, 2, \ldots, 6$); $\tilde{\alpha}_{11}$, $\tilde{\alpha}_{22}$, and $\tilde{\alpha}_{33}$ are the nonzero elements of the linear expansivity tensor.

Equations 14.1 through 14.4 establish relationships between the effective properties of a laminated composite and its local properties; they hold for various types of distributions of local properties, the best known examples being a piecewise-continuous and a piecewise-constant distribution, of which the latter is found most frequently in laminates. To cover all the possibilities, it proves advisable (Kolpakov and Rakin 1986) to introduce a set of functions defined by

$$W = \{f(t) \in L_\infty[0,1]: 1/f(t) \in L_\infty[0,1]\}$$

where for any $f(t)$ there exists a positive number $\xi(f)$ such that $f(t) > \xi(f)$ for almost all $t \in [0,1]$, $t = x_1/\varepsilon$. We will assume, from this point on, that the functions $c_v(t)$, $\lambda(t)$, $E(t)$, and $\alpha(t)$ belong to the set W, and we will treat Equations 14.1 through 14.4 as the mapping

$$(c_v(t), \quad \lambda(t), E(t), \quad \alpha(t)) \in W^4 \to R^{16}$$

of the set of the distributions of local properties onto the set of the values of its effective properties. Denoting by $I(W^4)$ the image of the set W^4 upon the mapping I, we can express as follows the problem of designing a laminated composite with prescribed effective properties:

First, obtain a composite material possessing the prescribed effective properties $[(\tilde{c}_v^{(0)}, \quad \tilde{\lambda}_{ij}^{(0)}, \quad \tilde{h}^{(0)}, \quad \tilde{\alpha}_{ij}^{(0)}) \in I(W^4)]$. Second, give a method for producing such a composite.

The problem is thus solved by first ascertaining whether or not a material with the prescribed properties may at all exist (at which stage a description of the set $I(W^4)$ is all the information needed) and then by finding a way for constructing a set of functions $(c_v(t), E(t), \lambda(t))$ such that $I(c_v, \lambda, E, \alpha) = x$ for $x \in R^{16}$. The second of these subproblems is thus that of synthesis for the image I (see Pontryagin et al. 1976, Alekseev et al. 1979). It should also be noted that Equations 14.1 through 14.4 are mutually independent and the effective properties they determine are in a one-to-one correspondence with the functionals of the form

$$\int_0^1 u_1(t)\, dt \qquad \int_0^1 \frac{dt}{u_1(t)} \qquad \int_0^1 u_2(t)\, dt \qquad \int_0^1 u_1(t)u_2(t)\, dt \quad (14.5)$$

where, for Equations 14.3 and 14.4, $u_1(t) = E(t)$, $u_2(t) = \alpha(t)$, and in Equations 14.1 and 14.2 the functionals of the first and second kinds are involved. It suffices, therefore, to solve the problem for Mappings 14.5 of the set W^2 onto R^4. We reproduce here only the final results of the analysis, referring the reader to the original paper, Kolpakov and Rakin (1986), for a treatment in detail.

First it is stated that, to within the boundaries of the sets involved, laminated structures composed of isotropic materials with equal Poisson ratios may have only the following effective properties.

1. Heat capacity:

$$\tilde{c}_v = X \qquad (X > 0) \tag{14.6}$$

2. Thermal conductivity tensor:

$$\tilde{\lambda}_{11} = Y \qquad \tilde{\lambda}_{22} = \tilde{\lambda}_{33} = Z \qquad \left(Z > 0, Y > \frac{1}{Z}\right) \qquad (14.7)$$

3. Young's moduli, Poisson ratios, and shear moduli, respectively:

$$\tilde{E}_1 = \frac{(1 - \nu)x}{(1 + \nu)(1 - 2\nu)xy + 2\nu^2} \qquad \tilde{E}_2 = \tilde{E}_3 = x$$

$$\tilde{\nu}_{12} = \tilde{\nu}_{13} = \frac{\nu(1 - \nu)}{(1 + \nu)(1 - 2\nu)xy + 2\nu^2} \qquad \tilde{\nu}_{23} = \nu \qquad (14.8)$$

$$\tilde{G}_{12} = \tilde{G}_{13} = \frac{2}{(1 + \nu)y} \qquad \tilde{G}_{23} = \frac{2x}{1 + \nu}$$

4. The (linear) expansivity tensor:

$$\tilde{\alpha}_{11} = \frac{1 + \nu}{1 - \nu}z - \frac{2\nu}{1 - \nu}\frac{t}{x} \qquad \tilde{\alpha}_{22} = \tilde{\alpha}_{33} = \frac{t}{x}$$

$$\left(x > 0, y > \frac{1}{x}, z > 0, t > 0\right) \qquad (14.9)$$

The variables X, Y, Z, x, y, z, and t involved in Equations 14.6 through 14.9 assume independent values in their respective domains, and any one of the allowable properties determined by these equations may characterize a laminated composite composed of no more than three dissimilar materials.

We can thus suggest the following procedure for designing a laminated composite with prescribed effective properties:

1. Equate the left-hand sides of Equations 14.5 through 14.8 to the (prescribed) values $\tilde{c}_v^{(0)}$, $\tilde{\lambda}_{ij}^{(0)}$, $\tilde{E}_i^{(0)}$, $\tilde{\nu}_{ij}^{(0)}$, $\tilde{G}_{ij}^{(0)}$, and $\tilde{\alpha}_{ij}^{(0)}$, and solve the resulting algebraic equations for X, Y, Z, x, y, z, and t.
2. If this system of equations turns out to be unsolvable or if its solutions invalidate the inequalities indicated in Equations 14.6 through 14.9, a material with the required properties cannot exist in the given class of composites.
3. If the system is solvable and the inequalities are satisfied, a material with the required properties does exist in the given class of composites and will be found by constructing functions $c_v(t)$, $\lambda(t)$, $E(t)$, $\alpha(t)) \in W^4$ such that $\langle c_v \rangle = X$, $\langle 1/\lambda \rangle = Y$, $\langle \lambda \rangle = Z$, $\langle E \rangle = x$, $\langle 1/E \rangle = y$, $\langle \alpha \rangle = z$, and $\langle E\alpha \rangle = t$.

As an interesting example of the application of these principles, it has been shown by Kolpakov and Rakin (1986) and Alekhin et al. (1988) that a laminate with a negative expansivity can be composed, in principle, of components with positive expansivities; in particular, the effective property $\tilde{\alpha}_{11}$ can be made negative by a proper choice of local properties. To see this, we repeat here the first of Equations 14.9,

$$\tilde{\alpha}_{11} = \frac{1+\nu}{1-\nu}\mathscr{Z} - \frac{2\nu}{1+\nu}\frac{t}{x} \qquad \left(x > 0,\ \mathscr{Z} > 0,\ t > 0,\ y > \frac{1}{x}\right) \quad (14.10)$$

which shows that the set of all possible values of $\tilde{\alpha}_{11}$ is $(-\infty, \infty)$ and hence there is a possibility for a composite with a negative value of $\tilde{\alpha}_{11}$ to exist. We restrict attention to a class of composites having no more than three constituents and denote by E_i, α_i, and ν_i, respectively, the Young's modulus, linear expansivity, and volume fraction of the ith material $(i = 1, 2, 3)$. All the possible values of $\tilde{\alpha}_{11}$ will then be given by Equation 14.10 with

$$x = \sum_{i=1}^{3} E_i \nu_i \qquad y = \sum_{i=1}^{3} \frac{\nu_i}{E_i}$$

$$z = \sum_{i=1}^{3} \dot{\alpha}_i \nu_i \qquad t = \sum_{i=1}^{3} E_i \alpha_i \nu_i$$

$$\left(0 \le \nu_i \le 1,\ \sum_{i=1}^{3} \nu_i = 1,\ E_i,\ \alpha_i > 0\right) \quad (14.11)$$

The synthesis problem that arises is finite-dimensional and leads, when stated in a discrete formulation, to the problem of exhaustive search, among a finite number of possibilities, for the sets E_i, α_i, and ν_i $(i = 1, 2, 3)$ that ensure the prescribed $\tilde{\alpha}_{11}$ value. The results obtained on a computer for a number of practical materials are summarized in Table 14.1, where binary compositions were formally treated as ternary ones with two components considered to be identical in their properties.

15. HOMOGENIZATION OF PIEZOELECTROELASTIC PROBLEMS

One way to describe the subject of piezoelectroelasticity is to say that it enables one to study the interaction between the electric field and deformable media of certain symmetry classes. In the context of composite

TABLE 14.1

Composition	i	$E_i \times 10^{10}$ (Pa)	$\alpha_i \times 10^{-6} \text{ K}^{-1}$	v_i	$\alpha_{11} \times 10^{-6} \text{ K}^{-1}$
Iridium	1	52.8	6.5	0.1	-3.79
Invar	2	13.5	0.2	0.9	
PTFE	1	9.8	220	0.1	-49.0
Hardened paper	2	1.2	20	0.9	
Iridium	1	52.8	6.5	0.05	
Tungsten	2	39	4.5	0.05	-2.79
Invar	3	13.5	0.2	0.9	

material behavior, the essential feature of the theory is that the electrostatic or electrodynamic Maxwell's equations of the problem must be averaged in some way over a medium with (generally) a periodic structure. The averaging of the equations of electrodynamics for a periodically nonhomogeneous medium has been a subject of extensive study in recent years (Sanchez-Palencia 1980, Caillerie and Lévy 1983, Bakhvalov and Panasenko 1984). In the framework of the zeroth-order approach discussed in the first of these references, it turns out that the averaging (or homogenization) procedure results in a system of integrodifferential equations with retardation; nonhomogeneities in microstructure thus impart to the homogenized media a property of memory with respect to electromagnetic wave propagation processes. Caillerie and Lévy (1983) applied the homogenization method to the electrodynamic analysis of a two-phase composite with a unit cell composed of a conductor (metal) and a dielectric (ceramic) and were able to derive homogenized macroscopic relations for electrostatic as well for magnetostatic problems.

Among the many piezoelectric materials currently available, polarized ceramic appears to be most widely used in the design of piezoelements for various applications. It should be noted, however, that polarized ceramic is weak against tensile loadings, a drawback it shares with all other ceramic materials; to remedy this, a number of methods have been put forward. In one of these, for example, tensile prestresses are produced in a piezoceramic element.

A method that is of more interest for us here depends on high-strength fiber reinforcement. To make this method practical, there are a number of questions that continuum mechanics and the theory of composites are expected to be able to answer, for example: Given reinforcement geometry, what is the fiber volume fraction needed to secure the desired reinforcing effect? What dielectric and mechanical properties must the fibers have if we wish to maximize the electromechanical coupling factor of the resulting piezocomposite?

A promising new direction in the field of electromechanical transducers is associated with the use of filled piezoelectric polymers. It is found (see

Shakhtakhtinskii et al. 1985, Shakhtakhtinskii et al. 1987) that filling a polymer with powder, fibers, or granules of a ferroelectric ceramic enhances the dielectric permittivity of the polymer and activates polarization processes in it, thereby producing a good material for designing high-performance piezoelectric device elements.

To evaluate the effective (or overall) properties of piezocomposites, it is customary to perform measurements on samples considered to be homogeneous and anisotropic in their material properties. The properties of a number of composite systems based on thermoplastic polymers or PZT ceramic have been studied experimentally by Furukawa et al. (1979). Numan and Keller (1984) describe methods for developing composite materials with different types of contact between piezoelectric particles within the matrix phase. The effects of polarization conditions and of the structure and material properties of constituents on the piezoelectric properties of a composite have been studied in the previously mentioned experimental work of Shakhtakhtinskii and co-workers (1985, 1987), who discovered, among other things, the occurrence of extrema in the dependences of the piezoelectric coefficients on the polarization temperature and electric field strength.

However, it is clear that, because of the large amount of input information needed, those concerned with optimum piezocomposite design cannot actually rely on experiment when evaluating the properties of the projected material for different reinforcement schemes; the only way to move forward is by developing theoretical models for piezoelectric composite materials, in terms of which both the averaged (effective) and the local properties of the medium might be determined. One possible approach to the coupled field analysis of piezocomposites is discussed by Vanin (1977, 1985) who performed a detailed examination of a piezoelectrically active, two-component periodic medium (unidirectional fiber composite) under antiplane strain conditions. The properties of the homogenized medium were determined by Vanin by equating the energy of the (nonhomogeneous!) unit cell to that of a certain fictitious medium described by effective properties. A similar energy approach to the determination of effective electromechanical properties was adopted by Ivanenko and Fil'shtinski (1986) in their study of a piezoceramic matrix strengthened by a doubly periodic system of anisotropic dielectric fibers; the relevant electroelasticity problem turned out to be reducible to the system of Fredholm integral equations of the second kind. Mention should also be made of a study by Miloserdova (1987) of coupled electroelastic fields in piezocomposites with a regular structure.

Suppose that we have a nonhomogeneous periodic piezoelectric medium with a characteristic nonhomogeneity dimension ε and that we use a rectangular Cartesian coordinate system x_i ($i = 1, 2, 3$) to describe points in it. Referring to Chapter 1 of this book (see also Parton and Kudryavtsev

1988), the electroelastic state of this medium is governed by the following system of equations.

1.   Equations of motion (neglecting body forces):

$$\frac{\partial \sigma_{ij}^{(\varepsilon)}}{\partial x_j} = \rho^{(\varepsilon)} \cdot \frac{\partial^2 u_i^{(\varepsilon)}}{\partial t^2} \tag{15.1}$$

2.   Maxwell's equations (in the quasistatic approximation):

$$\frac{\partial D_i^{(\varepsilon)}}{\partial x_i} = 0 \qquad E_i^{(\varepsilon)} = -\frac{\partial \varphi^{(\varepsilon)}}{\partial x_i} \tag{15.2}$$

3.   Constitutive relations:

$$\sigma_{ij}^{(\varepsilon)} = c_{ijkn}^{(\varepsilon)} \frac{\partial u_k^{(\varepsilon)}}{\partial x_n} - e_{ijk}^{(\varepsilon)} E_k^{(\varepsilon)}$$

$$\tag{15.3}$$

$$D_i^{(\varepsilon)} = e_{ikn}^{(\varepsilon)} \frac{\partial u_k^{(\varepsilon)}}{\partial x_n} + \epsilon_{ij}^{(\varepsilon)} E_j^{(\varepsilon)}$$

The quantities $c_{ijkn}^{(\varepsilon)}(x) = c_{ijkn}(y)$, $e_{ijk}^{(\varepsilon)}(x) = e_{ijk}(y)$, and $\epsilon_{ij}^{(\varepsilon)}(x) = \epsilon_{ij}(y)$ in this set of equations characterize the elastic, piezoelectric, and dielectric properties of the medium and are periodic functions of $y = x/\varepsilon$ ($y \in Y$, Y designating the unit cell of the medium).

In accordance with the two-scale expansion method, the displacements $u_k^{(\varepsilon)}$ and the electric potential $\varphi^{(\varepsilon)}$ are both expanded in power series of ε,

$$u_k^{(\varepsilon)} = u_k^{(0)}(x) + \varepsilon u_k^{(1)}(x, y) + \varepsilon^2 u_k^{(2)}(x, y) + \cdots \tag{15.4}$$

$$\varphi^{(\varepsilon)} = \varphi^{(0)}(x) + \varepsilon \varphi^{(1)}(x, y) + \varepsilon^2 \varphi^{(2)}(x, y) + \cdots \tag{15.5}$$

with functions $u_k^{(1)}, u_k^{(2)}, \varphi^{(1)}, \varphi^{(2)}, \ldots$ Y-periodic in y. Substitution into Equations 15.3 allows the latter to be written as

$$\sigma_{ij}^{(\varepsilon)} = \sigma_{ij}^{(0)}(x, y) + \varepsilon \sigma_{ij}^{(1)}(x, y) + \cdots \tag{15.6}$$

$$D_i^{(\varepsilon)} = D_i^{(0)}(x, y) + \varepsilon D_i^{(1)}(x, y) + \cdots \tag{15.7}$$

where

$$\sigma_{ij}^{(n)} = c_{ijkh}(y)\frac{\partial u_k^{(n)}}{\partial x_h} + e_{kij}(y)\frac{\partial \varphi^{(n)}}{\partial x_k} + c_{ijkh}(y)\frac{\partial u_k^{(n+1)}}{\partial y_h}$$

$$+ e_{kij}(y)\frac{\partial \varphi^{(n+1)}}{\partial y_k}$$

$$D_i^{(n)} = e_{ikh}(y)\frac{\partial u_k^{(n)}}{\partial x_h} - \epsilon_{ik}(y)\frac{\partial \varphi^{(n)}}{\partial x_k} + e_{ikh}(y)\frac{\partial u_k^{(n+1)}}{\partial y_h}$$

$$- \epsilon_{ik}(y)\frac{\partial \varphi^{(n+1)}}{\partial y_k} \qquad (n = 0, 1, 2, \dots)$$

(15.8)

Using Equations 15.6 and 15.7 in Equations 15.1 and 15.2 and equating terms of equal powers of ε yields

$$\frac{\partial \sigma_{ij}^{(0)}(x, y)}{\partial y_j} = 0 \qquad \frac{\partial D_i^{(0)}(x, y)}{\partial y_i} = 0 \qquad (15.9)$$

$$\frac{\partial \sigma_{ij}^{(0)}(x, y)}{\partial x_j} + \frac{\partial \sigma_{ij}^{(1)}(x, y)}{\partial y_j} = \rho(x)\frac{\partial^2 u_i^{(0)}(x)}{\partial t^2} \qquad (15.10)$$

$$\frac{\partial D_i^{(0)}(x, y)}{\partial x_i} + \frac{\partial D_i^{(1)}(x, y)}{\partial y_i} = 0 \qquad (15.11)$$

The equations of motion and equations of electroelastics of the homogenized piezoelectric medium now can be obtained by subjecting Equations 15.10 and 15.11 to the averaging operation

$$\langle \cdots \rangle = \frac{1}{|Y|}\int(\cdots)\, dy$$

and making use of the periodicity of $\sigma_{ij}^{(1)}(x, y)$ and $D_i^{(1)}(x, y)$ in y. The result is

$$\frac{\partial \tilde{\sigma}_{ij}^{(0)}}{\partial x_j} = \rho(x)\frac{\partial^2 u_i^{(0)}}{\partial t^2} \qquad (15.12)$$

$$\frac{\partial \tilde{D}_i^{(0)}}{\partial x_i} = 0 \qquad (15.13)$$

where

$$\tilde{\sigma}_{ij}^{(0)}(x) = \left\langle \sigma_{ij}^{(0)}(x,y) \right\rangle \quad \text{and} \quad \tilde{D}_i^{(0)}(x) = \left\langle D_i^{(0)}(x,y) \right\rangle$$

Now substitute Equations 15.8 into Equations 15.9, giving

$$\frac{\partial}{\partial y_i}\left[c_{ijkh}(y)\frac{\partial u_k^{(1)}(x,y)}{\partial y_h} + e_{kij}(y)\frac{\partial \varphi^{(1)}(x,y)}{\partial y_k} \right]$$

$$= -\frac{\partial c_{ijkh}(y)}{\partial y_i}\frac{\partial u_k^{(0)}(x)}{\partial x_h} - \frac{\partial e_{kij}(y)}{\partial y_i}\frac{\partial \varphi^{(0)}(x)}{\partial x_k} \qquad (15.14)$$

$$\times \frac{\partial}{\partial y_i}\left[e_{ikh}(y)\frac{\partial u_k^{(1)}(x,y)}{\partial y_h} - \epsilon_{ik}(y)\frac{\partial \varphi^{(1)}(x,y)}{\partial y_k} \right]$$

$$= -\frac{\partial e_{ikh}(y)}{\partial y_i}\frac{\partial u_k^{(0)}(x)}{\partial x_h} + \frac{\partial \epsilon_{ik}(y)}{\partial y_i}\frac{\partial \varphi^{(0)}(x)}{\partial x_k} \qquad (15.15)$$

We proceed by writing

$$u_n^{(1)}(x,y) = N_n^{kl}(y)\frac{\partial u_k^{(0)}(x,y)}{\partial x_l} + W_{nl}(y)\frac{\partial \varphi^{(0)}(x)}{\partial x_l} \qquad (15.16)$$

$$\varphi^{(1)}(x,y) = \Phi_{kl}(y)\frac{\partial u_k^{(0)}(x)}{\partial x_l} + \Psi_l(y)\frac{\partial \varphi^{(0)}(x)}{\partial x_l} \qquad (15.17)$$

and substituting into Equations 15.14 and 15.15 to obtain the following local problem for determining the (Y-periodic) functions $N_n^{kl}(y)$, $W_{nl}(y)$, $\Phi_{kl}(y)$, and $\Psi_l(y)$:

$$\frac{\partial \tau_{ij}^{kl}}{\partial y_j} = -\frac{\partial c_{ijkl}(y)}{\partial y_j}$$

$$\qquad (15.18)$$

$$\frac{\partial d_i^{kl}}{\partial y_i} = -\frac{\partial e_{ikl}(y)}{\partial y_i}$$

and

$$\frac{\partial \zeta_{ij}^l}{\partial y_j} = \frac{\partial e_{lij}(y)}{\partial y_j}$$

$$\frac{\partial g_{il}}{\partial y_i} = -\frac{\partial \epsilon_{il}(y)}{\partial y_i} \tag{15.19}$$

with

$$\tau_{ij}^{kl} = c_{ijnh}(y)\frac{\partial N_n^{kl}(y)}{\partial y_h} + e_{hij}\frac{\partial \Phi_{kl}(y)}{\partial y_h} \tag{15.20}$$

$$d_i^{kl} = e_{inh}(y)\frac{\partial N_n^{kl}(y)}{\partial y_h} - \epsilon_{ih}(y)\frac{\partial \Phi_{kl}(y)}{\partial y_h} \tag{15.21}$$

and

$$\zeta_{ij}^l = c_{ijkh}(y)\frac{\partial W_{kl}(y)}{\partial y_h} + e_{hij}(y)\frac{\partial \Psi_l(y)}{\partial y_h} \tag{15.22}$$

$$g_{il} = e_{ikh}(y)\frac{\partial W_{kl}(y)}{\partial y_h} - \epsilon_{ih}(y)\frac{\partial \Psi_l(y)}{\partial y_h} \tag{15.23}$$

Using Equations 15.16 and 15.17 in Equations 15.18, the latter become

$$\sigma_{ij}^{(0)} = \left[c_{ijkl}(y) + \tau_{ij}^{kl}(y)\right]\frac{\partial u_k^{(0)}(x)}{\partial x_l} + \left[e_{lij} + \zeta_{ij}^l(y)\right]\frac{\partial \varphi^{(0)}(x)}{\partial x_l} \tag{15.24}$$

$$D_i^{(0)} = \left[e_{ipl}(y) + d_i^{pl}(y)\right]\frac{\partial u_p^{(0)}(x)}{\partial x_l} - \left[\epsilon_{il}(y) - g_{il}(y)\right]\frac{\partial \varphi^{(0)}(x)}{\partial x_l} \tag{15.25}$$

which when averaged over the unit cell volume yield the constitutive relations of the homogenized piezoelectric medium:

$$\tilde{\sigma}_{ij} \cong \tilde{\sigma}_{ij}^{(0)} = \tilde{c}_{ijkl}\frac{\partial u_k^{(0)}}{\partial x_l} + \tilde{e}_{lij}\frac{\partial \varphi^{(0)}}{\partial x_l} \tag{15.26}$$

$$\tilde{D}_i \cong \tilde{D}_i^{(0)} = \tilde{e}_{ipl}'\frac{\partial u_p^{(0)}}{\partial x_l} - \tilde{\epsilon}_{il}\frac{\partial \varphi^{(0)}}{\partial x_l} \tag{15.27}$$

where the quantities

$$\tilde{c}_{ijkl} = \left\langle c_{ijkl}(y) + \tau_{ij}^{kl}(y) \right\rangle$$

$$\tilde{e}_{lij} = \left\langle e_{lij}(y) + \zeta_{ij}^{l}(y) \right\rangle$$

$$\tilde{e}'_{ipl} = \left\langle e_{ipl}(y) + d_{i}^{pl}(y) \right\rangle$$

$$\tilde{\epsilon}_{il} = \left\langle \epsilon_{il}(y) + g_{il}(y) \right\rangle \tag{15.28}$$

represent the effective moduli of the medium and are to be found from the solutions of the local problems given by Equations 15.8 and 15.9.

The effective properties of a polarized ceramic type of piezoelectric medium are expressed in terms of the local functions τ_{ij}^{kl}, d_{i}^{kl}, ζ_{ij}^{l}, and g_{il}, whose explicit forms are as follows (where the tensor components c_{ijkl} and e_{ijk} are given in two-index notation; the structure of the material constant matrices is given in Chapter 1):

$$\tau_{11}^{kl} = c_{11}(y)\frac{\partial N_1^{kl}}{\partial y_1} + c_{12}(y)\frac{\partial N_2^{kl}}{\partial y_2} + c_{13}(y)\frac{\partial N_3^{kl}}{\partial y_3} + e_{13}(y)\frac{\partial \Phi_{kl}}{\partial y_3}$$

$$\tau_{22}^{kl} = c_{12}(y)\frac{\partial N_1^{kl}}{\partial y_1} + c_{11}(y)\frac{\partial N_2^{kl}}{\partial y_2} + c_{13}(y)\frac{\partial N_3^{kl}}{\partial y_3} + e_{13}(y)\frac{\partial \Phi_{kl}}{\partial y_3}$$

$$\tau_{33}^{kl} = c_{13}(y)\left(\frac{\partial N_1^{kl}}{\partial y_1} + \frac{\partial N_2^{kl}}{\partial y_2}\right) + c_{33}(y)\frac{\partial N_3^{kl}}{\partial y_3} + e_{33}(y)\frac{\partial \Phi_{kl}}{\partial y_3}$$

$$\tau_{23}^{kl} = c_{44}(y)\left(\frac{\partial N_3^{kl}}{\partial y_2} + \frac{\partial N_2^{kl}}{\partial y_3}\right) + e_{51}(y)\frac{\partial \Phi_{kl}}{\partial y_2}$$

$$\tau_{13}^{kl} = c_{44}(y)\left(\frac{\partial N_3^{kl}}{\partial y_1} + \frac{\partial N_1^{kl}}{\partial y_3}\right) + e_{51}(y)\frac{\partial \Phi_{kl}}{\partial y_1}$$

$$\tau_{12}^{kl} = c_{66}(y)\left(\frac{\partial N_1^{kl}}{\partial y_2} + \frac{\partial N_2^{kl}}{\partial y_1}\right) \tag{15.29}$$

$$g_1^{kl} = e_{51}(y)\left(\frac{\partial N_3^{kl}}{\partial y_1} + \frac{\partial N_1^{kl}}{\partial y_3}\right) - \epsilon_{11}(y)\frac{\partial \Phi_{kl}}{\partial y_1}$$

$$g_2^{kl} = e_{51}(y)\left(\frac{\partial N_3^{kl}}{\partial y_2} + \frac{\partial N_2^{kl}}{\partial y_3}\right) - \epsilon_{11}(y)\frac{\partial \Phi_{kl}}{\partial y_2} \qquad (15.30)$$

$$g_3^{kl} = e_{13}(y)\left(\frac{\partial N_1^{kl}}{\partial y_1} + \frac{\partial N_2^{kl}}{\partial y_2}\right) + e_{33}(y)\frac{\partial N_3^{kl}}{\partial y_3} - \epsilon_{33}(y)\frac{\partial \Phi_{kl}}{\partial y_3}$$

$$g_{1l} = e_{51}(y)\left(\frac{\partial W_{3l}}{\partial y_1} + \frac{\partial W_{1l}}{\partial y_3}\right) - \epsilon_{11}(y)\frac{\partial \Psi_l}{\partial y_1}$$

$$g_{2l} = e_{51}(y)\left(\frac{\partial W_{3l}}{\partial y_2} - \frac{\partial W_{2l}}{\partial y_3}\right) - \epsilon_{11}(y)\frac{\partial \Psi_l}{\partial y_2}$$

$$g_{3l} = e_{13}(y)\left(\frac{\partial W_{1l}}{\partial y_1} + \frac{\partial W_{2l}}{\partial y_2}\right) + e_{33}(y)\frac{\partial W_{3l}}{\partial y_3} - \epsilon_{33}(y)\frac{\partial \Psi_l}{\partial y_3} \qquad (15.31)$$

$$\zeta_{11}^l = c_{11}(y)\frac{\partial W_{1l}}{\partial y_1} + c_{12}(y)\frac{\partial W_{2l}}{\partial y_2} + c_{13}(y)\frac{\partial W_{3l}}{\partial y_3} + e_{13}(y)\frac{\partial \Psi_l}{\partial y_3}$$

$$\zeta_{22}^l = c_{12}(y)\frac{\partial W_{1l}}{\partial y_1} + c_{11}(y)\frac{\partial W_{2l}}{\partial y_2} + c_{13}(y)\frac{\partial W_{3l}}{\partial y_3} + e_{13}(y)\frac{\partial \Psi_l}{\partial y_3}$$

$$\zeta_{33}^l = c_{13}(y)\left(\frac{\partial W_{1l}}{\partial y_1} + \frac{\partial W_{2l}}{\partial y_2}\right) + c_{33}(y)\frac{\partial W_{3l}}{\partial y_3} + e_{33}(y)\frac{\partial \Psi_l}{\partial y_3}$$

$$\zeta_{23}^l = c_{44}(y)\left(\frac{\partial W_{3l}}{\partial y_2} + \frac{\partial W_{2l}}{\partial y_3}\right) + e_{51}(y)\frac{\partial \Psi_l}{\partial y_2}$$

$$\zeta_{13}^l = c_{44}(y)\left(\frac{\partial W_{3l}}{\partial y_1} + \frac{\partial W_{1l}}{\partial y_3}\right) + e_{51}(y)\frac{\partial \Psi_l}{\partial y_1}$$

$$\zeta_{12}^l = c_{66}(y)\left(\frac{\partial W_{2l}}{\partial y_1} + \frac{\partial W_{1l}}{\partial y_2}\right) \qquad \left[c_{66} = \frac{1}{2}(c_{11} - c_{12})\right] \qquad (15.32)$$

16. EFFECTIVE PROPERTIES OF PIEZOELECTRIC COMPOSITES OF LAMELLAR AND UNIDIRECTIONAL FIBER TYPES

We show here that in the case of periodic laminate composites the local problems given by Equations 15.18 and 15.19 are integrated in an elementary fashion, thereby allowing analytical expressions for effective properties as functions of the constituent properties. We assume, without any loss of generality, that the periodically repeating "motif" of the composite consists of two layers, one piezoceramic and the other dielectric, and that the piezoceramic layer is polarized in the x_3 direction. The one-dimensional unit cell of such a composite may be visualized as comprising two parts: $Y = Y_p \cup Y_d$, the indices p and d referring to the piezoceramic and dielectric parts, respectively. The material properties of the problem depend on a single coordinate, remain constant within each layer, and undergo a discontinuity at the interface, thus

$$c_{ijkl}(y) = \begin{cases} c_{ijkl}^{(p)} & (y \in Y_p) \\ c_{ijkl}^{(d)} & (y \in Y_d) \end{cases}$$

$$e_{kij}(y) = \begin{cases} e_{kij} & (y \in Y_p) \\ 0 & (y \in Y_d) \end{cases} \qquad (16.1)$$

$$\epsilon_{ij}(y) = \begin{cases} \epsilon_{ij}^{(p)} & (y \in Y_p) \\ \epsilon_{ij}^{(d)} & (y \in Y_d) \end{cases}$$

We will consider two cases of piezoceramic polarization in the following analysis, one parallel and the other normal to the interface, as shown in Figures 16.1 and 16.2, respectively.

For the parallel case we have

$$N_i^{kl} = N_i^{kl}(y_1) \qquad \Phi_{kl} = \Phi_{kl}(y_1)$$

$$W_{ik} = W_{ik}(y_1) \qquad \Psi_k = \Psi_k(y_1)$$

(where $i, k, l = 1, 2, 3$), and Equations 15.18 and 15.19 take the form

$$\frac{d}{dy_1} \tau_{i1}^{kl} = -\frac{d}{dy_1} c_{i1kl}$$

$$\frac{d}{dy_1} d_1^{kl} = -\frac{d}{dy_1} e_{1kl} \qquad (16.2)$$

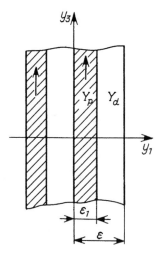

FIGURE 16.1. Piezoelectric laminar composite; piezoceramic polarization direction parallel to the layer interface.

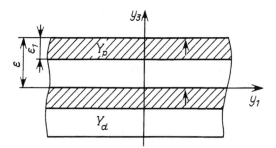

FIGURE 16.2. Piezoelectric laminar composite; piezoceramic polarization direction perpendicular to the layer interface.

and

$$\frac{d}{dy_1}\zeta_{i1}^{kl} = \frac{d}{dy_1}e_{li1}$$

$$\frac{d}{dy_1}g_{1l} = -\frac{d}{dy_1}\epsilon_{1l}$$

(16.3)

whereas Relations 15.20 through 15.23 may be rewritten as

$$\tau_{11}^{kl} = c_{11}(y_1)\frac{d}{dy_1}N_1^{kl}(y_1) \qquad \tau_{22}^{kl} = c_{12}(y_1)\frac{d}{dy_1}N_1^{kl}(y_1)$$

$$\tau_{33}^{kl} = c_{13}(y_1)\frac{d}{dy_1}N_1^{kl}(y_1) \qquad \tau_{23}^{kl} = 0$$

$$\tau_{13}^{kl} = c_{44}(y_1)\frac{d}{dy_1}N_3^{kl}(y_2) + e_{51}(y_1)\frac{d}{dy_1}\Phi_{kl}(y_1)$$

$$\tau_{12}^{kl} = \frac{1}{2}(c_{11}(y_1) - c_{12}(y_1))\frac{d}{dy_1}N_2^{kl}(y_1) \tag{16.4}$$

$$d_1^{kl} = e_{15}(y_1)\frac{d}{dy_1}N_3^{kl}(y_1) - \epsilon_{11}(y_1)\frac{d}{dy_1}\Phi_{kl}(y_1)$$

$$d_2^{kl} = 0 \qquad d_3^{kl} = e_{13}(y_1)\frac{d}{dy_1}N_1^{kl}(y_1) \tag{16.5}$$

$$g_{1l} = e_{15}(y_1)\frac{d}{dy_1}W_{3l}(y_1) - \epsilon_{11}(y_1)\frac{d}{dy_1}\Psi_l(y_1)$$

$$g_{2l} = 0 \qquad g_{3l} = e_{13}(y_1)\frac{d}{dy_1}W_{1l}(y_1) \tag{16.6}$$

$$\zeta_{11}^l = c_{11}(y_1)\frac{d}{dy_1}W_{1l}(y_1) \qquad \zeta_{22}^l = c_{12}(y_1)\frac{d}{dy_1}W_{1l}(y_1)$$

$$\zeta_{33}^l = c_{13}(y_1)\frac{d}{dy_1}W_{1l}(y_1) \qquad \zeta_{23}^l = 0$$

$$\zeta_{13}^l = c_{44}(y_1)\frac{d}{dy_1}W_{3l}(y_1) + l_{15}(y_1)\frac{d}{dy_1}\Psi_l(y_1)$$

$$\zeta_{12}^l = \frac{1}{2}(c_{11}(y_1) - c_{12}(y_1))\frac{d}{dy_1}W_{2l}(y_1) \tag{16.7}$$

Having solved the local problems given by Equations 16.2 and 16.3, the matrices of the effective material constants of the piezoelectric composite of Figure 16.1 are found to be

$$
\tilde{e}'_{i\alpha} =
\begin{pmatrix}
0 & 0 & 0 & 0 & \tilde{e}'_{15} & 0 \\
0 & 0 & 0 & \tilde{e}'_{24} & 0 & 0 \\
\tilde{e}'_{31} & \tilde{e}'_{32} & \tilde{e}'_{33} & 0 & 0 & 0
\end{pmatrix}
\tag{16.8}
$$

$$
\tilde{c}_{\alpha\beta} =
\begin{pmatrix}
\tilde{c}_{11} & \tilde{c}_{12} & \tilde{c}_{13} & 0 & 0 & 0 \\
\tilde{c}_{12} & \tilde{c}_{22} & \tilde{c}_{23} & 0 & 0 & 0 \\
\tilde{c}_{13} & \tilde{c}_{23} & \tilde{c}_{33} & 0 & 0 & 0 \\
0 & 0 & 0 & \tilde{c}_{44} & 0 & 0 \\
0 & 0 & 0 & 0 & \tilde{c}_{55} & 0 \\
0 & 0 & 0 & 0 & 0 & \tilde{c}_{66}
\end{pmatrix}
\tag{16.9}
$$

$$
\tilde{e}_{\alpha i} =
\begin{pmatrix}
0 & 0 & \tilde{e}_{13} \\
0 & 0 & \tilde{e}_{23} \\
0 & 0 & \tilde{e}_{33} \\
0 & \tilde{e}_{42} & 0 \\
\tilde{e}_{51} & 0 & 0 \\
0 & 0 & 0
\end{pmatrix}
\qquad
\tilde{\epsilon}_{ij} =
\begin{pmatrix}
\tilde{\epsilon}_{11} & 0 & 0 \\
0 & \tilde{\epsilon}_{22} & 0 \\
0 & 0 & \tilde{\epsilon}_{33}
\end{pmatrix}
\tag{16.10}
$$

where

$$
\tilde{c}_{11} = \varepsilon c_{11}^{(p)} \frac{1}{\Delta_{11}}
$$

$$
\tilde{c}_{12} = \left[\varepsilon_1 c_{12}^{(p)} + (\varepsilon - \varepsilon_1) \frac{c_{12}^{(d)} c_{11}^{(p)}}{c_{11}^{(d)}} \right] \frac{1}{\Delta_{11}}
$$

$$
\tilde{c}_{13} = \left[\varepsilon_1 c_{13}^{(p)} + (\varepsilon - \varepsilon_1) \frac{c_{13}^{(d)} c_{11}^{(p)}}{c_{11}^{(d)}} \right] \frac{1}{\Delta_{11}}
$$

$$
\tilde{c}_{22} = \frac{\varepsilon_1}{\varepsilon} \left\{ c_{11}^{(p)} - \frac{\left(c_{12}^{(p)}\right)^2}{c_{11}^{(p)}} + \left[\varepsilon_1 \frac{\left(c_{12}^{(p)}\right)^2}{c_{11}^{(p)}} \right. \right.
$$

$$+ (\varepsilon - \varepsilon_1) \frac{c_{12}^{(p)} c_{12}^{(d)}}{c_{11}^{(d)}} \Bigg] \frac{1}{\Delta_{11}} \Bigg\}$$

$$+ \frac{(\varepsilon - \varepsilon_1)}{\varepsilon} \Bigg\{ c_{11}^{(d)} - \frac{\left(c_{12}^{(d)}\right)^2}{c_{11}^{(d)}} + \Bigg[\varepsilon_1 \frac{c_{12}^{(p)} c_{12}^{(d)}}{c_{11}^{(d)}}$$

$$+ (\varepsilon - \varepsilon_1) \frac{\left(c_{12}^{(d)}\right)^2 c_{11}^{(p)}}{\left(c_{11}^{(d)}\right)^2} \Bigg] \frac{1}{\Delta_{11}} \Bigg\}$$

$$\tilde{c}_{23} = \frac{\varepsilon_1}{\varepsilon} \Bigg\{ c_{13}^{(p)} - \frac{c_{13}^{(p)} c_{12}^{(p)}}{c_{11}^{(p)}} + \Bigg[\varepsilon_1 \frac{c_{12}^{(p)} c_{13}^{(p)}}{c_{11}^{(p)}}$$

$$+ (\varepsilon - \varepsilon_1) \frac{c_{13}^{(d)} c_{12}^{(p)}}{c_{11}^{(d)}} \Bigg] \frac{1}{\Delta_{11}} \Bigg\}$$

$$+ \frac{(\varepsilon - \varepsilon_1)}{\varepsilon} \Bigg\{ c_{11}^{(d)} - \frac{c_{13}^{(d)} c_{12}^{(d)}}{c_{11}^{(d)}} + \Bigg[\varepsilon_1 \frac{c_{12}^{(d)} c_{13}^{(d)}}{c_{11}^{(d)}}$$

$$+ (\varepsilon - \varepsilon_1) \frac{c_{12}^{(d)} c_{13}^{(d)} c_{11}^{(p)}}{\left(c_{11}^{(d)}\right)^2} \Bigg] \frac{1}{\Delta_{11}} \Bigg\}$$

$$\tilde{c}_{33} = \frac{\varepsilon_1}{\varepsilon} \Bigg\{ c_{33}^{(p)} - \frac{c_{13}^{(p)}}{c_{11}^{(p)}} + \Bigg[\varepsilon_1 \frac{\left(c_{13}^{(p)}\right)^2}{c_{11}^{(p)}}$$

$$+ (\varepsilon - \varepsilon_1) \frac{c_{13}^{(p)} c_{13}^{(d)}}{c_{11}^{(d)}} \Bigg] \frac{1}{\Delta_{11}} \Bigg\}$$

$$+ \frac{(\varepsilon - \varepsilon_1)}{\varepsilon} \Bigg\{ c_{33}^{(d)} - \frac{c_{13}^{(d)}}{c_{11}^{(d)}} + \Bigg[\varepsilon_1 \frac{c_{13}^{(p)} c_{13}^{(d)}}{c_{11}^{(d)}}$$

$$+ (\varepsilon - \varepsilon_1) \frac{\left(c_{13}^{(d)}\right)^2 c_{11}^{(p)}}{\left(c_{11}^{(d)}\right)^2} \Bigg] \frac{1}{\Delta_{11}} \Bigg\}$$

$$\tilde{c}_{44} = \frac{\varepsilon_1}{\varepsilon} c_{44}^{(p)} + \frac{(\varepsilon - \varepsilon_1)}{\varepsilon} c_{44}^{(d)}$$

$$\tilde{c}_{55} = \left[\varepsilon \varepsilon_1 c_{44}^{(p)} c_{44}^{(d)} \varepsilon_{11}^{(d)} + \varepsilon(\varepsilon - \varepsilon_1) c_{44}^{(d)} \Delta_{12} \right] \frac{1}{\Sigma_1}$$

$$\tilde{c}_{66} = \varepsilon \frac{c_{66}^{(p)}}{\left[\varepsilon_1 + (\varepsilon - \varepsilon_1) c_{66}^{(p)}/c_{66}^{(d)} \right]} \qquad (16.11)$$

$$\tilde{e}_{13} = \tilde{e}_{31} = \varepsilon_1 e_{13} \frac{1}{\Delta_{11}}$$

$$\tilde{e}_{42} = \tilde{e}_{24}' = \frac{\varepsilon_1 e_{51}}{\varepsilon}$$

$$\tilde{e}_{23} = \tilde{e}_{32}' = \frac{\varepsilon_1}{\varepsilon} \left\{ e_{33} - e_{13} \frac{c_{12}^{(p)}}{c_{11}^{(p)}} \right.$$

$$\left. + \left[\varepsilon_1 e_{13} \frac{c_{12}^{(p)}}{c_{11}^{(p)}} + (\varepsilon - \varepsilon_1) \frac{c_{12}^{(d)} e_{13}}{c_{11}^{(d)}} \right] \frac{1}{\Delta_{11}} \right\}$$

$$\tilde{e}_{33} = \tilde{e}_{33}' = \frac{\varepsilon_1}{\varepsilon} \left\{ e_{33} - e_{13} \frac{c_{12}^{(p)}}{c_{11}^{(p)}} \right.$$

$$\left. + \left[\varepsilon_1 e_{13} \frac{c_{12}^{(p)}}{c_{11}^{(p)}} + (\varepsilon - \varepsilon_1) \frac{c_{12}^{(d)} e_{13}}{c_{11}^{(d)}} \right] \frac{1}{\Delta_{11}} \right\}$$

$$\tilde{e}_{51} = \tilde{e}_{15}' = \varepsilon \varepsilon_1 e_{51} c_{44}^{(d)} \varepsilon_{11}^{(d)} \frac{1}{\Sigma_1} \qquad (16.12)$$

$$\tilde{\epsilon}_{22} = \frac{\varepsilon_1}{\varepsilon} \epsilon_{11}^{(p)} + \frac{(\varepsilon - \varepsilon_1)}{\varepsilon} \epsilon_{11}^{(d)}$$

$$\tilde{\epsilon}_{11} = \left[\varepsilon \varepsilon_1 \epsilon_{11}^{(p)} c_{44}^{(d)} \varepsilon_{11}^{(d)} + \varepsilon(\varepsilon - \varepsilon_1) \epsilon_{11}^{(d)} \Delta_{12} \right] \frac{1}{\Sigma_1}$$

$$\tilde{\epsilon}_{33} = \frac{\varepsilon_1}{\varepsilon} \left\{ \epsilon_{33}^{(p)} + \frac{(e_{13})^2}{c_{11}^{(p)}} - \varepsilon_1 \frac{(e_{13})^2}{c_{11}^{(p)} \Delta_{11}} \right\} + \frac{\varepsilon - \varepsilon_1}{\varepsilon} \epsilon_{33}^{(d)} \qquad (16.13)$$

In Equations 16.11 through 16.13,

$$\Delta_{11} = \varepsilon_1 + (\varepsilon - \varepsilon_1)\frac{c_{11}^{(p)}}{c_{11}^{(d)}}$$

$$\Delta_{12} = c_{44}^{(p)}\epsilon_{11}^{(p)} + (e_{15})^2$$

$$\Sigma_1 = \varepsilon_1^2 c_{44}^{(d)}\epsilon_{11}^{(d)} + \varepsilon_1(\varepsilon - \varepsilon_1)(c_{44}^{(p)}\epsilon_{11}^{(d)} + \epsilon_{11}^{(p)}c_{44}^{(d)}) + (\varepsilon - \varepsilon_1)^2\Delta_{12}$$

An important point to be made about the structure of these matrices is the difference between the symmetry of the polarized ceramic (hexagonal class 6*mm*) and that of the homogenized piezoelectric medium obtained from the laminar piezoceramic composite of Figure 16.1 (orthorhombic system, class *mm*2).

The analysis proceeds along much the same line when the laminate has its layers polarized normal to the interface (Figure 16.2). In this case

$$N_i^{kl} = N_i^{kl}(y_3) \qquad W_{ik} = W_{ik}(y_3) \qquad \Phi_{kl} = \Phi_{kl}(y_3) \qquad \Psi_k = \Psi_k(y_3)$$

(where $i, k, l = 1, 2, 3$), and the local problems given by Equations 15.8 and 15.9 take the form

$$\frac{d}{dy_3}\tau_{i3}^{kl}(y_3) = -\frac{d}{dy_3}c_{i3kl}(y_3)$$

$$\frac{d}{dy_3}d_3^{kl}(y_3) = -\frac{d}{dy_3}e_{3kl}(y_3) \qquad (16.14)$$

$$\frac{d}{dy_3}\zeta_{i3}^l(y_3) = \frac{d}{dy_3}e_{l3i}(y_3)$$

$$\frac{d}{dy_3}g_{3l}(y_3) = -\frac{d}{dy_3}\epsilon_{3l}(y_3) \qquad (16.15)$$

The corresponding local functions can be expressed as

$$\tau_{11}^{kl} = c_{13}(y_3)\frac{d}{dy_3}N_3^{kl}(y_3) + e_{13}(y_3)\frac{d}{dy_3}\Phi_{kl}(y_3)$$

$$\tau_{22}^{kl} = \tau_{11}^{kl} \qquad \tau_{12}^{kl} = 0$$

$$\tau_{33}^{kl} = c_{33}(y_3)\frac{d}{dy_3}N_3^{kl}(y_3) + e_{33}(y_3)\frac{d}{dy_3}\Phi_{kl}(y_3)$$

$$\tau_{23}^{kl} = c_{44}(y_3)\frac{d}{dy_3}N_2^{kl}(y_3) \qquad \tau_{13}^{kl} = c_{44}(y_3)\frac{d}{dy_3}N_1^{kl}(y_3) \quad (16.16)$$

$$d_1^{kl} = e_{51}(y_3)\frac{d}{dy_3}N_1^{kl}(y_3) \qquad d_2^{kl} = e_{51}(y_3)\frac{d}{dy_3}N_2^{kl}(y_3)$$

$$d_3^{kl} = e_{33}(y_3)\frac{d}{dy_3}N_3^{kl}(y_3) - \epsilon_{33}(y_3)\frac{d}{dy_3}\Phi_{kl}(y_3) \qquad (16.17)$$

$$g_{1l} = e_{51}(y_3)\frac{d}{dy_3}W_{l3}(y_3) \qquad g_{2l} = g_{51}(y_3)\frac{d}{dy_3}W_{2l}(y_3)$$

$$g_{3l} = e_{33}(y_3)\frac{d}{dy_3}W_{3l}(y_3) - \epsilon_{33}(y_3)\frac{d}{dy_3}\Psi_l(y_3) \qquad (16.18)$$

$$\zeta_{11}^l = c_{13}(y_3)\frac{d}{dy_3}W_{3l}(y_3) + e_{13}(y_3)\frac{d}{dy_3}\Psi_l(y_3)$$

$$\zeta_{22}^l = \zeta_{11}^l \qquad \zeta_{12}^l = 0$$

$$\zeta_{33}^l = c_{33}(y_3)\frac{d}{dy_3}W_{3l}(y_3) + e_{13}(y_3)\frac{d}{dy_3}\Psi_l(y_3)$$

$$\zeta_{23}^l = c_{44}(y_3)\frac{d}{dy_3}W_{2l}(y_3) \qquad \zeta_{13}^l = c_{44}(y_3)\frac{d}{dy_3}W_{1l}(y_3) \quad (16.19)$$

Again, having solved Equations 16.14 and 16.15, the effective property matrices of the laminar composite are evaluated. In this case we have

$$\tilde{c}_{\alpha\beta} = \begin{pmatrix} \tilde{c}_{11} & \tilde{c}_{12} & \tilde{c}_{13} & 0 & 0 & 0 \\ \tilde{c}_{12} & \tilde{c}_{11} & \tilde{c}_{13} & 0 & 0 & 0 \\ \tilde{c}_{13} & \tilde{c}_{13} & \tilde{c}_{33} & 0 & 0 & 0 \\ 0 & 0 & 0 & \tilde{c}_{44} & 0 & 0 \\ 0 & 0 & 0 & 0 & \tilde{c}_{44} & 0 \\ 0 & 0 & 0 & 0 & 0 & \tilde{c}_{66} \end{pmatrix} \qquad (16.20)$$

$$\tilde{e}_{\alpha i} = \begin{pmatrix} 0 & 0 & \tilde{e}_{13} \\ 0 & 0 & \tilde{e}_{13} \\ 0 & 0 & \tilde{e}_{33} \\ 0 & \tilde{e}_{51} & 0 \\ \tilde{e}_{51} & 0 & 0 \\ 0 & 0 & 0 \end{pmatrix} \qquad \tilde{\epsilon}_{ij} = \begin{pmatrix} \tilde{\epsilon}_{11} & 0 & 0 \\ 0 & \tilde{\epsilon}_{11} & 0 \\ 0 & 0 & \tilde{\epsilon}_{33} \end{pmatrix} \qquad (16.21)$$

where

$$\tilde{c}_{11} = \frac{\varepsilon_1}{\varepsilon} \left[c_{11}^{(p)} + (A_{11} - c_{13}^{(p)}) \frac{\Delta_{21}}{\Delta_{23}} - (D_{11} - e_{13}) \frac{\Delta_{22}}{\Delta_{23}} \right]$$

$$+ \frac{(\varepsilon - \varepsilon_1)}{\varepsilon} \left[c_{11}^{(d)} + (A_{11} - c_{13}^{(d)}) \frac{c_{13}^{(d)}}{c_{33}^{(d)}} \right]$$

$$\tilde{c}_{12} = \frac{\varepsilon_1}{\varepsilon} \left[c_{12}^{(p)} + (A_{11} - c_{13}^{(p)}) \frac{\Delta_{21}}{\Delta_{23}} - (D_{11} - e_{13}) \frac{\Delta_{22}}{\Delta_{23}} \right]$$

$$+ \frac{(\varepsilon - \varepsilon_1)}{\varepsilon} \left[c_{12}^{(d)} + (A_{11} - c_{13}^{(d)}) \frac{c_{13}^{(d)}}{c_{33}^{(d)}} \right]$$

$$\tilde{c}_{13} = \frac{\varepsilon_1}{\varepsilon} \left[A_{33} \frac{\Delta_{21}}{\Delta_{23}} + D_{33} \frac{\Delta_{22}}{\Delta_{23}} \right] + \frac{(\varepsilon - \varepsilon_1)}{\varepsilon} A_{33} \frac{c_{13}^{(d)}}{c_{33}^{(d)}}$$

$$\tilde{c}_{33} = A_{33} \qquad \tilde{c}_{44} = \varepsilon \frac{c_{44}^{(p)}}{\left[\varepsilon_1 + (\varepsilon - \varepsilon_1) c_{44}^{(p)} / c_{44}^{(d)} \right]}$$

$$\tilde{c}_{66} = \frac{\varepsilon_1}{\varepsilon} c_{66}^{(p)} + \frac{(\varepsilon - \varepsilon_1)}{\varepsilon} c_{66}^{(d)} = \frac{1}{2} (\tilde{c}_{11} - \tilde{c}_{12}) \qquad (16.22)$$

$$\tilde{e}_{13} = D_{11} \qquad \tilde{e}_{33} = D_{33} \qquad \tilde{e}_{15} = -B_1 \qquad (16.23)$$

$$\tilde{\epsilon}_{11} = \frac{\varepsilon_1}{\varepsilon} \left[\epsilon_{11}^{(p)} + \frac{e_{51}^2}{c_{44}^{(p)}} + \frac{e_{51}}{c_{44}^{(p)}} B_1 \right] + \frac{(\varepsilon - \varepsilon_1)}{\varepsilon} \epsilon_{11}^{(d)}$$

$$\tilde{\epsilon}_{33} = \left[\varepsilon \varepsilon_1 \epsilon_{33}^{(p)} c_{33}^{(d)} \epsilon_{33}^{(d)} \Delta_{23} + \varepsilon(\varepsilon - \varepsilon_1) \epsilon_{33}^{(d)} (\Delta_{23})^2 \right] \frac{1}{\Sigma_2} \qquad (16.24)$$

In Equations 16.22 through 16.24,

$$A_{11} = \left\{ \varepsilon_1^2 c_{13}^{(p)} \Delta_{23} \Delta_{24} + \varepsilon_1 (\varepsilon - \varepsilon_1) \left[c_{33}^{(d)} \Delta_{21} + c_{33}^{(p)} c_{13}^{(d)} \epsilon_{33}^{(d)} \right] \Delta_{23} \right.$$

$$\left. + (\varepsilon - \varepsilon_1) c_{13}^{(p)} \Delta_{23} \right\} \frac{1}{\Sigma_2}$$

$$D_{11} = \left\{ \varepsilon_1^2 \left[e_{13}\Delta_{23} + 2c_{13}^{(p)}\epsilon_{33}^{(p)}e_{33} \right] \right.$$

$$\left. + \varepsilon_1(\varepsilon - \varepsilon_1)\left[c_{13}^{(d)}\epsilon_{33}^{(d)}e_{33} + \epsilon_{33}^{(d)}c_{33}^{(p)}e_{13} + \epsilon_{33}^{(p)}c_{13}^{(p)}e_{33} \right]\Delta_{23} \right\} \frac{1}{\Sigma_2}$$

$$A_{33} = \left[\varepsilon\varepsilon_1 c_{33}^{(p)}\Delta_{24}\Delta_{23} - \varepsilon(\varepsilon - \varepsilon_1)c_{33}^{(d)}(\Delta_{23})^2 \right]\frac{1}{\Sigma_2}$$

$$D_{33} = \left[\varepsilon\varepsilon_1 e_{33}\Delta_{24}\Delta_{23} \right]\frac{1}{\Sigma_2}$$

$$B_1 = \frac{\varepsilon_1 e_{15}}{\left[\varepsilon_1 + (\varepsilon - \varepsilon_1)c_{44}^{(p)}/c_{44}^{(d)} \right]}$$

$$\Sigma_2 = \varepsilon_1^2 e_{33}^2 \Delta_{24}\Delta_{23} + \varepsilon_1(\varepsilon - \varepsilon_1)(\epsilon_{33}^{(p)}c_{33}^{(d)} + \epsilon_{33}^{(d)}c_{33}^{(p)})\Delta_{23} + (\varepsilon - \varepsilon_1)^2(\Delta_{23})^2$$

$$\Delta_{21} = c_{13}^{(p)}\epsilon_{33}^{(p)} + e_{33}e_{13} \qquad \Delta_{22} = c_{33}^{(p)}e_{13} - c_{13}^{(p)}e_{33}$$

$$\Delta_{23} = c_{33}^{(p)}\epsilon_{33}^{(p)} + e_{33}^2 \qquad \Delta_{24} = \epsilon_{33}^{(d)}c_{33}^{(d)}$$

Note that, unlike the preceding case, the symmetry of piezoceramic (hexagonal class 6*mm*) is retained in the homogenized piezoelectric medium.

Turning now to the case of a unidirectional fiber composite with piezoactive constituents, we must consider the local problems given by Equations 15.18 and 15.19 as two-dimensional ones and may treat them, in principle, by means of the analytical methods of the theory of doubly periodic functions of a complex variable. However, we remarked earlier that, unless the fibers are circular in cross section, analytical methods are actually of no use—and must be replaced by finite-element techniques or similar numerical methods—when dealing with two-dimensional local problems. An important thing to bear in mind in doing this is that the two-dimensional local problems given by Equations 15.18 and 15.19 must be solved in the class of doubly periodic functions and therefore may not be considered as unit cell boundary value problems unless rendered such by imposing symmetry and periodicity requirements on the boundary of the unit cell. In particular, for a square unit cell $|y_1| < \varepsilon/2$, $|y_2| < \varepsilon/2$ (shown in Figure 16.3 for circular and square fibers), the reader is referred to Figure 16.4 for the boundary conditions to be satisfied by the functions N_i^{kl} and Φ_{kl} ($i, k, l = 1, 2, 3$).

We reproduce here some of the results obtained by Miloserdova (1987) from the finite-element analysis of the foregoing local problems for fiber composites with a piezoceramic constituent; the polarization of this latter

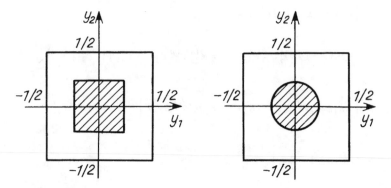

FIGURE 16.3. Unit cells of fiber composites: (left) geometry 1, square fiber cross section; (right) geometry 2, circular fiber cross section.

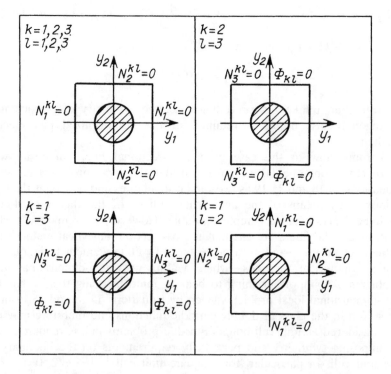

FIGURE 16.4. Boundary conditions on the contour of the unit cell for the functions N_i^{kl} and Φ_{kl} ($i, k, l = 1, 2, 3$).

TABLE 16.1

	Ceramic	Glass	Epoxy	Al–B–Si
Elastic moduli (N/m^2)				
$c_{11} \times 10^{-10}$	13.9	3.36	0.55	8.47
$c_{12} \times 10^{-10}$	7.78	2.8	0.296	2.53
$c_{13} \times 10^{-10}$	7.43	2.8	0.296	2.53
$c_{33} \times 10^{-10}$	11.5	3.36	0.55	8.47
$c_{44} \times 10^{-10}$	2.56	0.28	0.127	2.97
Piezoelectric constants (C/m^2)				
e_{31}	−5.2			
e_{33}	15.1			
e_{15}	12.7			
Dielectric permittivities (F/m)				
$\epsilon_{11} \times 10^{10}$	64.5	0.531	0.2655	0.4425
$\epsilon_{33} \times 10^{10}$	56.2	0.531	0.2655	0.4425

was always taken to be in the fiber direction. Specifically, the piezoceramic was chosen to be PZT-4, and the three compositions investigated were as follows:

(a) glass fibers in a piezoceramic matrix;
(b) Al–B–Si fibers in a piezoceramic matrix;
(c) piezoceramic fibers in an epoxy matrix.

Computations were carried out for the two cell types shown in Figure 16.3 for the fiber volume fractions $\gamma = 0.2, 0.4, 0.6, 0.7$ ($\gamma = S_F/S$, where S_F denotes the area of the fiber cross section and S denotes that of the unit cell). The values of the material constants employed in the calculations are listed in Table 16.1, and the effective properties computed are given in Tables 16.2 through 16.5 for the respective values of γ. In particular, we see that, qualitatively, the models with circular cross section and those with square cross section show the same dependences of the effective properties on the fiber volume fraction.

17. HOMOGENIZATION OF PERIODICALLY NONHOMOGENEOUS PIEZOELECTRIC MEDIA INCLUDING ELECTRICAL CONDUCTIVITY

We consider here a nonhomogeneous, periodic piezoelectric medium whose nonhomogeneity dimension ε is a small parameter characterizing

TABLE 16.2
Computed Effective Properties
for Compositions (a)–(c) and Geometries 1 and 2; $\gamma = 0.2$

	(a)		(b)		(c)	
	1	2	1	2	1	2
Elastic moduli (N/m^2)						
$c_{11} \times 10^{-11}$	1.152	1.0756	1.265	1.2516	0.204	0.163
$c_{12} \times 10^{-10}$	6.60	6.16	6.567	6.432	1.06	0.881
$c_{13} \times 10^{-10}$	6.637	6.01	6.301	6.176	1.07	0.867
$c_{33} \times 10^{-11}$	0.979	0.959	1.076	1.064	0.229	0.214
$c_{44} \times 10^{-10}$	0.202	0.204	2.6416	2.588	0.478	0.505
Piezoelectric constants (C/m^2)						
$-e_{31}$	4.015	3.61	4.0017	3.87	0.564	0.416
e_{33}	12.16	12.43	12.21	12.38	3.332	3.419
e_{15}	10.08	10.4	10.15	10.21	2.427	2.218
Dielectric permittivities (F/m)						
$\epsilon_{11} \times 10^8$	0.516	0.534	0.518	0.5213	0.132	0.131
$\epsilon_{33} \times 10^8$	0.4513	0.4552	0.4516	0.4542	0.116	0.116

the size of the unit cell of the problem; the unit cell Y is taken to be composed of two parts, a piezoelectric Y_p and a finite conductivity dielectric Y_d.

The system of equations describing the elastic and electromagnetic fields in the medium include the equations of motion,

$$\frac{\partial \sigma_{ij}^{(\varepsilon)}(x,t)}{\partial x_j} = \rho^{(\varepsilon)} \cdot \frac{\partial^2 u_i^{(\varepsilon)}(x,t)}{\partial t^2} \tag{17.1}$$

Maxwell's equations,

$$\text{curl } \mathbf{H}^{(\varepsilon)} = \frac{\partial \mathbf{D}^{(\varepsilon)}}{\partial t} + \mathbf{J}^{(\varepsilon)} \tag{17.2}$$

$$\text{curl } \mathbf{E}^{(\varepsilon)} = -\frac{\partial \mathbf{B}^{(\varepsilon)}}{\partial t} \tag{17.3}$$

TABLE 16.3
Computed Effective Properties
for Compositions (a)–(c) and Geometries 1 and 2; $\gamma = 0.4$

	(a)		(b)		(c)	
	1	2	1	2	1	2
Elastic moduli (N/m^3)						
$c_{11} \times 10^{-11}$	0.940	0.8311	1.152	1.129	0.364	0.2815
$c_{12} \times 10^{-11}$	0.547	0.4865	0.547	0.524	0.184	0.1481
$c_{13} \times 10^{-11}$	0.5398	0.4894	0.528	0.507	0.188	0.1478
$c_{33} \times 10^{-11}$	0.814	0.7838	1.011	0.992	0.406	0.3792
$c_{44} \times 10^{-11}$	0.1562	0.1786	0.236	0.264	0.1150	0.1014
Piezoelectric constants (C/m^2)						
$-e_{31}$	2.917	2.352	2.92	2.692	1.157	0.861
e_{33}	9.193	9.527	9.257	9.462	6.685	6.897
e_{15}	7.47	7.613	7.583	7.915	4.276	4.5335
Dielectric permittivities (F/m)						
$\epsilon_{11} \times 10^8$	0.393	0.399	0.411	0.402	0.258	0.268
$\epsilon_{33} \times 10^8$	0.341	0.344	0.341	0.343	0.231	0.233

and the constitutive relations for a finite conductivity piezoelectric–dielectric system

L06

$$\sigma_{ij}^{(\varepsilon)} = c_{ijkl}^{(\varepsilon)}(x)\frac{\partial u_k^{(\varepsilon)}}{\partial x_l} - e_{lij}^{(\varepsilon)}(x)E_k^{(\varepsilon)} \tag{17.4}$$

$$D_i^{(\varepsilon)} = e_{ikl}^{(\varepsilon)}(x)\frac{\partial u_k^{(\varepsilon)}}{\partial x_l} + \epsilon_{ik}^{(\varepsilon)}(x)E_k^{(\varepsilon)} \tag{17.5}$$

$$B_i^{(\varepsilon)} = \mu_{ik}^{(\varepsilon)}(x)H_k^{(\varepsilon)} \qquad J_i^{(\varepsilon)} = \kappa_{ik}^{(\varepsilon)}(x)E_k^{(\varepsilon)} \tag{17.6}$$

where $E_k^{(\varepsilon)}$, $H_k^{(\varepsilon)}$, $D_i^{(\varepsilon)}$, $B_i^{(\varepsilon)}$, and $J_i^{(\varepsilon)}$ are the components of electric field, magnetic field, electric displacement, magnetic flux density, and conduction current density, respectively.

TABLE 16.4
Computed Effective Properties
for Compositions (a)–(c) and Geometries 1 and 2; $\gamma = 0.6$

	(a)		(b)		(c)	
	1	2	1	2	1	2
Elastic moduli (N/m²)						
$c_{11} \times 10^{-11}$	0.726	0.6046	1.044	1.016	0.539	0.411
$c_{12} \times 10^{-11}$	0.4401	0.382	0.442	0.415	0.264	0.205
$c_{13} \times 10^{-11}$	0.4440	0.3906	0.4301	0.405	0.276	0.211
$c_{33} \times 10^{-11}$	0.6489	0.6172	0.951	0.928	0.587	0.544
$c_{44} \times 10^{-11}$	0.1184	0.1216	0.279	0.278	0.152	0.148
Piezoelectric constants (C/m²)						
$-e_{31}$	1.843	1.243	1.88	1.614	1.795	1.324
e_{33}	6.182	6.542	6.224	6.470	9.968	10.31
e_{15}	5.093	5.217	5.47	5.59	7.269	7.9337
Dielectric permittivities (F/m)						
$\epsilon_{11} \times 10^{8}$	0.262	0.251	0.261	0.275	0.39	0.411
$\epsilon_{33} \times 10^{8}$	0.2296	0.234	0.231	0.232	0.345	0.347

As before, all coefficients in Equations 17.4 through 17.6 are considered to be periodic functions of $y = x/\varepsilon$ and are defined as follows:

L-03

$$c_{ijkl}^{(\varepsilon)}(x) = c_{ijkl}(y) = \begin{cases} c_{ijkl}^{(p)} & (y \in Y_p) \\ c_{ijkl}^{(d)} & (y \in Y_d) \end{cases}$$

$$e_{kij}^{(\varepsilon)}(x) = e_{kij}(y) = \begin{cases} e_{kij} & (y \in Y_p) \\ 0 & (y \in Y_d) \end{cases}$$

$$\epsilon_{ik}^{(\varepsilon)}(x) = \varepsilon_{ik}^{(\varepsilon)}(y) = \begin{cases} \epsilon_{ik}^{(p)} & (y \in Y_p) \\ \epsilon_{ik}^{(d)} & (y \in Y_d) \end{cases}$$

$$\kappa_{ij}^{(\varepsilon)}(x) = \kappa_{ij}(y) = \begin{cases} 0 & (y \in Y_p) \\ \kappa_{ij} & (y \in Y_d) \end{cases}$$

$$\mu_{ik}^{(\varepsilon)}(x) = \mu_{ik}(y) = \begin{cases} \mu_{ik}^{(p)} & (y \in Y_p) \\ \mu_{ik}^{(d)} & (y \in Y_d) \end{cases} \qquad (17.7)$$

TABLE 16.5
Computed Effective Properties
for Compositions (a)–(c) and Geometries 1 and 2; $\gamma = 0.7$

	(a)		(b)		(c)	
	1	2	1	2	1	2
Elastic moduli (N/m^2)						
$c_{11} \times 10^{-11}$	0.624	0.483	0.9915	0.9623	0.640	0.485
$c_{12} \times 10^{-11}$	0.395	0.3355	0.3922	0.3629	0.309	0.231
$c_{13} \times 10^{-11}$	0.400	0.341	0.3832	0.3563	0.326	0.246
$c_{33} \times 10^{-11}$	0.569	0.5334	0.9225	0.8978	0.682	0.6282
$c_{44} \times 10^{-11}$	0.101	0.0993	0.2833	0.2828	0.1867	0.2008
Piezoelectric constants (C/m^2)						
$-e_{31}$	1.35	0.678	1.382	1.097	2.16	1.575
e_{33}	4.646	5.06	4.685	4.9596	11.584	12.005
e_{15}	3.974	3.986	3.8	4.032	8.768	8.761
Dielectric permittivities (F/m)						
$\epsilon_{11} \times 10^8$	0.195	0.2026	0.1979	0.2081	0.4501	0.4577
$\epsilon_{33} \times 10^8$	0.1734	0.1783	0.1731	0.1763	0.4015	0.4051

In accordance with the asymptotic two-expansion method, the mechanical displacement and the electric and magnetic fields are represented in the respective forms

$$u_k^{(\varepsilon)} = u_k^{(0)}(x,t) + \varepsilon u_k^{(1)}(x,y,t) + \cdots \tag{17.8}$$

$$E_k^{(\varepsilon)} = E_k^{(0)}(x,y,t) + \varepsilon E_k^{(1)}(x,y,t) + \cdots \tag{17.9}$$

$$H_k^{(\varepsilon)} = H_k^{(0)}(x,y,t) + \varepsilon H_k^{(1)}(x,y,t) + \cdots \tag{17.10}$$

Equations 17.4 through 17.6 then yield

$$\sigma_{ij}^{(\varepsilon)} = \sigma_{ij}^{(0)}(x,y,t) + \varepsilon \sigma_{ij}^{(1)}(x,y,t) + \cdots$$

$$D_i^{(\varepsilon)} = D_i^{(0)}(x,y,t) + \varepsilon D_i^{(1)}(x,y,t) + \cdots$$

$$B_i^{(\varepsilon)} = B_i^{(0)}(x,y,t) + \varepsilon B_i^{(1)}(x,y,t) + \cdots \tag{17.11}$$

$$J_i^{(\varepsilon)} = J_i^{(0)}(x,y,t) + \varepsilon J_i^{(1)}(x,y,t) + \cdots$$

where we have denoted

$$\sigma_{ij}^{(n)} = c_{ijkl}(y)\frac{\partial u_k^{(n)}}{\partial x_l} - e_{kij}(y)E_k^{(n)} + c_{ijkl}(y)\frac{\partial u_k^{(n+1)}}{\partial y_l} \qquad (17.12)$$

$$D_i^{(n)} = e_{ikl}(y)\frac{\partial u_k^{(n)}}{\partial x_l} + \epsilon_{ik}(y)E_k^{(n)} + e_{ikl}(y)\frac{\partial u_k^{(n+1)}}{\partial y_l} \qquad (17.13)$$

$$B_i^{(n)} = \mu_{ik}(y)H_k^{(n)} \qquad J_i^{(n)} = \kappa_{ij}(y)E_j^{(n)} \qquad (17.14)$$

(where $n = 0, 1, \dots$).

Substituting Equations 17.11 into Equations 17.1 through 17.3 and equating to zero the terms $O(\varepsilon^{-1})$ and $O(\varepsilon^0)$ we obtain

$$\frac{\partial \sigma_{ij}^{(0)}}{\partial y_j} = 0 \qquad \text{curl}_y\, \mathbf{H}^{(0)} = 0 \qquad \text{curl}_y\, \mathbf{E}^{(0)} = 0 \qquad (17.15)$$

$$\frac{\partial \sigma_{ij}^{(0)}}{\partial x_j} + \frac{\partial \sigma_{ij}^{(1)}}{\partial y_j} = \rho(y)\frac{\partial^2 u_i^{(0)}}{\partial t^2} \qquad (17.16)$$

$$\text{curl}_x\, \mathbf{H}^{(0)} + \text{curl}_y\, \mathbf{H}^{(1)} = \frac{\partial \mathbf{D}^{(0)}}{\partial t} + \mathbf{J}^{(0)} \qquad (17.17)$$

$$\text{curl}_x\, \mathbf{E}^{(0)} + \text{curl}_y\, \mathbf{E}^{(1)} = -\frac{\partial \mathbf{B}^{(0)}}{\partial t} \qquad (17.18)$$

where the subscripts on "curl" indicate the variables acted upon by the operator.

Equations 17.16 through 17.18 are now subjected to the averaging operation

$$\langle f \rangle = \frac{1}{|Y|}\int_Y f\, dy \qquad \langle f \rangle \equiv \tilde{f}$$

and the periodicity of $\sigma_{ij}^{(1)}$, $H_i^{(0)}$, and $E_k^{(0)}$ in y used to give the mechanical and electrodynamical equations of the problem,

$$\frac{\partial \tilde{\sigma}_{ij}^{(0)}}{\partial x_j} = \tilde{\rho}\frac{\partial^2 u_i^{(0)}}{\partial t^2} \qquad (17.19)$$

$$\text{curl}_x\, \tilde{\mathbf{H}}^{(0)} = \frac{\partial \tilde{\mathbf{D}}^{(0)}}{\partial t} + \tilde{\mathbf{J}}^{(0)} \qquad \text{curl}_x\, \tilde{\mathbf{E}}^{(0)} = -\frac{\partial \tilde{\mathbf{B}}^{(0)}}{\partial t} \qquad (17.20)$$

From the second and third of Equations 17.15 it follows that

$$E_k^{(0)} = \tilde{E}_k^{(0)}(x,t) - \frac{\partial \varphi(x,y,t)}{\partial y_k}$$

$$H_k^{(0)} = \tilde{H}_k(x,t) - \frac{\partial \psi(x,y,t)}{\partial y_k}$$

(17.21)

where the functions φ and ψ are Y-periodic in y and can be determined by means of the relations

$$\mathrm{div}\left(\frac{\partial \mathbf{D}^{(\varepsilon)}}{\partial t} + \mathbf{J}^{(\varepsilon)}\right) = 0 \quad \text{and} \quad \mathrm{div}\left(\frac{\partial \mathbf{B}^{(\varepsilon)}}{\partial t}\right) = 0 \qquad (17.22)$$

that result from the application of the operation div to Maxwell's equations, Equations 17.2 and 17.3. Substituting Equations 17.11 into Equations 17.22 results in the equations

$$\mathrm{div}_y\left(\frac{\partial \mathbf{D}^{(0)}}{\partial t} + \mathbf{J}^{(0)}\right) = 0 \quad \text{and} \quad \mathrm{div}_y\left(\frac{\partial \mathbf{B}^{(0)}}{\partial t}\right) = 0 \qquad (17.23)$$

which, together with the first of Equations 17.15, form local problems on the unit cell of the component. Using Equations 17.21 and 17.14, the local problems take the form

$$\frac{\partial}{\partial y_j}\left(c_{ijkl}(y)\frac{\partial u_k^{(1)}}{\partial y_l} + e_{kij}(y)\frac{\partial \varphi}{\partial y_k}\right) = -\frac{\partial c_{ijkl}}{\partial y_j}\frac{\partial u_k^{(0)}}{\partial x_l} + \frac{\partial e_{kij}}{\partial y_j}\tilde{E}_k^{(0)} \quad (17.24)$$

$$\frac{\partial}{\partial y_j}\left(e_{ikl}(y)\frac{\partial}{\partial t}\frac{\partial u_k^{(1)}}{\partial y_l} - \epsilon_{ik}(y)\frac{\partial}{\partial t}\frac{\partial \varphi}{\partial y_k} - \kappa_{ik}(y)\frac{\partial \varphi}{\partial y_k}\right)$$

$$= -\frac{\partial e_{ikl}(y)}{\partial y_i}\frac{\partial}{\partial t}\frac{\partial u_k^{(0)}}{\partial x_l} - \frac{\partial \epsilon_{ik}(y)}{\partial y_i}\frac{\partial}{\partial t}\tilde{E}_k^{(0)} - \frac{\partial \kappa_{ik}(y)}{\partial y_i}\tilde{E}_k^{(0)} \quad (17.25)$$

$$\frac{\partial}{\partial y_i}\left(\mu_{ik}(y)\frac{\partial \psi}{\partial y_k}\right) = \frac{\partial \mu_{ik}(y)}{\partial y_i}\tilde{H}_k^{(0)}(x,t) \qquad (17.26)$$

Taking the Laplace transforms of Equations 17.24 and 17.25 with respect to t, using the zero initial conditions, now gives

$$\frac{\partial}{\partial y_j}\left(c_{ijkl}(y)\frac{\partial \hat{u}_k^{(1)}}{\partial y_l} + e_{kij}(y)\frac{\partial \hat{\varphi}}{\partial y_k}\right) = -\frac{\partial c_{ijkl}(y)}{\partial y_j}\frac{\partial \hat{u}_k^{(0)}}{\partial x_l} + \frac{\partial e_{kij}(y)}{\partial y_j}\hat{E}_k^{(0)}$$

(17.27)

$$\frac{\partial}{\partial y_j}\left(pe_{ikl}(y)\frac{\partial \hat{u}_k^{(1)}}{\partial y_l} - p\epsilon_{ik}(y)\frac{\partial \hat{\varphi}}{\partial y_k} - \kappa_{ik}(y)\frac{\partial \hat{\varphi}}{\partial y_k}\right)$$

$$= -p\frac{\partial e_{ikl}(y)}{\partial y_i}\frac{\partial \hat{u}_k^{(0)}}{\partial x_l} - \frac{\partial}{\partial y_i}(p\epsilon_{ik}(y) + \kappa_{ik}(y))\hat{E}_k^{(0)} \quad (17.28)$$

where the caret (^) denotes a Laplace transform with p the transform parameter.

Now if we write

$$\hat{u}_n^{(1)} = \hat{N}_n^{kl}(y,p)\frac{\partial \hat{u}_k^{(0)}(x,p)}{\partial x_l} + \hat{W}_{nk}(y,p)\hat{E}_k^{(0)}(x,p) \quad (17.29)$$

$$\hat{\varphi} = \hat{\Phi}_{kl}(y,p)\frac{\partial \hat{u}_k^{(0)}(x,p)}{\partial x_l} + \hat{\Psi}_k(y,p)\hat{E}_k^{(0)}(x,p) \quad (17.30)$$

$$\psi = V_k(y)\tilde{H}_k^{(0)}(x,t) \quad (17.31)$$

then, upon substitution into Equations 17.26 through 17.28, we obtain the following system of equations for determining the Y-periodic functions $\hat{N}_n^{kl}(y,p)$, $\hat{W}_{nk}(y,p)$, $\hat{\Phi}_{kl}(y,p)$, $\Psi_k(y,p)$, and $V_k(y)$:

$$\frac{\partial \hat{\tau}_{ij}^{kl}}{\partial y_j} = -\frac{\partial c_{ijkl}(y)}{\partial y_j} \quad (17.32)$$

$$\frac{\partial \hat{d}_i^{kl}}{\partial y_i} = -\frac{\partial e_{ikl}}{\partial y_i} \quad (17.33)$$

$$\frac{\partial \hat{\bar{\tau}}_{ij}^k}{\partial y_j} = \frac{\partial e_{kij}(y)}{\partial y_j} \tag{17.34}$$

$$\frac{\partial \hat{\bar{d}}_i^k}{\partial y_i} = -\frac{\partial}{\partial y_i}\left(\epsilon_{ik}(y) + \frac{1}{p}\kappa_{ik}(y)\right)$$

$$(i, k, l = 1, 2, 3) \tag{17.35}$$

and

$$\frac{\partial}{\partial y_i}\left(\mu_{ih}(y)\frac{\partial V_k(y)}{\partial y_h}\right) = \frac{\partial \mu_{ik}(y)}{\partial y_i} \qquad (k = 1, 2, 3) \tag{17.36}$$

where

$$\hat{\tau}_{ij}^{kl} = c_{ijnh}(y)\frac{\partial \hat{N}_n^{kl}(y, p)}{\partial y_h} + e_{hij}(y)\frac{\partial \hat{\Phi}_{kl}(y, p)}{\partial y_h} \tag{17.37}$$

$$\hat{d}_i^{kl} = e_{inh}(y)\frac{\partial \hat{N}_n^{kl}(y, p)}{\partial y_h} - \left(\epsilon_{ih}(y) + \frac{1}{p}\kappa_{ih}(y)\right)\frac{\partial \hat{\Phi}_{kl}(y, p)}{\partial y_h} \tag{17.38}$$

$$\hat{\bar{\tau}}_{ij}^k = c_{ijnh}(y)\frac{\partial \hat{W}_{nk}(y, p)}{\partial y_h} + e_{hij}(y)\frac{\partial \hat{\Psi}_k(y, p)}{\partial y_h} \tag{17.39}$$

$$\hat{\bar{d}}_i^k = e_{inh}(y)\frac{\partial \hat{W}_{nk}(y, p)}{\partial y_h} - \left(\epsilon_{ih}(y) + \frac{1}{p}\kappa_{ih}(y)\right)\frac{\partial \hat{\Psi}_k(y, p)}{\partial y_h} \tag{17.40}$$

Using Equations 17.29 through 17.31 and setting $n = 0$ in Equations 17.12 through 17.14 yields

$$\hat{\sigma}_{ij}^{(0)}\left(c_{ijkl}(y) + \hat{\tau}_{ij}^{kl}(y, p)\right)\frac{\partial \hat{u}_k^{(0)}(x, p)}{\partial x_l} - \left(e_{kij}(y)\right.$$

$$\left. - \hat{\bar{\tau}}_{ij}^k(y, p)\right)\hat{\bar{E}}_k^{(0)}(x, p) \tag{17.41}$$

$$\hat{D}_i^{(0)} = \left(e_{ikl}(y) + \hat{d}_i^{kl}(y,p) + \frac{1}{p}\hat{\eta}^{kl}(y,p)\right)\frac{\partial\hat{u}_k^{(0)}(x,p)}{\partial x_l}$$

$$+ \left(\epsilon_{ik}(y) + \hat{\bar{d}}_i^k(y,p) + \frac{1}{p}\hat{\bar{\eta}}_i^k(y,p)\right)\hat{E}_k^{(0)}(x,p) \quad (17.42)$$

$$\hat{J}_i^{(0)} = -\eta_i^{kl}(y,p)\frac{\partial\hat{u}_k^{(0)}(x,p)}{\partial x_l} + \left(\kappa_{ik}(y) - \hat{\bar{\eta}}_i^k(y,p)\right)\hat{E}_k^{(0)}(x,p) \quad (17.43)$$

$$B_i^{(0)} = \left(\mu_{ik}(y) - \mu_{ih}(y)\frac{\partial V_k(y)}{\partial y_h}\right)H_k^{(0)}(x,t) \quad\quad (17.44)$$

where we have defined

$$\hat{\eta}_i^{kl}(y,p) = \kappa_{ih}(y)\frac{\partial\hat{\Phi}_{kl}(y,p)}{\partial y_h}$$

$$(17.45)$$

$$\hat{\bar{\eta}}_i^k(y,p) = \kappa_{ih}(y)\frac{\partial\hat{\Psi}_k(y,p)}{\partial y_h}$$

Taking a volume average over the unit cell of Equations 17.41 through 17.44 now yields the (averaged) Laplace transforms of the stress, electrical displacement, and current density components,

$$\hat{\bar{\sigma}}_{ij}^{(0)} = \hat{\bar{c}}_{ijkl}(p)\frac{\partial\hat{u}_k^{(0)}(x,p)}{\partial x_l} - \hat{\bar{e}}_{kij}(p)\hat{E}_k^{(0)}(x,p) \quad\quad (17.46)$$

$$\hat{\bar{D}}_i^{(0)} = \hat{\bar{e}}_{ikl}'(p)\frac{\partial\hat{u}_k^{(0)}(x,p)}{\partial x_l} + \hat{\bar{\epsilon}}_{ik}(p)\hat{E}_k^{(0)}(x,p) \quad\quad (17.47)$$

$$\hat{\bar{J}}_i^{(0)} = -\hat{\bar{\psi}}_{ikl}(p)\frac{\partial\hat{u}_k^{(0)}(x,p)}{\partial x_l} + \hat{\bar{\kappa}}_{ik}(p)\hat{E}_k^{(0)}(x,p) \quad\quad (17.48)$$

and the relation for the magnetic flux density,

$$\tilde{B}_i^{(0)} = \tilde{\mu}_{ik}\tilde{H}_k^{(0)}(x,t) \quad\quad (17.49)$$

where

$$\hat{\tilde{c}}_{ijkl}(p) = \langle c_{ijkl}(y) + \hat{\tau}_{ij}^{kl}(y,p)\rangle$$

$$\hat{\tilde{e}}_{ijk}(p) = \langle e_{ijk}(y) - \hat{\tau}_{ij}^{k}(y,p)\rangle$$

$$\hat{\tilde{e}}'_{ikl}(p) = \left\langle e_{ikl}(y) + \hat{d}_i^{kl}(y,p) + \frac{1}{p}\hat{\eta}_i^{kl}(y,p)\right\rangle$$

$$\hat{\tilde{\chi}}_{ikl}(p) = \langle \hat{\eta}_i^{kl}(y,p)\rangle \tag{17.50}$$

$$\hat{\tilde{\epsilon}}_{ik}(p) = \left\langle \epsilon_{ik}(y) + \hat{d}_i^{k}(y,p) + \frac{1}{p}\hat{\eta}_i^{k}(y,p)\right\rangle$$

$$\hat{\tilde{\kappa}}_{ik}(p) = \langle \kappa_{ik}(y) - \hat{\eta}_i^{k}(y,p)\rangle$$

$$\tilde{\mu}_{ik} = \left\langle \mu_{ik}(y) - \mu_{ih}(y)\frac{\partial V_k(y)}{\partial y_h}\right\rangle$$

The constitutive relations of the piezoelectric composite are found at this point through the transform inversion of Equations 17.46 through 17.48, and they turn out to exhibit memory properties as a consequence of the finite electrical conductivity in one of the constituents:

$$\tilde{\sigma}_{ij}^{(0)} = \int_0^t \tilde{c}_{ijkl}(t-\tau)\frac{\partial u_k^{(0)}(x,\tau)}{\partial x_l}\,d\tau - \int_0^t \tilde{e}_{kij}(t-\tau)\tilde{E}_k^{(0)}(x,\tau)\,d\tau$$

$$\tilde{D}_i^{(0)} = \int_0^t \tilde{e}'_{ikl}(t-\tau)\frac{\partial u_k^{(0)}(x,\tau)}{\partial x_l}\,d\tau + \int_0^t \tilde{e}_{kij}(t-\tau)\tilde{E}_k^{(0)}(x,\tau)\,d\tau$$

$$\tilde{J}_i^{(0)} = -\int_0^t \tilde{\psi}_{ikl}(t-\tau)\frac{\partial u_k^{(0)}(x,\tau)}{\partial x_l}\,d\tau \tag{17.51}$$

$$+ \int_0^t \tilde{\kappa}_{ik}(t-\tau)\tilde{E}_k^{(0)}(x,\tau)\,d\tau$$

$$\tilde{B}_i^{(0)} = \tilde{\mu}_{ik}\tilde{H}_k^{(0)}(x,t)$$

Note that the kernels involved in the integrals in Equations 17.51 are determined by solving the local problems given by Equations 17.32 through 17.35 in the space of Laplace transforms and subsequently inverting the transforms so found. A noteworthy feature of the constitutive equations (Equations 17.51) of the homogenized piezoelectric medium is the strain dependence shown by the electrical current density of the system.

Consider next a laminar composite whose unit cell is composed of a finite-conductivity, isotropic dielectric layer and a piezoceramic layer polarized parallel to the interface between the layers (see Figure 16.2). For the piezoceramic, the structure of the material constant matrices is given in Chapter 1. For the isotropic dielectric, the only nonzero components in these matrices are

$$c_{11}^{(d)} = c_{22}^{(d)} = c_{33}^{(d)} = \lambda + 2G \qquad c_{12}^{(d)} = c_{13}^{(d)} = \lambda$$

$$c_{44}^{(d)} = c_{55}^{(d)} = c_{66}^{(d)} = G \qquad \epsilon_{ij}^{(d)} = \epsilon^{(d)}\delta_{ij} \qquad \kappa_{ij}^{(d)} = \kappa\delta_{ij}$$

where λ and G are Lamé constants and δ_{ij} is the Kronecker symbol.

Using the structure of the matrices $c_{\alpha\beta}$, $e_{\alpha i}$, ϵ_{ij}, and κ_{ij} and noting that the only space variable of the problem is y_1, Equations 17.37 through 17.40 can be put into the form

$$\hat{\tau}_{11}^{kl} = c_{11}(y_1)\frac{\partial}{\partial y_1}\hat{N}_1^{kl}(y_1, p) \qquad \hat{\tau}_{22}^{kl} = c_{12}(y_1)\frac{\partial}{\partial y_1}\hat{N}_1^{kl}(y_1. p)$$

$$\hat{\tau}_{33}^{kl} = c_{13}(y_1)\frac{\partial}{\partial y_1}\hat{N}_1^{kl}(y_1, p) \qquad \hat{\tau}_{12}^{kl} = c_{66}(y_1)\frac{\partial}{\partial y_1}\hat{N}_2^{kl}(y_1, p)$$

$$\hat{\tau}_{13}^{kl} = c_{44}(y_1)\hat{N}_3^{kl}(y_1, p) + e_{15}(y_1)\frac{\partial}{\partial y_1}\hat{\Phi}_{kl}(y_1, p)$$

$$\hat{\tau}_{23}^{kl} = 0 \tag{17.52}$$

$$d_1^{kl} = e_{15}(y_1)\frac{\partial}{\partial y_1}\hat{N}_3^{kl}(y_1, p) - (\epsilon_{11}(y_1) + p^{-1}\kappa_{11}(y_1))\frac{\partial}{\partial y_1}\hat{\Phi}_{kl}(y_1, p)$$

$$d_2^{kl} = 0 \qquad d_3^{kl} = e_{13}(y_1)\frac{\partial}{\partial y_1}\hat{N}_1^{kl}(y_1, p) \tag{17.53}$$

$$\hat{\tilde{\tau}}_{11}^l = c_{11}(y_1)\frac{\partial}{\partial y_1}\hat{W}_1^l(y_1, p) \qquad \hat{\tilde{\tau}}_{22}^l = c_{12}(y_1)\frac{\partial}{\partial y_1}\hat{W}_1^l(y_1, p)$$

$$\hat{\tilde{\tau}}_{33}^l = c_{13}(y_1)\frac{\partial}{\partial y_1}\hat{W}_1^l(y_1, p) \qquad \hat{\tilde{\tau}}_{12}^l = c_{66}(y_1)\frac{\partial}{\partial y_1}\hat{W}_2^l(y_1, p)$$

$$\hat{\tilde{\tau}}_{13}^l = c_{44}(y_1)\frac{\partial}{\partial y_1}\hat{W}_3^l(y_1, p) + e_{15}(y_1)\frac{\partial}{\partial y_1}\hat{\Psi}_l$$

$$\hat{\tilde{\tau}}_{23}^l = 0 \tag{17.54}$$

$$\hat{d}_1^l = e_{15}(y_1)\frac{\partial}{\partial y_1}\hat{W}_3^l(y_1, p) - (\epsilon_{11}(y_1) + p^{-1}\kappa_{11}(y_1))\frac{\partial}{\partial y_1}\hat{\Psi}_l(y_1, p)$$

$$\hat{d}_2^l = 0 \qquad \hat{d}_3^l = e_{15}(y_1)\frac{\partial}{\partial y_1}\hat{W}_1^l(y_1, p) \tag{17.55}$$

Finally, the functions $\hat{N}_n^{kl}$, $\hat{\Phi}_{kl}$, $\hat{W}_n^l$, and $\hat{\psi}_l$ are determined by solving the local problems given by Equations 17.32 through 17.35, after some algebra, the matrices of the effective properties of the homogenized medium are found to be

$$\tilde{c}_{\alpha\beta}(t) = \begin{pmatrix} \tilde{c}_{11}(t) & \tilde{c}_{12}(t) & \tilde{c}_{13}(t) & 0 & 0 & 0 \\ \tilde{c}_{12}(t) & \tilde{c}_{22}(t) & \tilde{c}_{23}(t) & 0 & 0 & 0 \\ \tilde{c}_{13}(t) & \tilde{c}_{23}(t) & \tilde{c}_{33}(t) & 0 & 0 & 0 \\ 0 & 0 & 0 & \tilde{c}_{44}(t) & 0 & 0 \\ 0 & 0 & 0 & 0 & \tilde{c}_{55}(t) & 0 \\ 0 & 0 & 0 & 0 & 0 & \tilde{c}_{66}(t) \end{pmatrix}$$

$$\tilde{e}_{\alpha i}(t) = \begin{pmatrix} 0 & 0 & \tilde{e}_{13}(t) \\ 0 & 0 & \tilde{e}_{23}(t) \\ 0 & 0 & \tilde{e}_{33}(t) \\ 0 & \tilde{e}_{42}(t) & 0 \\ \tilde{e}_{51}(t) & 0 & 0 \\ 0 & 0 & 0 \end{pmatrix} \tag{17.56}$$

$$\tilde{e}'_{i\alpha}(t) = \begin{pmatrix} 0 & 0 & 0 & 0 & \tilde{e}_{15}(t) & 0 \\ 0 & 0 & 0 & \tilde{e}'_{24}(t) & 0 & 0 \\ \tilde{e}'_{31}(t) & \tilde{e}'_{32}(t) & \tilde{e}'_{33}(t) & 0 & 0 & 0 \end{pmatrix}$$

$$\tilde{\varepsilon}_{ij}(t) = \begin{pmatrix} \tilde{\varepsilon}_{11}(t) & 0 & 0 \\ 0 & \tilde{\varepsilon}_{22}(t) & 0 \\ 0 & 0 & \tilde{\varepsilon}_{33}(t) \end{pmatrix} \tag{17.57}$$

$$\tilde{\chi}_{i\alpha}(t) = \begin{pmatrix} 0 & 0 & 0 & 0 & \tilde{\chi}_{15} & 0 \\ 0 & 0 & 0 & 0 & 0 & 0 \\ 0 & 0 & 0 & 0 & 0 & 0 \end{pmatrix}$$

$$\tilde{\kappa}_{ij}(t) = \begin{pmatrix} \tilde{\kappa}_{11}(t) & 0 & 0 \\ 0 & \tilde{\kappa}_{22}(t) & 0 \\ 0 & 0 & \tilde{\kappa}_{33}(t) \end{pmatrix} \tag{17.58}$$

where

$$\tilde{c}_{11}(t) = \varepsilon c_{11}^{(p)} [\varepsilon_1 + (\varepsilon - \varepsilon_1)c'_{11}]^{-1} \delta(t)$$

$$\tilde{c}_{12}(t) = A_1 \delta(t) \qquad \tilde{c}_{13}(t) = A_2 \delta(t)$$

$$\tilde{c}_{22}(t) = \left\{ \varepsilon_1 \varepsilon^{-1} \left[c_{11}^{(p)} - (c_{12}^{(p)} - A_1)\frac{c_{12}^{(p)}}{c_{11}^{(p)}} \right] \right.$$

$$\left. + (\varepsilon - \varepsilon_1)\varepsilon^{-1} \left[c_{11}^{(d)} - (c_{12}^{(d)} - A_1)\frac{c_{12}^{(d)}}{c_{11}^{(d)}} \right] \right\} \delta(t)$$

$$\tilde{c}_{33}(t) = \left\{ \varepsilon_1 \varepsilon^{-1} \left[c_{33}^{(p)} - (c_{13}^{(p)} - A_2)\frac{c_{13}^{(p)}}{c_{11}^{(p)}} \right] \right.$$

$$\left. + (\varepsilon - \varepsilon_1)\varepsilon^{-1} \left[c_{33}^{(d)} - (c_{13}^{(d)} - A_2)\frac{c_{13}^{(d)}}{c_{11}^{(d)}} \right] \right\} \delta(t)$$

$$\tilde{c}_{23}(t) = \left\{ \varepsilon_1 \varepsilon^{-1} \left[c_{12}^{(p)} - (c_{13}^{(p)} - A_2)\frac{c_{12}^{(p)}}{c_{11}^{(p)}} \right] \right.$$

$$\left. + (\varepsilon - \varepsilon_1)\varepsilon^{-1} \left[c_{13}^{(d)} - (c_{13}^{(d)} - A_2)\frac{c_{12}^{(d)}}{c_{11}^{(d)}} \right] \right\} \delta(t)$$

$$\tilde{c}_{44}(t) = \left[\varepsilon_1\varepsilon^{-1}c_{44}^{(p)} + (\varepsilon - \varepsilon_1)\varepsilon^{-1}c_{44}^{(d)}\right]\delta(t)$$

$$\tilde{c}_{66}(t) = \varepsilon c_{66}^{(p)}\left[\varepsilon_1 + (\varepsilon - \varepsilon_1)\frac{c_{66}^{(d)}}{c_{66}^{(p)}}\right]^{-1}\delta(t)$$

$$\tilde{c}_{55}(t) = \left\{2\varepsilon_1\varepsilon^{-1}(c_{44}^{(p)} - c_{44}^{(d)}) + 2c_{44} - 2\varepsilon_1\varepsilon^{-1}c_{44}\right.$$

$$\times \frac{b_3}{b_2}(e_{15}^2 + c_{44}^{(p)}\epsilon_{11}^{(p)})^{-1} + \left.\frac{b_1}{b_2}\right\}\delta(t)$$

$$- \frac{1}{b_2^2}\left\{a_1 b_2 + b_1 a_2 - 2\varepsilon_1\varepsilon^{-1}c_{44}^{(p)}e_{15}(a_3 b_2 - b_3 a_2)\right.$$

$$\times (e_{15}^2 + c_{44}^{(p)}\epsilon_{11}^{(p)})^{-1}\Big\}e^{-\lambda_2 t}$$

$$\tilde{e}_{32}'(t) = \left[\varepsilon_1\varepsilon e_{13} - \varepsilon^{-1}\frac{c_{12}^{(p)}}{c_{11}^{(p)}} - \varepsilon^{-1}\frac{A_1}{c_{11}^{(p)}}\right]\delta(t)$$

$$\tilde{e}_{31}'(t) = \varepsilon e_{31}[\varepsilon_1 + (\varepsilon - \varepsilon_1)c_{11}']^{-1}\delta(t)$$

$$\tilde{e}_{33}'(t) = \left\{\varepsilon_1\varepsilon^{-1}\left[e_{33} - \frac{e_{13}(c_{13}^{(p)} - A_2)}{c_{11}^{(p)}}\right]\right\}\delta(t)$$

$$\tilde{e}_{15}'(t) = \frac{b_3}{b_2}\delta(t) + \frac{(a_3 b_2 - b_3 a_2)}{b_2^2}$$

$$\times \left[1 + (\varepsilon - \varepsilon_1)\varepsilon^{-1}\lambda_1(\lambda_1 - \lambda_2)^{-1}\right]e^{-\lambda_2 t}$$

$$- (\varepsilon - \varepsilon_1)\varepsilon\lambda_1 b_2^{-2}\left[b_3 b_2 + (a_3 b_2 - b_3 a_2)\right.$$

$$\times (\lambda_1 - \lambda_2)^{-1}\Big]e^{-\lambda_1 t}$$

$$\tilde{e}_{24}'(t) = \varepsilon_1\varepsilon^{-1}e_{15}\delta(t) \qquad \tilde{e}_{42}'(t) = \varepsilon_1\varepsilon^{-1}e_{15}\delta(t)$$

$$\tilde{e}_{13}(t) = \varepsilon_1 e_{13}$$

$$\times [\varepsilon_1 + (\varepsilon - \varepsilon_1)c_{11}']\delta(t)$$

$$\tilde{e}_{23}(t) = \left\{ \varepsilon_1 \varepsilon^{-1}\left(e_{13} - e_{13}\frac{c_{12}^{(p)}}{c_{11}^{(p)}} \right) \right.$$

$$+ \left[\varepsilon_1 \varepsilon^{-1}\frac{c_{12}^{(p)}}{c_{11}^{(p)}} + (\varepsilon - \varepsilon_1)\varepsilon^{-1}\frac{c_{12}^{(d)}}{c_{11}^{(d)}} \right]\varepsilon_1 e_{13}$$

$$\left. \times [\varepsilon_1 + (\varepsilon - \varepsilon_1)c_{11}']^{-1} \right\}\delta(t)$$

$$\tilde{e}_{51}(t) = b_3 b_2^{-1}\delta(t) + (a_3 b_2 - b_3 a_2)b_2^{-2}e^{-\lambda_2 t}$$

$$\tilde{e}_{33}(t) = \left\{ \varepsilon_1 \varepsilon^{-1}\left(e_{33} - c_{13}^{(p)}\frac{e_{31}}{c_{11}^{(p)}} \right) \right.$$

$$+ \left[\varepsilon_1 \varepsilon^{-1}\frac{c_{13}^{(p)}}{c_{11}^{(p)}} + (\varepsilon - \varepsilon_1)\varepsilon^{-1}\frac{c_{13}^{(d)}}{c_{11}^{(d)}} \right]$$

$$\left. \times \varepsilon_1 e_{31}[\varepsilon_1 + (\varepsilon - \varepsilon_1)c_{11}']^{-1} \right\}\delta(t)$$

$$\tilde{\epsilon}_{11}(t) = \frac{b_4}{b_2}\delta(t) + \frac{1}{b_2^2}\left[(a_4 b_2 - b_4 a_2)(\lambda_1 - \lambda_2)^{-1} - b_4 b_2 \lambda_1 \right]e^{-\lambda_1 t}$$

$$+ \frac{1}{b_2^2}(a_4 b_2 - b_4 a_2)\left[1 - (\lambda_1 - \lambda_2)^{-1} \right]e^{-\lambda_2 t}$$

$$\tilde{\epsilon}_{22}(t) = \left[\varepsilon_1 \varepsilon^{-1}\epsilon_{11}^{(p)} + (\varepsilon - \varepsilon_1)\varepsilon^{-1}\epsilon_{11}^{(d)} \right]\delta(t)$$

$$\tilde{\epsilon}_{33}(t) = \varepsilon_1 \varepsilon^{-1}\left(\frac{e_{13}^2}{c_{11}^{(p)}} \right)\left[1 - \varepsilon_1(\varepsilon_1 + (\varepsilon - \varepsilon_1)c_{11}')^{-1} \right]\delta(t)$$

$$\tilde{\kappa}_{11}(t) = (\varepsilon - \varepsilon_1)\varepsilon^{-1}\left\{ b_1 b_2^{-1}\delta(t) - \lambda_1^2(a_1 - b_1\lambda_1)(a_2 - b_2\lambda_1)^{-1}e^{-\lambda_1 t} \right.$$

$$\left. + \lambda_1 a_2 b_2^{-2}(a_1 b_2 - b_1 a_2)(a_2 - b_2\lambda_1)^{-1}e^{-\lambda_2 t} \right\}$$

$$\tilde{\kappa}_{22}(t) = \tilde{\kappa}_{33}(t) = (\varepsilon - \varepsilon_1)\varepsilon^{-1}\kappa$$

and

$$\tilde{\chi}_{15}(t) = -\varepsilon_1^2 e_{15} c_{44}^{(d)} b_2^{-1}\{\delta(t) - \lambda_2 e^{-\lambda_2 t}\}$$

in which

$$A_1\left[\varepsilon_1 c_{12}^{(p)} + (\varepsilon - \varepsilon_1)c_{12}^{(d)}c_{11}'\right]\left[\varepsilon_1 + (\varepsilon - \varepsilon_1)c_{11}'\right]^{-1}$$

$$A_2 = \left[\varepsilon_1 c_{13}^{(p)} + (\varepsilon - \varepsilon_1)c_{13}^{(d)}c_{11}'\right]\left[\varepsilon_1 + (\varepsilon - \varepsilon_1)c_{11}'\right]^{-1}$$

$$a_1 = \varepsilon_1 \varepsilon c_{44}^{(p)}c_{44}^{(d)}\kappa$$

$$a_2 = \varepsilon_1 \kappa\left[\varepsilon_1 c_{44}^{(d)} + (\varepsilon - \varepsilon_1)c_{44}^{(p)}\right]$$

$$a_3 = \varepsilon_1 \varepsilon e_{15} c_{44}^{(d)}\kappa \qquad a_4 = \lambda_1 b_4$$

$$b_1 = \varepsilon_1 \varepsilon c_{44}^{(d)}c_{44}^{(p)}\epsilon^{(d)} + \varepsilon(\varepsilon - \varepsilon_1)c_{44}^{(d)}(e_{15}^2 + c_{44}^{(p)}\epsilon_{11}^{(p)})$$

$$b_2 = \varepsilon_1^2 c_{44}^{(d)}\epsilon^{(d)} + (\varepsilon - \varepsilon_1)^2(e_{15}^2 + c_{44}^{(p)}\epsilon_{11}^{(p)})$$

$$+ \varepsilon_1(\varepsilon - \varepsilon_1)c_{44}^{(p)}\epsilon^{(d)} + \varepsilon_1(\varepsilon - \varepsilon_1)c_{44}^{(d)}$$

$$b_3 = \varepsilon_1 \varepsilon e_{15} c_{44}^{(d)}\epsilon^{(d)}$$

$$b_4 = \varepsilon\epsilon^{(d)}\left[\varepsilon_1 \epsilon_{11}^{(p)}c_{44}^{(d)} + (\varepsilon - \varepsilon_1)(e_{15}^2 + c_{44}^{(p)}\epsilon_{11}^{(d)})\right]$$

$$c_{11}' = \frac{c_{11}^{(p)}}{c_{11}^{(d)}} \qquad \lambda_1 = \frac{\kappa}{\epsilon^{(d)}} \qquad \lambda_2 = \frac{a_2}{b_2}$$

and $\delta(t)$ is Dirac's delta function.

18. POROUS ELASTIC BODY FILLED WITH A VISCOUS FLUID

There are a wide variety of problems, apart from those concerned with composite materials, that also involve periodic deformable media and are amenable to treatment in a manner discussed in this book. Problems of this kind arise, for example, in the study of compound media with complex structure and in the special case of elastic-solid–viscous-fluid combinations, a detailed linear perturbation dynamic analysis for a wide range of viscosities is available (Sanchez-Palencia 1980). Our aim in this section is to derive the homogenized equations for the dynamics of a connected

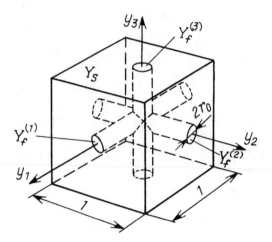

FIGURE 18.1. Unit cell for a body with a set of fluid-filled channels.

elastic solid body pierced with a number of channels filled with a viscous barotropic fluid. As is customary (Sanchez-Palencia 1980), we assume that the body has a periodic structure and that each period Y in the space of variables y_1 is composed of a solid part Y_s separated by a smooth boundary Γ^s from a fluid part Y_f; the unions of all the Y_s and Y_f regions are assumed to form connected sets. A simple example of such a structure is a system of fluid-filled, mutually orthogonal rectilinear channels of circular cross section (shown in Figure 18.1); the unit cell—that is, the periodically repeating building block of this medium—is a cube of dimension ε in the space of variables $x_i = \varepsilon y_i$. We denote by the symbol ε the dimensionless small parameter of our problem.

We begin the solution by deriving equations of motion for a viscous compressible fluid in a stiff nondeformable medium with a periodic structure (Bakhvalov and Panasenko 1984). The low-viscosity, linear form of the Navier–Stokes equations for a noncompressible barotropic fluid is (cf. Sanchez-Palencia 1980)

$$\rho_0 \frac{\partial \mathbf{v}}{\partial t} = \operatorname{grad} p + \nu\varepsilon^2 \nabla^2 \mathbf{v} + \eta\varepsilon^2 \operatorname{grad} \operatorname{div} \mathbf{v} \qquad (18.1)$$

$$\frac{\partial p}{\partial t} + \rho_0 c^2 \operatorname{div} \mathbf{v} = 0 \qquad (18.2)$$

where $\nu\varepsilon^2$ $\eta\varepsilon^2$ are the coefficients of viscosity, p and $\mathbf{v}$ are the pressure and velocity perturbations, ρ_0 is the unperturbed mass density, c is the sound velocity, and ∇^2 is the Laplace operator.

Following the two-scale asymptotic expansion procedure, we write (Sanchez-Palencia 1980, Bakhvalov and Panasenko 1984)

$$\mathbf{v} = \mathbf{v}^{(0)}(x, y, t) + \varepsilon \mathbf{v}^{(1)}(x, y, t) + \cdots \tag{18.3}$$

$$p = p^{(0)}(x, t) + \varepsilon p^{(1)}(x, y, t) + \cdots \tag{18.4}$$

where $y = x/\varepsilon$, $x = (x_1, x_2, x_3)$, and the functions $p^{(1)}$, $\mathbf{v}^{(0)}$, and $\mathbf{v}^{(1)}$ are periodic with period 1 in $y = (y_1, y_2, y_3)$. Substituting into Equations 18.1 and 18.2 and retaining terms $O(\varepsilon^0)$ gives

$$\rho_0 \frac{\partial \mathbf{v}^{(0)}}{\partial t} = -\operatorname{grad}_y p^{(1)} - \operatorname{grad}_x p^{(0)} + \nu \nabla_y^2 \mathbf{v}^{(0)} \tag{18.5}$$

$$\operatorname{div}_y \mathbf{v}^{(0)} = 0 \tag{18.6}$$

$$\frac{\partial p^{(0)}}{\partial t} + \rho_0 c^2 \left(\operatorname{div}_x \mathbf{v}^{(0)} + \operatorname{div}_y \mathbf{v}^{(1)} \right) = 0 \tag{18.7}$$

where the subscripts x and y refer to the variables acted upon by the corresponding operator.

Equations 18.5 and 18.6 form a unit cell local problem whose solution, $\mathbf{v}^{(0)}$, must satisfy zero initial conditions, zero boundary conditions on Γ, and periodicity conditions with respect to variables y.

We next introduce the averaging operator with respect to the local variables,

$$\langle u \rangle = \frac{1}{|Y|} \int_{Y_f} u \, dy \tag{18.8}$$

where $|Y|$ is the unit cell volume, and apply it to Equation 18.7 to obtain the following macroscopic acoustic equation for the fluid within the stiff porous solid:

$$\pi_0 \frac{\partial p^{(0)}(x, t)}{\partial t} + \rho_0 c^2 \operatorname{div}_x \langle \mathbf{v}^{(0)}(x, t) \rangle = 0 \tag{18.9}$$

where the y-periodicity of $\mathbf{v}^{(1)}$ and the "adherence condition" on Γ, $\mathbf{v}^{(1)} = 0$, have been satisfied and where $\pi_0 = |Y_f|/|Y|$ is, obviously, the porosity of the medium.

Taking the Laplace transform of Equations 18.5 and 18.6 with respect to t and using the zero initial conditions for $\mathbf{v}^{(0)}$ and p, results in

$$\rho_0 s \hat{\mathbf{v}}^{(0)}(x, y, s) = -\text{grad}_y\, \hat{p}^{(1)}(x, y, s) - \text{grad}_y\, \hat{p}^{(0)}(x, s)$$

$$+ \nu \nabla_y \hat{\mathbf{v}}^{(0)}(x, y, s) \tag{18.10}$$

and

$$\text{div}_y\, \hat{\mathbf{v}}^{(0)}(x, y, s) = 0 \tag{18.11}$$

where

$$\hat{\mathbf{v}}^{(0)}(x, y, s) = \int_0^\infty e^{-st} \mathbf{v}^{(0)}(x, y, t)\, dt$$

$$\hat{p}^{(1)}(x, y, s) = \int_0^\infty e^{-st} p^{(1)}(x, y, t)\, dt$$

We can now represent the functions $v_i^{(0)}$ and $p^{(1)}$ in the respective forms

$$\hat{v}_i^{(0)}(x, y, s) = \hat{V}_{in}(y, s) \frac{\partial \hat{p}^{(0)}(x, s)}{\partial x_n}$$

$$\hat{p}^{(1)}(x, y, s) = \hat{P}_n(y, s) \frac{\partial \hat{p}^{(0)}(x, s)}{\partial x_s} \qquad (i, n = 1, 2, 3) \tag{18.12}$$

which reduces Equations 18.10 and 18.11 to the following transformed local problem:

$$\rho_0 s \hat{V}_{in}(y, s) = -\frac{\partial \hat{P}_n(y, s)}{\partial y_i} - \delta_{in} + \nu \nabla_y^2 \hat{V}_{in}(y, s) \tag{18.13}$$

$$\frac{\partial \hat{V}_{in}(y, s)}{\partial y_i} = 0 \quad \text{in } Y_s \tag{18.14}$$

Note that the functions $\hat{V}_{in}(y, s)$ and $\hat{P}_n(y, s)$ are 1-periodic in y and $\hat{V}_{in} = 0$ on Γ.

Having solved the problem given by Equations 18.13 and 18.14, Equation 18.12 gives the transformed velocity vector $\hat{v}_i^{(0)}(x, y, s)$, which then is reversed to its original, is averaged, and is inserted into the macroscopic equation (Equation 18.9) to give

$$\pi_0 \frac{\partial p^{(0)}(x,t)}{\partial t} + \rho_0 c^2 \int_0^t \langle V_{in}(t-\tau) \rangle \frac{\partial^2 p^{(0)}(x,\tau)}{\partial x_i \, \partial x_n} \, d\tau = 0 \quad (18.15)$$

where

$$\langle V_{in}(t) \rangle = \frac{1}{|Y|} \int_{Y_f} V_{in}(y,t) \, dy$$

The integrodifferential equation, Equation 18.15, is the basic governing equation in the study of acoustic waves in a viscous fluid filling a porous medium. Once the averaged pressure $p^{(0)}$ is known, the velocity and the pressure are approximated by

$$v_i(x,t) \cong v_i^{(0)}(x,t) = \int_0^t V_{in}(y,t-\tau) \frac{\partial p^{(0)}(x,\tau)}{\partial x_n} \, d\tau \quad (18.16)$$

$$p(x,t) \cong p^{(0)}(x,t) + \varepsilon \int_0^t P_n(y,t-\tau) \frac{\partial p^{(0)}(x,\tau)}{\partial x_n} \, d\tau \quad (18.17)$$

in the limit of vanishing ε.

To illustrate the construction of the kernel for the integrodifferential equation, Equation 18.15, consider the local space problem given by Equations 18.13 and 18.14 for a stiff undeformable solid containing a system of capillary tubes of circular cross section filled with a viscous fluid (see Figure 18.1 earlier in this section). We are assuming the radius of a capillary tube to be small as compared to the cell dimension ($r_0 \ll 1$) in order to be able to employ the splitting principle due to Zobnin et al. (1988) in our discussion.

For a capillary tube $Y_f^{(3)} = \{y: |y_3| < \frac{1}{2}, \ y_1^2 + y_2^2 \le r_0^2\}$, current lines may be considered to be parallel to the y_3 axis everywhere except for the immediate vicinity of an intersection with other tubes, with a consequence that only the components v_{3k} ($k = 1,2,3$) are nonzero. It is obvious from Equation 18.4 that $V_{3k} = V_{3k}(y_1, y_2, s)$, and from Equation 18.13, noting the periodicity in y_3, it follows that

$$\hat{P}_\alpha = -y_\alpha \quad (\alpha = 1,2) \qquad \hat{P}_3 = 0 \qquad \rho_0 s \hat{V}_{3k} + \delta_{3k} = \nu \nabla_y^2 V_{3k} \quad (18.18)$$

which, together with the zero boundary conditions on the walls of the capillary tubes, imply that $\hat{V}_{31} = \hat{V}_{32} = 0$ and that the function $\hat{V}_{33}$ satisfies the equation

$$\rho_0 s \hat{V}_{33} + 1 = \nu \frac{1}{r} \frac{d}{dr} \left(r \frac{d\hat{V}_{33}}{dr} \right) \tag{18.19}$$

with zero boundary condition at $r = r_0$. The solution of Equation 18.19 is

$$V_{33}(r, s) = -\frac{1}{\rho_0 s} \left(1 - \frac{I_0(\gamma)}{I_0(\gamma_0)} \right) \tag{18.20}$$

where $I_0(\gamma)$ is the modified Bessel function, $\gamma = r(\rho_0 s/\nu)^{1/2}$, $\gamma_0 = r_0(\rho_0 s/\nu)^{1/2}$, and $r = (y_1^2 + y_2^2)^{1/2}$, and because this solution applies equally well to the capillary tubes $Y_f^{(1)}$ and $Y_f^{(2)}$ by reasons of symmetry, we may formulate as follows the solution of the unit cell problem for everywhere outside the intersection zones:

> $\hat{V}_{ii}$ ($i = 1, 2, 3$; no summation convention) is given by Equation 18.20 if $y \in Y_f^{(i)}$, and it is zero otherwise; for $i = j$, $\hat{V}_{ij}$ is everywhere zero.

Averaging Equation 18.20 now yields

$$\langle \hat{V}_{ii} \rangle = -\frac{\pi_0}{3\rho_0 s} \left(1 - \frac{2I_1(\gamma_0)}{\gamma_0 I_0(\gamma_0)} \right) \tag{18.21}$$

and reverting to originals gives the following nonzero kernels of the integrodifferential equation, Equation 18.15:

$$\langle V_{11}(t) \rangle = \langle V_{22}(t) \rangle = \langle V_{33}(t) \rangle$$

$$= -\frac{4\pi_0}{3\rho_0} \sum_{k=0}^{\infty} \frac{1}{\alpha_k^2} \exp\left(-\frac{\nu \alpha_k^2}{\rho_0 r_0^2} t \right) \tag{18.22}$$

where α_k are the zeros of the Bessel function $J_0(x)$ and $\pi_0 = 3\pi r_0^2$ is the porosity of the medium.

We now proceed to examine the motion of a compressible fluid in a deformable elastic medium with a periodic structure. Assuming displacements to be small in both the solid (Y_s) and fluid (Y_f) phases of the unit cell, the equations of motion and the constitutive relations of the problem are

$$\rho^{(s)}\frac{\partial^2 u_i}{\partial t^2} = \frac{\partial \sigma_{ij}^{(s)}}{\partial x_j} \tag{18.23}$$

$$\sigma_{ij}^{(s)} = c_{ijkh}^{(s)}\frac{\partial u_k}{\partial x_h} \quad \text{in } Y_s \tag{18.24}$$

and

$$\rho^{(f)}\frac{\partial^2 u_i}{\partial t^2} = \frac{\partial \sigma_{ij}^{(f)}}{\partial x_j}$$

$$\sigma_{ij}^{(f)} = -p\delta_{ij} + \delta_{ij}\delta_{kh}\eta\frac{\partial}{\partial t}\left(\frac{\partial u_k}{\partial x_h}\right) + (\delta_{ik}\delta_{jh} + \delta_{ih}\delta_{jk}) \tag{18.25}$$

$$\mu\frac{\partial}{\partial t}\left(\frac{\partial u_k}{\partial x_h}\right) \quad \text{in } Y_f \tag{18.26}$$

where η and μ are the viscosity coefficients ($\mu > 0$, $\eta/\mu > -\frac{2}{3}\alpha$, $0 < \alpha < 1$), $p = c^2\rho$, c is the state-of-rest sound velocity, ρ is the mass density perturbation, $c_{ijkh}^{(s)}$ is the elastic tensor of the solid phase, and $\rho^{(f)}$ and $\rho^{(s)}$ are the state-of-rest mass densities of the fluid and solid phases, respectively.

Now the (linearized) continuity condition for Y_f may be written as

$$\frac{\partial \rho}{\partial t} + \rho^{(f)}\frac{\partial}{\partial t}\left(\frac{\partial u_k}{\partial x_k}\right) = 0 \tag{18.27}$$

which when integrated with respect to t yields

$$\rho = \rho^{(f)}\frac{\partial u_k}{\partial x_k} \tag{18.28}$$

Hence

$$p = -c^2\rho^{(f)}\frac{\partial u_k}{\partial x_k} \quad \text{in } Y_f \tag{18.29}$$

in Equation 18.26, which becomes

$$\sigma_{ij}^{(f)} = \delta_{ij}\delta_{kh}c^2\rho^{(f)}\left(\frac{\partial u_k}{\partial x_h}\right) + \left[\delta_{ij}\delta_{kh}\eta + (\delta_{ik}\delta_{jh} + \delta_{ih}\delta_{jk})\mu\right]\frac{\partial}{\partial t}\left(\frac{\partial u_k}{\partial x_h}\right)$$

(18.30)

and is combined with Equation 18.24 to give

$$\sigma_{ij} = \left(c_{ijkh} + b_{ijkh}\frac{\partial}{\partial t}\right)\frac{\partial u_k}{\partial x_h}$$

(18.31)

where

$$c_{ijkh} = \begin{cases} c_{ijkh}^{(s)} & \text{in } Y_s \\ c^2\rho^{(f)}\delta_{ij}\delta_{kh} & \text{in } Y_f \end{cases}$$

(18.32)

$$b_{ijkh} = \begin{cases} 0 & \text{in } Y_s \\ \delta_{ij}\delta_{kh}\eta + (\delta_{ik}\delta_{jh} + \delta_{ih}\delta_{jk})\mu & \text{in } Y_f \end{cases}$$

(18.33)

The equations of motion in the region $Y = Y_s \cup Y_f$ take the form

$$\frac{\partial \sigma_{ij}}{\partial x_j} = \rho\frac{\partial^2 u_i}{\partial t^2}$$

(18.34)

where

$$\rho = \begin{cases} \rho^{(s)} & \text{in } Y_s \\ \rho^{(f)} & \text{in } Y_f \end{cases}$$

(18.35)

and on the solid–fluid interface the conditions

$$[u_i] = 0 \qquad [\sigma_{ij}n_j] = 0 \quad \text{on } \Gamma$$

(18.36)

must be satisfied, n_i being the unit vector normal to Γ.

Proceeding now to the homogenization process, we define the coefficients

$$\rho^{(\varepsilon)}(x) = \rho\left(\frac{x}{\varepsilon}\right) = \begin{cases} \rho^{(s)} & \text{in } Y_s \\ \rho^{(f)} & \text{in } Y_f \end{cases} \tag{18.37}$$

$$c_{ijkh}^{(\varepsilon)}(x) = c_{ijkh}\left(\frac{x}{\varepsilon}\right) = \begin{cases} c_{ijkh}^{(s)} & \text{in } Y_s \\ c^2\rho^{(f)}\delta_{ij}\delta_{kh} & \text{in } Y_f \end{cases} \tag{18.38}$$

$$b_{ijkh}^{(\varepsilon)}(x) = b_{ijkh}\left(\frac{x}{\varepsilon}\right) = \begin{cases} 0 & \text{in } Y_s \\ \delta_{ij}\delta_{kh}\eta + (\delta_{ik}\delta_{jh} + \delta_{ih}\delta_{jk})\mu & \text{in } Y_f \end{cases} \tag{18.39}$$

and consider the problem

$$\frac{\partial \sigma_{ij}^{(\varepsilon)}}{\partial x_j} = \rho^{(\varepsilon)}(x)\frac{\partial^2 u_i^{(\varepsilon)}}{\partial t^2}$$

$$\tag{18.40}$$

$$\sigma_{ij}^{(\varepsilon)} = \left(c_{ijkh}(y) + b_{ijkh}(y)\frac{\partial}{\partial t}\right)\frac{\partial u_k^{(\varepsilon)}}{\partial x_h}$$

assuming its solution to be of the form

$$u_k^{(\varepsilon)} = u_k^{(0)}(x,t) + \varepsilon u_k^{(1)}(x,y,t) + \cdots \tag{18.41}$$

Then

$$\sigma_{ij}^{(\varepsilon)} = \sigma_{ij}^{(0)}(x,y,t) + \varepsilon\sigma_{ij}^{(1)}(x,y,t) + \cdots \tag{18.42}$$

$$\sigma_{ij}^{(0)} = \left(c_{ijkh}(y) + b_{ijkh}(y)\frac{\partial}{\partial t}\right)\left(\frac{\partial u_k^{(0)}}{\partial x_h} + \frac{\partial u_k^{(1)}}{\partial y_h}\right) \tag{18.43}$$

$$\sigma_{ij}^{(1)} = \left(c_{ijkh}(y) + b_{ijkh}(y)\frac{\partial}{\partial t}\right)\left(\frac{\partial u_k^{(1)}}{\partial x_h} + \frac{\partial u_k^{(2)}}{\partial y_h}\right) \tag{18.44}$$

If we substitute Equations 18.41 and 18.42 into Equation 18.40 and keep terms $O(\varepsilon^{-1})$ and $O(\varepsilon^0)$ in the resulting expansions, we find

$$\frac{\partial \sigma_{ij}^{(0)}}{\partial y_j} = 0 \qquad (18.45)$$

$$\frac{\partial \sigma_{ij}^{(0)}}{\partial x_j} + \frac{\partial \sigma_{ij}^{(1)}}{\partial y_j} = \rho(y) \frac{\partial^2 u_i^{(0)}}{\partial t^2} \qquad (18.46)$$

Using Equation 18.43, the unit cell local problem expressed by Equation 18.45 can be rewritten as

$$\frac{\partial}{\partial y_j} \left[\left(c_{ijkh}(y) + b_{ijkh}(y) \frac{\partial}{\partial t} \right) \frac{\partial u_k^{(1)}}{\partial y_h} \right]$$

$$= -\frac{\partial}{\partial y_j} \left[\left(c_{ijkh}(y) + b_{ijkh}(y) \frac{\partial}{\partial t} \right) \right] \frac{\partial u_k^{(0)}}{\partial x_h} \qquad (18.47)$$

or, after Laplace transforming with respect to t,

$$\frac{\partial}{\partial y_j} \left[\left(c_{ijkh}(y) + s b_{ijkh}(y) \right) \frac{\partial \hat{u}_k^{(1)}}{\partial y_h} \right]$$

$$= -\frac{\partial}{\partial y_j} \left[\left(c_{ijkh}(y) + s b_{ijkh}(y) \right) \right] \frac{\partial \hat{u}_k^{(0)}}{\partial x_h} \qquad (18.48)$$

where $u_k^{(0)}$ and $u_k^{(1)}$ are the Laplace transforms of the respective noncareted variables.

As usual, the solution of Equation 18.48 is assumed to be of the form

$$\hat{u}_k^{(1)} = \hat{N}_k^{mn}(y, s) \frac{\partial \hat{u}_m^{(0)}(x, s)}{\partial x_n} \qquad (18.49)$$

where the functions $\hat{N}_k^{mn}$ are Y-periodic in y, solve

$$\frac{\partial}{\partial y_j} \left[\left(c_{ijkh}(y) + s b_{ijkh}(y) \right) \frac{\partial \hat{N}_k^{mn}(y, s)}{\partial y_h} \right]$$

$$= -\frac{\partial}{\partial y_j} \left(c_{ijmn}(y) + s b_{ijmn}(y) \right) \qquad (i, m, n = 1, 2, 3) \quad (18.50)$$

in region Y and, together with the quantities

$$\left(c_{ijmn}(y) + sb_{ijmn}(y)\right) + \left(c_{ijkh}(y) + sb_{ijkh}(y)\right)\left(\frac{N_k^{mn}(y,s)}{y_n}\right)$$

and

$$\left[\left(c_{ijmn}(y) + sb_{ijmn}(y)\right) + \left(c_{ijkh}(y) + sb_{ijkh}(y)\right)\left(\frac{\partial N_k^{mn}(y,s)}{\partial y_n}\right)\right]n_j$$

(where $m, n, i = 1, 2, 3$) are continuous on the solid–fluid interface Γ.
The Laplace-transformed stresses are found to be

$$\hat{\sigma}_{ij}^{(0)} = \left[\left(c_{ijmn}(y) + sb_{ijmn}(y)\right) + \left(c_{ijkh}(y)\right.\right.$$

$$\left.\left. + sb_{ijkh}(y)\right)\frac{\partial \hat{N}_k^{mn}(y)}{\partial y_h}\right]\frac{\partial \hat{u}_m^{(0)}}{\partial x_n} \tag{18.51}$$

using Equation 18.49, and reverting to originals we obtain

$$\sigma_{ij}^{(0)} = c_{ijmn}(y)\frac{\partial u_m^{(0)}}{\partial x_n} + b_{ijmn}(y)\frac{\partial}{\partial t}\left(\frac{\partial u_m^{(0)}}{\partial x_n}\right)$$

$$+ c_{ijkh}(y)\int_0^t \frac{\partial u_m^{(0)}(x,\tau)}{\partial x_n}\frac{\partial}{\partial y_h}\left[\hat{N}_k^{mn}(y, t - \tau)\right]d\tau$$

$$+ b_{ijkh}(y)\int_0^t \frac{\partial}{\partial \tau}\left(\frac{\partial u_m^{(0)}(x,\tau)}{\partial x_n}\right)\frac{\partial}{\partial y_h}\left[\hat{N}_k^{mn}(y, t - \tau)\right]d\tau \tag{18.52}$$

To derive the macroscopic equation of motion of the medium, average
Equation 18.46 over the unit cell Y, giving

$$\frac{\partial \tilde{\sigma}_{ij}^{(0)}}{\partial x_j} = \tilde{\rho}\frac{\partial^2 u_i^{(0)}}{\partial t^2} \tag{18.53}$$

where

$$\tilde{\rho} = \frac{1}{|Y|} \int_Y \rho \, dy$$

and

$$\tilde{\sigma}_{ij}^{(0)} = \frac{1}{|Y|} \int \sigma_{ij}^{(0)} \, dy$$

or, by Equation 18.52,

$$\tilde{\sigma}_{ij}^{(0)} = \tilde{c}_{ijmn} \frac{\partial u_m^{(0)}}{\partial x_n} + \tilde{b}_{ijmn} \frac{\partial}{\partial t} \left(\frac{\partial u_m^{(0)}}{\partial x_n} \right) + \int_0^t \tilde{c}'_{ijmn}(t - \tau) \frac{\partial u_m^{(0)}(x, \tau)}{\partial x_n} \, d\tau$$

$$+ \int_0^t \tilde{b}'_{ijmn}(t - \tau) \frac{\partial}{\partial \tau} \left(\frac{\partial u_m^{(0)}(x, \tau)}{\partial x_n} \right) d\tau \qquad (18.54)$$

where

$$\tilde{c}_{ijmn} = \frac{1}{|Y|} \int_Y c_{ijmn}(y) \, dy$$

$$\tilde{c}'_{ijmn}(t) = \frac{1}{|Y|} \int_Y c_{ijkh}(y) \frac{\partial \hat{N}_k^{mn}(y, t)}{\partial y_h} \, dy$$

$$\qquad (18.55)$$

$$\tilde{b}_{ijmn} = \frac{1}{|Y|} \int_Y b_{ijmn}(y) \, dy$$

$$\tilde{b}'_{ijmn}(t) = \frac{1}{|Y|} \int_Y b_{ijkh}(y) \frac{\partial \hat{N}_k^{mn}(y, t)}{\partial y_h} \, dy$$

We thus see that the homogenized macroscopic equations of motion of a porous medium filled with a viscous fluid are in fact equations of linear viscoelasticity with constitutive relations given by Equation 18.54.

5

General Homogenization Models for Composite Plates and Shells with Rapidly Varying Thickness

Today, the preponderance of uses for composite materials is in the form of plate and shell structural members whose strength and reliability, combined with reduced weight and concomitant material savings, offer the designer very impressive possibilities in some commercial applications. A fact that is of interest for us here is that whether the reinforcing effect comes from "real" elements (say, fibers) or "formal" ones (such as voids, holes, and similar design features), it often happens that these form a regular structure with a period much smaller than the characteristic dimension of the structural member; consequently, the asymptotic homogenization analysis becomes applicable. The homogenized models of plates with periodic inhomogeneities in tangential coordinate(s) have been developed in this way by Duvaut (1976,1977), Duvaut and Metellus (1976), and Artola and Duvaut (1977) and more recently by Andrianov and Manevich (1983), Andrianov et al. (1985), and some other workers (see Kalamkarov et al. [1987a] for a review). It should be noted, however, that the asymptotic homogenization method cannot be applied to a two-dimensional plate-and-shell theory if the space inhomogeneities of the material vary on a scale comparable with the small thickness of the three-dimensional body under consideration. A refined approach developed by Caillerie (1981a, 1981b) in his heat conduction studies consists of applying the two-scale formalism directly to the three-dimensional problem of a thin nonhomogeneous layer. Accordingly, Caillerie introduces two sets of "rapid" coordinates. One of these, in the tangential directions, is associated with rapid

215

periodic oscillations in the composite properties or layer thickness; in the transverse—thickness—direction, there is no periodicity and the corresponding "rapid" variable is associated with the small thickness of the layer. The two small parameters that arise from this approach, ε and δ, are determined by the period of the coefficients of the pertinent equations and the layer thickness, respectively, and may or may not be of the same order of magnitude. Kohn and Vogelius (1984, 1985, 1986) adopted this approach in their study of the pure bending of a linearly elastic, thin homogeneous plate whose thickness was taken to be described by a rapidly oscillating function with a period $\varepsilon = \delta^a$ $(0 < a < \infty)$, where δ denotes the mean plate thickness.

Of special practical interest, however, is the case of proportionality between the two small parameters,

$$\varepsilon = h\delta \qquad (\delta \ll 1, h \sim 1)$$

which has been treated by Caillerie (1982, 1984, 1987) in his linearly elastic asymptotic analysis of a periodically nonhomogeneous plate. For a still more special case of $\varepsilon = \delta$, Panasenko and Reztsov (1987) have been able to provide necessary justifications for the three-dimensional linear elasticity solutions they obtained for a thin periodically nonhomogeneous plate.

In this chapter, the modified $\varepsilon = h\delta$ homogenization method is applied to the study of a curved thin composite plate with a regular structure and wavy surfaces in the contexts of, successively, linear elasticity theory (Parton et al. 1986a, Kalamkarov et al. 1987b, Parton et al. 1989a, Parton et al. 1989d) and thermoelasticity (Parton et al. 1986b, Kalamkarov et al. 1987c, Parton and Kalamkarov 1988a). The starting point in each particular case is the exact three-dimensional formulation of the corresponding problem, without resorting to the Kirchhoff–Love hypotheses or any similar simplifying assumptions. Because of the presence of the small parameter δ, each original three-dimensional problem then proves to be amenable to a rigorous asymptotic analysis in which an asymptotic three-to-two dimensional process is combined with a homogenization composite-material–homogeneous-material process. The general homogenized models so obtained have many practical applications in the design of stiffened plates and shells possessing a regular structure.

19. ELASTICITY PROBLEM FOR A SHELL OF REGULARLY NONHOMOGENEOUS MATERIAL WITH WAVY SURFACES

19.1. Formulation in an orthogonal coordinate system

Let $(\alpha_1, \alpha_2, \gamma)$ be an orthogonal coordinate system such that the coordinate lines α_1 and α_2 coincide with the main curvature lines of the

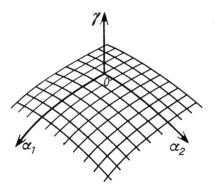

FIGURE 19.1. Orthogonal coordinate system $\alpha_1, \alpha_2, \gamma$.

midsurface of the shell and the γ axis is normal to the midsurface (Figure 19.1). All three coordinates are assumed to have been made dimensionless by dividing by a certain characteristic dimension of the body, L.

The metric tensor of this coordinate system is

$$g_{ij} = \begin{pmatrix} H_1^2 & 0 & 0 \\ 0 & H_2^2 & 0 \\ 0 & 0 & 1 \end{pmatrix}$$

and the Lamé coefficients H_1 and H_2 are given by

$$H_1 = A_1(1 + k_1\gamma) \quad \text{and} \quad H_2 = A_2(1 + k_2\gamma)$$

where $A_1(\alpha_1, \alpha_2)$ and $A_2(\alpha_1, \alpha_2)$ are the coefficients of the first quadratic form and $k_1(\alpha_1, \alpha_2)$ and $k_2(\alpha_1, \alpha_2)$ are the main curvatures of the midsurface. The unit cell of the problem (see Figure 19.2) is defined by the inequalities

$$\left\{ -\frac{\delta h_1}{2} < \alpha_1 < \frac{\delta h_1}{2}, \ -\frac{\delta h_2}{2} < \alpha_2 < \frac{\delta h_2}{2}, \gamma^- < \gamma < \gamma^+ \right\}$$

$$\gamma^{\pm} = \pm\frac{\delta}{2} \pm \delta F^{\pm}\left(\frac{\alpha_1}{\delta h_1}, \frac{\alpha_2}{\delta h_2}\right)$$

in which the dimensionless small parameter δ determines the thickness of the shell; the parameters h_1 and h_2 characterize the ratio of the tangential dimensions to the thickness of the unit cell; the functions F^+ and F^- describe the profiles of the upper and lower surfaces of the plate (respectively, S^+ and S^-) and are generally 1-periodic in $\alpha_1/\delta h_1$ and $\alpha_2/\delta h_2$.

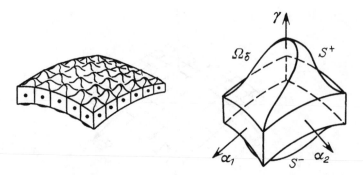

FIGURE 19.2. Curvilinear thin regularly nonhomogeneous (composite) layer with wavy surfaces; unit cell Ω_δ.

With the expressions given previously for H_1 and H_2, the Cauchy relations for the physical strain and displacement components are

$$e_{11} = \frac{1}{H_1}\frac{\partial u_1}{\partial \alpha_1} + \frac{1}{H_1 A_2}\frac{\partial A_1}{\partial \alpha_2}u_2 + \frac{A_1 k_1}{H_1}u_3$$

$$e_{22} = \frac{1}{H_2}\frac{\partial u_2}{\partial \alpha_2} + \frac{1}{H_2 A_1}\frac{\partial A_2}{\partial \alpha_1}u_1 + \frac{A_2 k_2}{H_2}u_3$$

$$e_{33} = \frac{\partial u_3}{\partial \gamma}$$

$$2e_{12} = \frac{1}{H_2}\frac{\partial u_1}{\partial \alpha_2} - \frac{1}{H_2 A_1}\frac{\partial A_2}{\partial \alpha_1}u_2 + \frac{1}{H_1}\frac{\partial u_2}{\partial \alpha_1} - \frac{1}{H_1 A_2}\frac{\partial A_1}{\partial \alpha_2}u_1$$

$$2e_{13} = \frac{\partial u_1}{\partial \gamma} + \frac{1}{H_1}\frac{\partial u_3}{\partial \alpha_1} - \frac{A_1 k_1}{H_1}u_1$$

$$2e_{23} = \frac{\partial u_2}{\partial \gamma} + \frac{1}{H_2}\frac{\partial u_3}{\partial \alpha_2} - \frac{A_2 k_2}{H_2}u_2$$

(19.1)

where (and throughout this chapter) the subscripts 1, 2, and 3 refer to α_1, α_2, and γ, respectively, and where the Gauss–Codazzi relations

$$\frac{1}{H_1}\frac{\partial H_2}{\partial \alpha_1} = \frac{1}{A_1}\frac{\partial A_2}{\partial \alpha_1} \quad \text{and} \quad \frac{1}{H_2}\frac{\partial H_1}{\partial \alpha_2} = \frac{1}{A_2}\frac{\partial A_1}{\partial \alpha_2}$$

(19.2)

have been used.

The equilibrium equations are given by

$$\frac{\partial(H_2\sigma_{11})}{\partial\alpha_1} + \frac{\partial(H_1\sigma_{12})}{\partial\alpha_2} + \frac{\partial(H_1H_2\sigma_{13})}{\partial\gamma} - \frac{H_1}{A_1}\frac{\partial A_2}{\partial\alpha_1}\sigma_{22}$$

$$+ \frac{H_2}{A_2}\frac{\partial A_1}{\partial\alpha_2}\sigma_{21} + H_2A_1k_1\sigma_{31} + H_1H_2P_1 = 0$$

$$\frac{\partial(H_2\sigma_{21})}{\partial\alpha_1} + \frac{\partial(H_1\sigma_{22})}{\partial\alpha_2} + \frac{\partial(H_1H_2\sigma_{23})}{\partial\gamma} - \frac{H_2}{A_2}\frac{\partial A_1}{\partial\alpha_2}\sigma_{11}$$

$$+ \frac{H_1}{A_1}\frac{\partial A_2}{\partial\alpha_1}\sigma_{12} + H_1A_2k_2\sigma_{32} + H_1H_2P_2 = 0 \qquad (19.3)$$

$$\frac{\partial(H_2\sigma_{31})}{\partial\alpha_1} + \frac{\partial(H_1\sigma_{32})}{\partial\alpha_2} + \frac{\partial(H_1H_2\sigma_{33})}{\partial\gamma} - H_2A_1k_1\sigma_{11}$$

$$- H_1A_2k_2\sigma_{22} + H_1H_2P_3 = 0$$

where P_i $(i = 1, 2, 3)$ are the body force components.

The physical stresses and strains are related by the generalized Hooke's law

$$\sigma_{ij} = c_{ijmn}e_{mn} \qquad (19.4)$$

(We shall always assume, unless otherwise stated, that Latin indices range from 1 to 3 and Greek indices from 1 to 2).

The surfaces of the shall are subjected to the conditions

$$\sigma^{ij}n_j = p^i$$

where σ^{ij} is the stress tensor, n_j is the unit vector in the outward direction to the surface, and p^i is the vector of the outer surface force.

Now the physical components of interest are given by (cf. Sedov 1972)

$$\sigma_{11} = \sigma_{\alpha_1\alpha_1} = \sigma^{11}g_{11} = \sigma^{11}H_1^2$$

$$\sigma_{12} = \sigma_{\alpha_1\alpha_2} = \sigma^{12}\sqrt{g_{11}}\sqrt{g_{22}} = \sigma^{12}H_1H_2$$

$$\sigma_{22} = \sigma_{\alpha_2\alpha_2} = \sigma^{22}g_{22} = \sigma^{22}H_2^2$$

$$\sigma_{31} = \sigma_{\gamma\alpha_1} = \sigma^{31}\sqrt{g_{11}} = \sigma^{31}H_1$$

$$\sigma_{32} = \sigma_{\gamma\alpha_2} = \sigma^{32}\sqrt{g_{22}} = \sigma^{32}H_2$$

$$\sigma_{33} = \sigma_{\gamma\gamma} = \sigma^{33}\sqrt{g_{33}} = \sigma^{33}$$

$$p_1 = p_{\alpha_1} = p^1\sqrt{g_{11}} = p^1 H_1$$

$$p_2 = P_{\alpha_2} = p^2\sqrt{g_{22}} = p^2 H_2$$

$$p_3 = p_\gamma = p^3\sqrt{g_{33}} = p^3$$

and substituting these into the surface boundary conditions we obtain

$$\frac{\sigma_{11}}{H_1}n_1 + \frac{\sigma_{12}}{H_2}n_2 + \sigma_{13}n_3 = p_1$$

$$\frac{\sigma_{12}}{H_1}n_1 + \frac{\sigma_{22}}{H_2}n_2 + \sigma_{23}n_3 = p_2$$

$$\frac{\sigma_{13}}{H_1}n_1 + \frac{\sigma_{23}}{H_2}n_2 + \sigma_{33}n_3 = p_3$$

which, after multiplication by $H_1 H_2$, leads to the result

$$H_2\sigma_{i1}n_1^\pm + H_1\sigma_{i2}n_2^\pm + H_1 H_2\sigma_{i3}n_3^\pm = \pm H_1 H_2 p_i^\pm \qquad (i = 1,2,3) \quad (19.5)$$

for $\gamma = \gamma^\pm$, where n_i^+ (n_i^-) is the outward (inward) unit normal vector (hence the minus sign on the right-hand side of Equation 19.5 for $\gamma = \gamma^-$).

The components of the normal vector are defined in a curvilinear coordinate system by (see, e.g., Berdichevskii 1983)

$$n_i = \frac{1}{\sqrt{a}}\,\varepsilon_{ijm}r_1^j r_2^m$$

where ε_{ijm} (the Levi–Civita symbols) have the property that

$$\varepsilon_{ijm} = \begin{cases} \sqrt{g} & \text{if all three indices are different} \\ & \text{and are obtained by an even number} \\ & \text{of interchanges from 123} \\ -\sqrt{g} & \text{if all three indices are different} \\ & \text{and are obtained by an odd number} \\ & \text{of interchanges from 123} \\ 0 & \text{if any two of the three indices} \\ & \text{are the same} \end{cases}$$

For the surfaces $\gamma = \gamma^{\pm}$ we find

$$\mathbf{n}^{\pm} = \left\{ -\frac{\partial \gamma^{\pm}}{\partial \alpha_1}, -\frac{\partial \gamma^{\pm}}{\partial \alpha_2}, 1 \right\} \left[1 \frac{1}{H_1^2} \left(\frac{\partial \gamma^{\pm}}{\partial \alpha_1} \right)^2 + \frac{1}{H_2^2} \left(\frac{\partial \gamma^{\pm}}{\partial \alpha_2} \right)^2 \right]^{-1/2} \tag{19.6}$$

The static (quasistatic) formulation of the foregoing elasticity problem will be complete by adjoining the proper specification of boundary conditions at the lateral surfaces of the shell. The usual practice is to make some requirements on stress or displacement fields on these surfaces or, in the case of closed shells, to impose periodicity conditions on the corresponding coordinates.

19.2. Asymptotic analysis

We begin by introducing the "rapid" coordinates of the problem,

$$y_1 = \frac{\alpha_1}{\delta h_1} \qquad y_2 = \frac{\alpha_2}{\delta h_2} \qquad z = \frac{\gamma}{\delta}$$

in terms of which the unit cell Ω is defined by the inequalities

$$\{ -\tfrac{1}{2} < y_1, y_2 < \tfrac{1}{2}, z^- < z < z^+ \} \qquad z^{\pm} = \pm \tfrac{1}{2} \pm F^{\pm}(y)$$

and the expression for the normal vector (Equation 19.6) is rewritten as

$$\mathbf{n}^{\pm} = \left\{ \mp \frac{1}{h_1} \frac{\partial F^{\pm}}{\partial y_1}, \mp \frac{1}{h_2} \frac{\partial F^{\pm}}{\partial y_2}, 1 \right\}$$

$$\times \left[1 + \frac{1}{H_1^2 h_1^2} \left(\frac{\partial F^{\pm}}{\partial y_1} \right) + \frac{1}{H_2^2 h_2^2} \left(\frac{\partial F^{\pm}}{\partial y_2} \right) \right]^{-1/2} \tag{19.7}$$

The regular inhomogeneity of the material is modeled mathematically by the requirement that the functions $c_{ijmn}(y, z)$, $y = (y_1, y_2)$, be periodic with unit cell Ω in the coordinates y_1 and y_2.

The solution of the problem will be represented as the asymptotic expansion

$$u_i = u_i^{(0)}(\alpha) + \delta u_i^{(1)}(\alpha, y, z) + \delta^2 u_i^{(2)}(\alpha, y, z) + \cdots \tag{19.8}$$

where the functions $u_i^{(l)}(\alpha, y, z)$ ($l = 1, 2, \ldots$) are periodic in y_1, y_2 with unit cell Ω; $\alpha = (\alpha_1, \alpha_2)$.

Now let the radii of curvature of the shell midsurface ($\gamma = 0$) be much larger than the shell thickness. We are then justified in assuming the asymptotic forms

$$k_1 = \delta K_2(\alpha) \qquad k_2 = \delta K_1(\alpha) \qquad k_1 + k_2 = \delta K_3(\alpha) \qquad (19.9)$$

for the main curvatures of the shell. For the external forces we write

$$p_\nu^\pm = \delta^2 r_\nu^\pm(\alpha, y) \qquad p_3^\pm = \delta^3 q_3^\pm(\alpha, y) \qquad (19.10)$$

$$P_\nu = \delta f_\nu(\alpha, y, z) \qquad P_3 = \delta^2 g_3(\alpha, y, z) \qquad (19.11)$$

where all functions involved are again periodic in y_1 and y_2 with unit cell Ω.

To simplify the solution, we introduce the notation

$$\xi_1 = A_1 y_1 \quad \text{and} \quad \xi_2 = A_2 y_2$$

and we define the following differential operators to be applied to φ_{nm} components:

$$\mathfrak{B}_1(\varphi_{n\mu}) = \frac{1}{A_1 A_2} \left[\frac{\partial(A_2 \varphi_{11})}{\partial \alpha_1} + \frac{\partial(A_1 \varphi_{21})}{\partial \alpha_2} + \frac{\partial A_1}{\partial \alpha_2} \varphi_{12} - \frac{\partial A_2}{\partial \alpha_1} \varphi_{22} \right]$$

$$\mathfrak{B}_2(\varphi_{n\mu}) = \frac{1}{A_1 A_2} \left[\frac{\partial(A_2 \varphi_{12})}{\partial \alpha_1} + \frac{\partial(A_1 \varphi_{22})}{\partial \alpha_2} + \frac{\partial A_2}{\partial \alpha_1} \varphi_{21} - \frac{\partial A_1}{\partial \alpha_2} \varphi_{11} \right]$$

$$\mathfrak{B}_3(\varphi_{n\mu}) = \frac{1}{A_1 A_2} \left[\frac{\partial(A_2 \varphi_{31})}{\partial \alpha_1} + \frac{\partial(A_1 \varphi_{32})}{\partial \alpha_2} \right]$$

$$\mathfrak{R}_1(\varphi_{n\mu}) = K_2 \varphi_{31}$$

$$\mathfrak{R}_2(\varphi_{n\mu}) = K_1 \varphi_{32}$$

$$\mathfrak{R}_3(\varphi_{n\mu}) = -K_2 \varphi_{11} - K_1 \varphi_{22} \qquad (19.12)$$

It is essential in the asymptotic homogenization method to distinguish between "rapid" and "slow" variables when differentiating, that is,

$$\frac{\partial}{\partial \alpha_\mu} \to \frac{\partial}{\partial \alpha_\mu} + \frac{1}{\delta h_\mu} \frac{\partial}{\partial y_\mu} \quad \text{and} \quad \frac{\partial}{\partial y} = \frac{1}{\delta} \frac{\partial}{\partial z}$$

where μ assumes the values 1 and 2 and is not summed.

Using Equations 19.8 and 19.9 in Equations 19.1 we obtain

$$e_{ij} = e_{ij}^{(0)} + \delta e_{ij}^{(1)} + \delta^2 e_{ij}^{(2)} + \cdots \tag{19.13}$$

where

$$e_{11}^{(0)} = \frac{1}{A_1} \frac{\partial u_1^{(0)}}{\partial \alpha_1} + \frac{1}{h_1 A_1} \frac{\partial u_1^{(1)}}{\partial y_1} + \frac{1}{A_1 A_2} \frac{\partial A_1}{\partial \alpha_2} u_2^{(0)}$$

$$e_{22}^{(0)} = \frac{1}{A_2} \frac{\partial u_2^{(0)}}{\partial \alpha_2} + \frac{1}{h_2 A_2} \frac{\partial u_2^{(1)}}{\partial y_2} + \frac{1}{A_1 A_2} \frac{\partial A_2}{\partial \alpha_1} u_1^{(0)}$$

$$e_{33}^{(0)} = \frac{\partial u_3^{(1)}}{\partial z}$$

$$2e_{12}^{(0)} = \left[\left(\frac{1}{A_2} \frac{\partial u_1^{(0)}}{\partial \alpha_2} - \frac{1}{A_1 A_2} \frac{\partial A_2}{\partial \alpha_1} u_2^{(0)} \right) \right.$$

$$+ \left(\frac{1}{A_1} \frac{\partial u_2^{(0)}}{\partial \alpha_1} - \frac{1}{A_1 A_2} \frac{\partial A_1}{\partial \alpha_2} u_1^{(0)} \right) \tag{19.14}$$

$$\left. + \left(\frac{1}{h_2 A_2} \frac{\partial u_1^{(1)}}{\partial y_2} + \frac{1}{h_1 A_1} \frac{\partial u_2^{(1)}}{\partial y_1} \right) \right]$$

$$2e_{13}^{(0)} = \frac{\partial u_1^{(1)}}{\partial z} + \frac{1}{A_1} \frac{\partial u_3^{(0)}}{\partial \alpha_1} + \frac{1}{h_1 A_1} \frac{\partial u_3^{(1)}}{\partial y_1}$$

$$2e_{23}^{(0)} = \frac{\partial u_2^{(1)}}{\partial z} + \frac{1}{A_2} \frac{\partial u_3^{(0)}}{\partial \alpha_2} + \frac{1}{h_2 A_2} \frac{\partial u_3^{(1)}}{\partial y_2}$$

$$e_{11}^{(1)} = \frac{1}{A_1} \frac{\partial u_1^{(1)}}{\partial \alpha_1} + \frac{1}{A_1 A_2} \frac{\partial A_1}{\partial \alpha_2} u_2^{(1)} + K_2 u_3^{(0)} + \frac{1}{h_1 A_1} \frac{\partial u_1^{(2)}}{\partial y_1}$$

$$e_{22}^{(1)} = \frac{1}{A_2} \frac{\partial u_2^{(1)}}{\partial \alpha_2} + \frac{1}{A_1 A_2} \frac{\partial A_2}{\partial \alpha_1} u_1^{(1)} + K_1 u_3^{(0)} + \frac{1}{h_2 A_2} \frac{\partial u_2^{(2)}}{\partial y_2}$$

$$e_{33}^{(1)} = \frac{\partial u_3^{(2)}}{\partial z}$$

$$2e_{12}^{(1)} = \left[\left(\frac{1}{A_2} \frac{\partial u_1^{(1)}}{\partial \alpha_2} - \frac{1}{A_1 A_2} \frac{\partial A_2}{\partial \alpha_1} u_2^{(1)} \right) \right.$$

$$+ \left(\frac{1}{A_1} \frac{\partial u_2^{(1)}}{\partial \alpha_1} - \frac{1}{A_1 A_2} \frac{\partial A_1}{\partial \alpha_2} u_1^{(1)} \right)$$

$$\left. + \left(\frac{1}{h_2 A_2} \frac{\partial u_1^{(2)}}{\partial y_2} + \frac{1}{h_1 A_1} \frac{\partial u_2^{(2)}}{\partial y_1} \right) \right]$$

(19.15)

$$2e_{13}^{(1)} = \frac{\partial u_1^{(2)}}{\partial z} + \frac{1}{A_1} \frac{\partial u_3^{(1)}}{\partial \alpha_1} - K_2 u_1^{(0)} + \frac{1}{h_1 A_1} \frac{\partial u_3^{(2)}}{\partial y_1}$$

$$2e_{23}^{(1)} = \frac{\partial u_2^{(2)}}{\partial z} + \frac{1}{A_2} \frac{\partial u_3^{(1)}}{\partial \alpha_2} - K_1 u_2^{(0)} + \frac{1}{h_2 A_2} \frac{\partial u_3^{(2)}}{\partial y_2}$$

Substituting Equation 19.13 into Equation 19.4 results in

$$\sigma_{ij} = \sigma_{ij}^{(0)} + \delta \sigma_{ij}^{(1)} + \delta^2 \sigma_{ij}^{(2)} + \cdots$$

(19.16)

$$\sigma_{ij}^{(l)} = c_{ijmn}(y, z) e_{mn}^{(l)} \qquad (l = 0, 1, 2, \ldots)$$

From Equations 19.9 and 19.11 we substitute into the equilibrium equations (Equations 19.3) and then expand in powers of δ^1 ($l = -1, 0, 1, 2$) to get

$$\frac{1}{h_\mu} \frac{\partial \sigma_{i\mu}^{(0)}}{\partial \xi_\mu} + \frac{\partial \sigma_{i3}^{(0)}}{\partial z} = 0$$

(19.17)

$$\frac{1}{h_\mu} \frac{\partial \sigma_{i\mu}^{(1)}}{\partial \xi_\mu} + \frac{\partial \sigma_{i3}^{(1)}}{\partial z} + \mathcal{B}_i \left(\sigma_{n\mu}^{(0)} \right) = 0$$

(19.18)

$$\frac{1}{h_\mu} \frac{\partial}{\partial \xi_\mu} \left(\sigma_{i\mu}^{(2)} + z K_\mu \sigma_{i\mu}^{(0)} \right) + \frac{\partial}{\partial z} (\sigma_{i3}^{(2)} + z K_3 \sigma_{i3}^{(0)})$$

$$+ \mathcal{B}_i \left(\sigma_{n\mu}^{(1)} \right) + \mathcal{R}_i \left(\sigma_{n\mu}^{(0)} \right) + f_i = 0$$

(19.19)

$$\frac{1}{h_\mu} \frac{\partial}{\partial \xi_\mu} \left(\sigma_{i\mu}^{(3)} + z K_\mu \sigma_{i\mu}^{(1)} \right) + \frac{\partial}{\partial z} (\sigma_{i3}^{(3)} + z K_3 \sigma_{i3}^{(1)})$$

$$+ \mathcal{B}_i \left(\sigma_{n\mu}^{(2)} + z K_n \sigma_{n\mu}^{(0)} \right) + \mathcal{R}_i \left(\sigma_{n\mu}^{(1)} \right) + g_i = 0$$

(19.20)

where

$$\frac{\partial}{\partial \xi_1} = \frac{1}{A_1} \frac{\partial}{\partial y_1} \quad \text{and} \quad \frac{\partial}{\partial \xi_2} = \frac{1}{A_2} \frac{\partial}{\partial y_2}$$

and where Relations 19.2 have been used, which may be rewritten in the form

$$A_2 \frac{\partial K_1}{\partial \alpha_1} = (K_2 - K_1) \frac{\partial A_2}{\partial \alpha_1} \quad \text{and} \quad A_1 \frac{\partial K_2}{\partial \alpha_2} = (K_1 - K_2) \frac{\partial A_1}{\partial \alpha_2}$$

using Equations 19.9. Note that $f_3 = 0$ and $g_\mu = 0$ in accordance with Equations 19.11 and that index n is not summed in the third term in Equation 19.20.

Our next aim is to expand the boundary conditions given by Equation 19.5 in powers of δ. Consider first the components of the unit normal vector (Equation 19.7). From Equations 19.9 we see that they can be expanded as

$$n_i^\pm = n_i^{\pm (0)} + \delta^2 n_i^{\pm (2)} + O(\delta^4)$$

$$\mathbf{n}^{\pm (0)} = \left\{ \mp \frac{1}{h_1} \frac{\partial F^\pm}{\partial y_1}, \; \mp \frac{1}{h_2} \frac{\partial F^\pm}{\partial y_2}, 1 \right\}$$

$$\times \left[1 + \frac{1}{A_1^2 h_1^2} \left(\frac{\partial F^\pm}{\partial y_1} \right)^2 + \frac{1}{A_2^2 h_2^2} \left(\frac{\partial F^\pm}{\partial y_2} \right)^2 \right]^{-1/2}$$

$$n_i^{\pm (2)} = n_i^{\pm (0)} \cdot \zeta^\pm$$

$$\zeta^\pm = z^\pm \left[\frac{K_2}{A_1^2 h_1^2} \left(\frac{\partial F^\pm}{\partial y_1} \right)^2 + \frac{K_1}{A_2^2 h_2^2} \left(\frac{\partial F^\pm}{\partial y_2} \right)^2 \right]$$

$$\times \left[1 + \frac{1}{A_1^2 h_1^2} \left(\frac{\partial F^\pm}{\partial y_1} \right)^2 + \frac{1}{A_2^2 h_2^2} \left(\frac{\partial F^\pm}{\partial y_2} \right)^2 \right]^{-1}$$

(19.21)

and using this along with Equations 19.9 and 19.10, Conditions 19.5 are expanded in terms of δ^l ($l = 0, 1, 2, 3$) as follows:

$$\sigma_{ij}^{(l)} N_j^{\pm} = 0 \qquad (z = z^{\pm}, l = 0, 1) \quad (19.22)$$

$$\left(\sigma_{ij}^{(2)} + z^{\pm} K_j \sigma_{ij}^{(0)} \right) N_j^{\pm} = \pm \omega^{\pm} r_i^{\pm} \qquad (z = z^{\pm}) \qquad (19.23)$$

$$\left(\sigma_{ij}^{(3)} + z^{\pm} K_j \sigma_{ij}^{(1)} \right) N_j^{\pm} = \pm \omega^{\pm} q_i^{\pm} \qquad (z = z^{\pm}) \qquad (19.24)$$

where we have defined

$$\mathbf{N}^{\pm} = \left\{ \mp \frac{1}{h_1} \frac{\partial F^{\pm}}{\partial \xi_1}, \mp \frac{1}{h_2} \frac{\partial F^{\pm}}{\partial \xi_2}, 1 \right\} \qquad (19.25)$$

$$\omega^{\pm} = \left[1 + \frac{1}{h_1^2} \left(\frac{\partial F^{\pm}}{\partial \xi_1} \right)^2 + \frac{1}{h_2} \left(\frac{\partial F^{\pm}}{\partial \xi_2} \right)^2 \right]^{1/2} \qquad (19.26)$$

Note that, in view of Equations 19.10,

$$r_3^{\pm} = 0 \quad \text{and} \quad q_{\mu}^{\pm} = 0$$

in Equations 19.23 and 19.24.

Introducing the averaging procedure

$$\langle \varphi \rangle = \int_{\Omega} \varphi \, dy_1 \, dy_2 \, dz \qquad (19.27)$$

with respect to the coordinates y_1, y_2, and z over the unit cell Ω, and noting that this operation is interchangeable with differentiation with respect to α_1 and α_2, let us prove that if a function Q_1 is periodic with unit cell Ω in y_1 and y_2, the formula

$$\left\langle \frac{1}{h_{\mu}} \frac{\partial Q_{\mu}}{\partial \xi_{\mu}} + \frac{\partial Q_3}{\partial z} \right\rangle = \int_{-1/2}^{1/2} \int_{-1/2}^{1/2} (Q_i^+ N_i^+ - Q_i^- N_i^-) \, dy_1 \, dy_2 \quad (19.28)$$

holds, where $N_i^\pm$ are the components of the vectors given by Equation 19.25, and $Q_i^\pm$ are values that the function Q_i assumes at the surfaces $S^\pm$ defined by $z = z^\pm \equiv \pm \frac{1}{2} \pm F^\pm(y)$. Using the Green–Gauss theorem, we find first that

$$
\left\langle \frac{1}{h_\mu} \frac{\partial Q_\mu}{\partial \xi_\mu} + \frac{\partial Q_3}{\partial z} \right\rangle = \left\langle \frac{\partial}{\partial y_\mu} \left(\frac{Q_\mu}{h_\mu A_\mu} \right) + \frac{\partial Q_3}{\partial z} \right\rangle
$$

$$
= \int_{S^+} \left(\frac{Q_\mu^+}{h_\mu A_\mu} n_{y_\mu}^+ + Q_3^+ n_{y_3}^+ \right) dS_\Omega^+
$$

$$
- \int_{S^-} \left(\frac{Q_\mu^-}{h_\mu A_\mu} n_{y_\mu}^- + Q_3^- n_{y_3}^- \right) dS_\Omega^- \quad (19.29)
$$

Now the integrals over the lateral surfaces of Ω cancel because of the periodicity of the functions Q_i in y_1 and y_2; by the definition of the surfaces $S^\pm$ we have

$$
dS_\Omega^\pm = \left[1 + \left(\frac{\partial F^\pm}{\partial y_1} \right)^2 + \left(\frac{\partial F^\pm}{\partial y_2} \right)^2 \right]^{1/2} dy_1\, dy_2
$$

$$
(19.30)
$$

$$
\mathbf{n}_y^\pm = \left\{ \mp \frac{\partial F^\pm}{\partial y_1}, \mp \frac{\partial F^\pm}{\partial y_2}, 1 \right\} \left[1 + \left(\frac{\partial F^\pm}{\partial y_1} \right)^2 + \left(\frac{\partial F^\pm}{\partial y_2} \right)^2 \right]^{-1/2}
$$

where the minus sign before the second integral in Equation 19.29 is due to the fact that the normal vector n_y^- is an "inner" with respect to the volume Ω. The proof is completed by inserting Equations 19.30 into Equation 19.29 and recalling the notation given in Equation 19.25.

Using Equations 19.14 in Equations 19.16, the equation

$$
\sigma_{ij}^{(0)} = c_{ijn\mu} \frac{1}{h_\mu} \frac{\partial u_n^{(1)}}{\partial \xi_\mu} + c_{ijn3} \frac{\partial u_n^{(1)}}{\partial z} + c_{ijn\mu} \varepsilon_{n\mu}^{(0)} \quad (19.31)
$$

is obtained, in which

$$\varepsilon_{11}^{(0)} = \frac{1}{A_1} \frac{\partial u_1^{(0)}}{\partial \alpha_1} + \frac{1}{A_1 A_2} \frac{\partial A_1}{\partial \alpha_2} u_2^{(0)}$$

$$\varepsilon_{22}^{(0)} = \frac{1}{A_2} \frac{\partial u_2^{(0)}}{\partial \alpha_2} + \frac{1}{A_1 A_2} \frac{\partial A_2}{\partial \alpha_1} u_1^{(0)}$$

$$\varepsilon_{12}^{(0)} = \varepsilon_{21}^{(0)} = \frac{1}{2} \left[\frac{A_1}{A_2} \frac{\partial}{\partial \alpha_2} \left(\frac{u_1^{(0)}}{A_1} \right) + \frac{A_2}{A_1} \frac{\partial}{\partial \alpha_1} \left(\frac{u_2^{(0)}}{A_2} \right) \right]$$

(19.32)

$$\varepsilon_{31}^{(0)} = \frac{1}{A_1} \frac{\partial u_3^{(0)}}{\partial \alpha_1} \qquad \varepsilon_{32}^{(0)} = \frac{1}{A_2} \frac{\partial u_3^{(0)}}{\partial \alpha_2}$$

and introducing the notation

$$L_{ijn} = c_{ijn\mu} \frac{1}{h_\mu} \frac{\partial}{\partial \xi_\mu} + c_{ijn3} \frac{\partial}{\partial z} \qquad (19.33)$$

$$D_{in} = \frac{1}{h_\beta} \frac{\partial}{\partial \xi_\beta} L_{i\beta n} + \frac{\partial}{\partial z} L_{i3n} \qquad (19.34)$$

$$C_{in\mu}(\xi, z) = \frac{1}{h_\beta} \frac{\partial c_{i\beta n\mu}}{\partial \xi_\beta} + \frac{\partial c_{i3n\mu}}{\partial z} \qquad (19.35)$$

where $\xi = (\xi_1, \xi_2)$, and substituting Equation 19.31 into Equation 19.17, one has to consider

$$D_{im} u_m^{(1)} = -C_{in\mu} \varepsilon_{n\mu}^{(0)} \qquad (19.36)$$

The solution of Equation 19.36 must be periodic in ξ_1 and ξ_2 (with periods A_1 and A_2, respectively) as well as satisfying the boundary conditions in Equation 19.22 for $l = 0$, which can be rewritten as

$$\left(L_{ijm} u_m^{(1)} + c_{ijn\mu} \varepsilon_{n\mu}^{(0)} \right) N_j^\pm = 0 \qquad (z = z^\pm) \qquad (19.37)$$

following the notation just introduced. To meet these requirements we write

$$u_m^{(1)} = U_m^{n\mu}(\xi, z) \varepsilon_{n\mu}^{(0)}(\alpha) + v_m^{(1)}(\alpha) \qquad (19.38)$$

where the functions U are periodic in ξ_1 and ξ_2 and solve the problem

$$D_{im}U_m^{n\mu} = -C_{in\mu}$$

$$(L_{ijm}V_m^{n\mu} + c_{ijn\mu})N_j^{\pm} = 0 \qquad (z = z^{\pm})$$

(19.39)

For $n\mu = 31, 32$, the problem given by Equations 19.39 is solved exactly and we find that

$$U_1^{31} = -z \qquad U_2^{31} = U_3^{31} = 0$$

$$U_2^{32} = -z \qquad U_1^{32} = U_3^{32} = 0$$

(19.40)

This is easily verified by writing Equations 19.39 in the explicit form

$$\frac{1}{h_\beta} \frac{\partial}{\partial \xi_\beta} \left(\frac{c_{i\beta m\nu}}{h_\nu} \frac{\partial U_m^{n\mu}}{\partial \xi_\nu} + c_{i\beta m3} \frac{\partial U_m^{n\mu}}{\partial z} \right) + \frac{\partial}{\partial z} \left(\frac{c_{i3m\nu}}{h_\nu} \frac{\partial U_m^{n\mu}}{\partial \xi_\nu} + c_{i3m3} \frac{\partial U_m^{n\mu}}{\partial z} \right)$$

$$= -\frac{1}{h_\nu} \frac{\partial c_{i\beta n\mu}}{\partial \xi_\beta} - \frac{\partial c_{i3n\mu}}{\partial z}$$

$$\left(\frac{c_{ijm\nu}}{h_\nu} \frac{\partial U_m^{n\mu}}{\partial \xi_\nu} + c_{ijm3} \frac{\partial U_m^{n\mu}}{\partial z} + c_{ijn\mu} \right) N_j^{\pm} = 0 \qquad (z = z^{\pm})$$

and observing that they turn into identities after substituting Equations 19.40.

To proceed further, define

$$b_{ij}^{n\mu} = L_{ijm}U_m^{n\mu} + c_{ijn\mu}$$

(19.41)

and note that, by Equations 19.40,

$$b_{ij}^{3\mu} \equiv 0$$

(19.42)

Upon substituting Equation 19.38 into Equation 19.31 we then find

$$\sigma_{ij}^{(0)} = b_{ij}^{\beta\mu}\varepsilon_{\beta\mu}^{(0)}$$

(19.43)

Now the averaging of Equation 19.18, using Equation 19.28 and Conditions 19.22 for $l = 1$, yields

$$\mathcal{B}_i\left(\langle \sigma_{n\mu}^{(0)} \rangle\right) = 0$$

(19.44)

Using Equation 19.43, it is readily seen that the homogeneous Equations 19.44 subjected to Conditions 19.22 at $l = 0$ have the zero solution

$$\varepsilon_{11}^{(0)} = \varepsilon_{22}^{(0)} = \varepsilon_{12}^{(0)} = 0$$

From Equations 19.32, 19.38, and 19.40 it is found that

$$u_1^{(0)} = u_2^{(0)} = 0 \qquad\qquad u_3^{(0)} = w(\alpha)$$

$$u_\mu^{(1)} = v_\mu^{(1)}(\alpha) - z\frac{1}{A_\mu}\frac{\partial w}{\partial \alpha_\mu} \qquad u_3^{(1)} = v_3^{(1)}(\alpha) \tag{19.45}$$

where μ is not summed.

Now because

$$\sigma_{ij}^{(0)} = 0 \tag{19.46}$$

from Equation 19.43, we find, using Equations 19.15 and 19.16, that

$$\sigma_{ij}^{(1)} = L_{ijm}u_m^{(2)} + c_{ijm\nu}\varepsilon_{m\nu}^{(1)} + zc_{ij\mu\nu}\tau_{\mu\nu} \tag{19.47}$$

where

$$\varepsilon_{11}^{(1)} = \frac{1}{A_1}\frac{\partial v_1^{(1)}}{\partial \alpha_1} + \frac{1}{A_1 A_2}\frac{\partial A_1}{\partial \alpha_2}v_2^{(1)} + K_2 w$$

$$\varepsilon_{22}^{(1)} = \frac{1}{A_2}\frac{\partial v_2^{(1)}}{\partial \alpha_2} + \frac{1}{A_1 A_2}\frac{\partial A_2}{\partial \alpha_1}v_1^{(1)} + K_1 w$$

$$\varepsilon_{12}^{(1)} = \varepsilon_{21}^{(1)} = \frac{1}{2}\left[\frac{A_1}{A_2}\frac{\partial}{\partial \alpha_2}\left(\frac{v_1^{(1)}}{A_1}\right) + \frac{A_2}{A_1}\frac{\partial}{\partial \alpha_1}\left(\frac{v_2^{(1)}}{A_2}\right)\right] \tag{19.48}$$

$$\varepsilon_{31}^{(1)} = \frac{1}{A_1}\frac{\partial v_3^{(1)}}{\partial \alpha_1} \qquad \varepsilon_{32}^{(1)} = \frac{1}{A_2}\frac{\partial v_3^{(1)}}{\partial \alpha_2}$$

$$\tau_{11} = -\frac{1}{A_1}\frac{\partial}{\partial \alpha_1}\left(\frac{1}{A_1}\frac{\partial w}{\partial \alpha_1}\right) - \frac{1}{A_1 A_2^2}\frac{\partial A_1}{\partial \alpha_2}\frac{\partial w}{\partial \alpha_2}$$

$$\tau_{22} = -\frac{1}{A_2}\frac{\partial}{\partial \alpha_2}\left(\frac{1}{A_2}\frac{\partial w}{\partial \alpha_2}\right) - \frac{1}{A_1^2 A_2}\frac{\partial A_2}{\partial \alpha_1}\frac{\partial w}{\partial \alpha_1} \tag{19.49}$$

$$\tau_{12} = \tau_{21} = -\frac{1}{A_1 A_2}\left(\frac{\partial^2 w}{\partial \alpha_1 \partial \alpha_2} - \frac{1}{A_1}\frac{\partial A_1}{\partial \alpha_2}\frac{\partial w}{\partial \alpha_1} - \frac{1}{A_2}\frac{\partial A_2}{\partial \alpha_1}\frac{\partial w}{\partial \alpha_2}\right)$$

Substituting Equations 19.46 and 19.47 into Equations 19.18 and 19.22 at $l = 1$ now yields

$$D_{im}u_m^{(2)} = -C_{im\nu}\varepsilon_{m\nu}^{(1)} - (c_{i3\mu\nu} + zC_{i\mu\nu})\tau_{\mu\nu}$$

$$\left(L_{ijm}u_m^{(2)} + c_{ijm\nu}\varepsilon_{m\nu}^{(1)} + z^{\pm}c_{ij\mu\nu}\tau_{\mu\nu}\right)N_j^{\pm} = 0 \qquad (z = z^{\pm}) \tag{19.50}$$

The solution of Equations 19.50 is assumed to be periodic in ξ_1 and ξ_2 (with respective periods A_1 and A_2) and can be expressed in the form

$$u_m^{(2)} = U_m^{l\nu}\varepsilon_{l\nu}^{(1)} + V_m^{\mu\nu}\tau_{\mu\nu} \tag{19.51}$$

where the functions $U_m^{l\nu}(\xi, z)$ are the solutions of the problem posed by Equations 19.39, and the functions $V_m^{\mu\nu}(\xi, z)$ are periodic in ξ_1 and ξ_2 and must satisfy the equations

$$D_{im}V_m^{\mu\nu} = -c_{i3\mu\nu} - zC_{i\mu\nu}$$

$$\left(L_{ijm}V_m^{\mu\nu} + z^{\pm}c_{ij\mu\nu}\right)N_j^{\pm} = 0 \qquad (z = z^{\pm}) \tag{19.52}$$

Let

$$b_{ij}^{*\mu\nu} = L_{ijm}V_m^{\mu\nu} + zc_{ij\mu\nu} \tag{19.53}$$

Using Equations 19.41 and 19.42 along with Equation 19.51 then reduces Equation 19.47 to the form

$$\sigma_{ij}^{(1)} = b_{ij}^{\mu\nu}\varepsilon_{\mu\nu}^{(1)} + b_{ij}^{*\mu\nu}\tau_{\mu\nu} \tag{19.54}$$

Using Equation 19.42 in Equations 19.39 and 19.52, we obtain

$$\frac{1}{h_\beta}\frac{\partial}{\partial\xi_\beta}b_{i\beta}^{\mu\nu} + \frac{\partial}{\partial z}b_{i3}^{\mu\nu} = 0$$

$$b_{ij}^{\mu\nu}N_j^{\pm} = 0 \qquad (z = z^{\pm}) \tag{19.55}$$

and

$$\frac{1}{h_\beta}\frac{\partial}{\partial\xi_\beta}b_{i\beta}^{*\mu\nu} + \frac{\partial}{\partial z}b_{i3}^{*\mu\nu} = 0$$

$$b_{ij}^{*\mu\nu}N_j^{\pm} = 0 \qquad (z = z^{\pm}) \tag{19.56}$$

according to the previous notation.

Together with Equations 19.41 and 19.53, the problems expressed by Equations 19.55 and 19.56 (henceforth referred to as the unit cell local problems) provide the functions $U_m^{\mu\nu}(\xi, z)$ and $V_m^{\mu\nu}(\xi, z)$, periodic in ξ_1 (with period A_1) and ξ_2 (with period A_2) and determining, in turn, the coefficients $b_{ij}^{\mu\nu}$ and $b_{ij}^{*\mu\nu}$ in Equation 19.54. It should be noted that, unlike the unit cell problems of "classical" homogenization schemes (Sanchez-Palencia 1980, Bakhvalov and Panasenko 1984), those set by Equations 19.55 and 19.56 depend on the boundary conditions at $z = z^{\pm}$ rather than on periodicity in the z direction. Clearly, in this case also the solutions U and V are unique up to constant terms (Bakhvalov and Panasenko 1984), an ambiguity that may be removed by imposing the conditions

$$\langle U_m^{\mu\nu} \rangle_\xi = 0 \quad \text{and} \quad \langle V_m^{\mu\nu} \rangle_\xi = 0 \quad \text{for } z = 0 \qquad (19.57)$$

where $\langle \cdots \rangle_\xi$ indicates average with respect to ξ_1 and ξ_2 only. Combined with Equations 19.8, 19.45, and 19.51, these conditions make clearer the mechanical meaning of the functions $v_1^{(1)}(\alpha)$, $v_2^{(1)}(\alpha)$, and $w(\alpha)$ characterizing the displacements of the midsurface of the shell, $z = 0$.

For actually solving the local problems given by Equations 19.55 and 19.56, a more convenient form of the boundary conditions at $z = z^{\pm}$ may be derived. Recalling Equation 19.25 and using the expressions for the ξ_1-, ξ_2-, and z-components of the normal $n_i^{\pm}$ to these surfaces we find

$$\frac{1}{h_\beta} n_\beta^{\pm(\xi)} b_{i\beta}^{\mu\nu} + n_3^{\pm(\xi)} b_{i3}^{\mu\nu} = 0 \qquad (z = z^{\pm}) \qquad (19.58)$$

$$\frac{1}{h_\beta} n_\beta^{\pm(\xi)} b_{i\beta}^{*\mu\nu} + n_3^{\pm(\xi)} b_{i3}^{*\mu\nu} = 0 \qquad (z = z^{\pm}) \qquad (19.59)$$

An important point to be made in conclusion is that although the functions $c_{ijmn}(y, z)$ have been assumed to be smooth in the preceding discussion, we can readily generalize the local problems given by Equations 19.55 and 19.56 to the case of piecewise-smooth elastic constants c_{ijmn}, with discontinuities of the first kind on a finite number of nonintersecting contact surfaces between dissimilar constituents (such as matrix and fibers, binder and inclusions, etc.) This is effected by adding the continuity conditions on contact surfaces,

$$[U_m^{\mu\nu}] = 0 \qquad \left[\frac{1}{h_\beta} n_\beta^{(k)} b_{i\beta}^{\mu\nu} + n_3^{(k)} b_{i3}^{\mu\nu}\right] = 0 \qquad (19.60)$$

$$[V_m^{\mu\nu}] = 0 \qquad \left[\frac{1}{h_\beta} n_\beta^{(k)} b_{i\beta}^{*\mu\nu} + n_3^{(k)} b_{i3}^{*\mu\nu}\right] = 0 \qquad (19.61)$$

where square brackets denote the jump of a function on the contact surface, and $n_i^{(k)}$ (the component of the normal vector to this surface) is referred to the coordinate system (ξ_1, ξ_2, z). Conditions 19.60 and 19.61 express the discontinuity of the displacements (Equation 19.51) and stresses (Equation 19.54) across the contact surface or, in other words, describe an ideal contact (or bonding) between dissimilar components of the composite. Note also that if we rewrite these conditions with respect to the functions $Q_m(b_{im}^{\mu\nu} \to Q_m, b_{im}^{*\mu\nu} \to Q_m)$, a generalization of Formula 19.28 to the case of piecewise-smooth functions Q_m becomes possible.

19.3. Governing equations of the homogenized shell

To derive these equations we transform Equations 19.55 and 19.56 by multiplying by z and z^2 and averaging. Using Equations 19.28 and the boundary conditions on the surfaces $z = z^{\pm}$ this gives

$$\langle b_{i3}^{\mu\nu} \rangle = \langle z b_{i3}^{\mu\nu} \rangle = \langle b_{i3}^{*\mu\nu} \rangle = \langle z b_{i3}^{*\mu\nu} \rangle = 0 \qquad (19.62)$$

and averaging Equation 19.54 we then have

$$\langle \sigma_{i3}^{(1)} \rangle = \langle z \sigma_{i3}^{(1)} \rangle = 0 \qquad (19.63)$$

The averaging procedure is next applied to Equations 19.19 and 19.20. Again using Equation 19.28 and employing Conditions 19.23 and 19.24 along with Relations 19.46 and 19.63, we find

$$\mathcal{B}_\lambda\!\left(\langle \sigma_{\nu\mu}^{(1)} \rangle \right) + r_\lambda + \langle f_\lambda \rangle = 0 \qquad (19.64)$$

$$\mathcal{B}_\lambda\!\left(\langle \sigma_{\nu\mu}^{(2)} \rangle \right) = 0 \qquad (19.65)$$

$$\mathcal{B}_3\!\left(\langle \sigma_{3\mu}^{(2)} \rangle \right) + \aleph_3\!\left(\langle \sigma_{\nu\mu}^{(1)} \rangle \right) + q_3 + \langle g_3 \rangle = 0 \qquad (19.66)$$

where the function $r_\lambda(\alpha)$ is defined by the equations

$$r_\lambda(\alpha) = \int_{-1/2}^{1/2}\int_{-1/2}^{1/2} (\omega^+ r_\lambda^4 + \omega^- r_\lambda^-)\, dy_1\, dy_2$$

$$(19.67)$$

$$q_3(\alpha) = \int_{-1/2}^{1/2}\int_{-1/2}^{1/2} (\omega^+ q_3^+ + \omega^- q_3^-)\, dy_1\, dy_2$$

Now if Equation 19.19 is averaged after multiplying by z, then, by Equation 19.23, we have for $i = 1, 2$

$$\mathcal{B}_\mu\!\left(\langle z \sigma_{\nu\beta}^{(1)} \rangle \right) - \langle \sigma_{3\mu}^{(2)} \rangle + \rho_\mu + \langle z f_\mu \rangle = 0 \qquad (19.68)$$

where

$$\rho_\mu(\alpha) = \int_{-1/2}^{1/2} \int_{-1/2}^{1/2} (z^+ \omega^+ r_\mu^+ + z^- \omega^- r_\mu^-) \, dy_1 \, dy_2 \qquad (19.69)$$

Equations 19.66 and 19.68 may be combined to eliminate $\langle \sigma_{3\mu}^{(2)} \rangle$, giving (cf. Equations 19.12)

$$\frac{1}{A_1 A_2} \left\{ \frac{\partial}{\partial \alpha_1} \left[A_2 \left(\mathfrak{B}_1 \langle z \sigma_{\nu\beta}^{(1)} \rangle + \rho_1 + \langle z f_1 \rangle \right) \right] \right.$$

$$+ \frac{\partial}{\partial \alpha_2} \left[A_1 \left(\mathfrak{B}_2 \langle z \sigma_{\nu\beta}^{(1)} \rangle + \rho_2 + \langle z f_2 \rangle \right) \right] \Bigg\}$$

$$- K_2 \langle \sigma_{11}^{(1)} \rangle - K_1 \langle \sigma_{22}^{(1)} \rangle + q_3 + \langle g_3 \rangle = 0 \qquad (19.70)$$

Note that as far as the stresses are concerned, Equations 19.64 and 19.70 only contain the averaged quantities $\langle \sigma_{\mu\nu}^{(1)} \rangle$ and $\langle z \sigma_{\mu\nu}^{(1)} \rangle$, which may be represented in the respective forms

$$\langle \sigma_{\mu\nu}^{(1)} \rangle = \langle b_{\mu\nu}^{\beta\lambda} \rangle \varepsilon_{\beta\lambda}^{(1)} + \langle b_{\mu\nu}^{*\beta\lambda} \rangle \tau_{\beta\lambda} \qquad (19.71)$$

$$\langle z \sigma_{\mu\nu}^{(1)} \rangle = \langle z b_{\mu\nu}^{\beta\lambda} \rangle \varepsilon_{\beta\lambda}^{(1)} + \langle z b_{\mu\nu}^{*\beta\lambda} \rangle \tau_{\beta\lambda} \qquad (19.72)$$

The system of three governing equations for the unknown functions of the problem can now be written down by substituting Equations 19.71 and 19.72 into Equations 19.64 and 19.70 and noting that $\varepsilon_{\beta\lambda}^{(1)}$, $\tau_{\beta\alpha}$ are expressed in terms of $v_1^{(1)}(\alpha)$, $v_2^{(1)}(\alpha)$ and $w(\alpha)$ (see Equations 19.48 and 19.49).

We conclude that the original problem for the regularly nonhomogeneous (composite material) shell with rapidly oscillating thickness reduces to two simpler types of problem. One of these types involves in turn a pair of local problems, the first of which is set by Equations 19.55, 19.57, and 19.60 and yields the functions $V_m^{\mu\nu}(\xi_1, \xi_2, z)$, which are periodic in ξ_1 and ξ_2 with periods A_1 and A_2, respectively. Once these are known, the functions $b_{ij}^{\mu\nu}(\xi_1, \xi_2, z)$ can be determined by means of Equation 19.41. The second local problem is posed by Equations 19.56, 1957, and 19.61 and produces the functions $V_m^{\mu\nu}(\xi_1, \xi_2, z)$ (also periodic in ξ_1 and ξ_2) in terms of which the functions $b_{ij}^{*\mu\nu}(\xi_1, \xi_2, z)$ are calculated from Equation 19.53. Both local problems are in fact elasticity-theory problems formulated for a generally nonhomogeneous body (with unit cell Ω) subjected to certain artificial forces (surface and bulk ones) that are expressed in terms of the elastic coefficients and their derivatives, as clearly seen from Equations 19.39 and 19.52. As regards the boundary conditions, there are periodicity requirements along the ξ_1 and ξ_2 coordinates, and the conditions expressed by the problems given in Equations 19.55 and 19.56 or Relations 19.58 and 19.59 must be satisfied on the surfaces $z = z^{\pm}$.

The local problems having been solved, the functions $b_{ij}^{\mu\nu}$ and $b_{ij}^{*\mu\nu}$ are averaged by application of Equation 19.27, giving the effective stiffness moduli of the homogenized shell, $\langle b_{\mu\nu}^{\beta\lambda} \rangle$, $\langle zb_{\mu\nu}^{\beta\lambda} \rangle$, $\langle b_{\mu\nu}^{*\beta\lambda} \rangle$, and $\langle zb_{\mu\nu}^{*\beta\lambda} \rangle$, entering the elastic relations, Equations 19.71 and 19.72 as coefficients. One can then proceed to the second of the two previously mentioned types of problem, namely, the boundary value problem for the homogenized shell, expressed by relations 19.64, 19.70 through 19.72, 19.48, and 19.49 and yielding the functions $v_1^{(1)}(\alpha)$, $v_2^{(1)}(\alpha)$, and $w(\alpha)$. Note that the solution of the homogenized problem corresponds to the application of the effective modulus method (Pobedrya 1984). An important feature of the method discussed is that the solutions of the local problems and of the homogenized problem enable one to make very accurate predictions concerning the three-dimensional local structure of the displacement and stress fields. In particular, Equations 19.8, 19.45, and 19.51 yield

$$u_1 = \delta v_1^{(1)}(\alpha) - \gamma \frac{1}{A_1} \frac{\partial w}{\partial \alpha_1} + \delta U_1^{\mu\nu} \delta \varepsilon_{\mu\nu}^{(1)} + \delta^2 V_1^{\mu\nu} \tau_{\mu\nu} + \cdots$$

$$u_2 = \delta v_2^{(1)}(\alpha) - \gamma \frac{1}{A_2} \frac{\partial w}{\partial \alpha_2} + \delta U_2^{\mu\nu} \delta \varepsilon_{\mu\nu}^{(1)} + \delta^2 V_2^{\mu\nu} \tau_{\mu\nu} + \cdots \qquad (19.73)$$

$$u_3 = w(\alpha) + \delta U_3^{\mu\nu} \delta \varepsilon_{\mu\nu}^{(1)} + \delta^2 V_3^{\mu\nu} \tau_{\mu\nu} + \cdots$$

for the displacement components. Note that the function $v_3^{(1)}(\alpha)$ involved in Equations 19.38 and 19.45 can be set equal to zero without any loss of generality because we proved earlier that this particular function does not enter into the system of governing equations of the problem.

Similarly, it is found from Equations 19.16 and 19.46 that the stresses are given by

$$\sigma_{ij} = b_{ij}^{\mu\nu} \delta \varepsilon_{\mu\nu}^{(1)} + \delta b_{ij}^{*\mu\nu} \tau_{\mu\nu} + \cdots \qquad (19.74)$$

From the way they are stated, it is seen that the local problems are completely determined by the structure of the unit cell of the composite and are totally independent of the formulation of the global boundary value problem. It follows that the solutions of these problems—particularly the effective elastic moduli of the homogenized shell—are universal in their nature and, once found, may therefore be utilized in studying very different types of boundary value problems associated with the given composite material.

It should be noted, finally, that the coordinates ξ_1 and ξ_2 involved in the local problems are defined in terms of the quantities $A_1(\alpha)$ and $A_2(\alpha)$, so that if these latter are not constant they may be constant in the case of developable surfaces), the effective stiffness moduli will also depend on the "slow" coordinates α_1 and α_2. This means that even in the case of an originally homogeneous material we may come up with a structural nonhomogeneity after the homogenization procedure.

19.4. The symmetry of effective properties

When the properties of a material are periodic in all three coordinates, it can be shown (Pobedrya 1984) that the symmetry properties of the coefficients involved remain the same after the homogenization process. In the case we treat here, there is no periodicity along the thickness coordinate z of the shell, and the question of symmetry therefore is given special consideration.

Let us first come back to the definitions (Equations 19.41 and 19.53 for the quantities b and b^*. Modifying them by allowing all the indices to range from 1 to 3, and using Relations 19.33 through 19.35, 19.39, and 19.52, we find that

$$b_{ij}^{mn} = b_{ji}^{mn} = b_{ij}^{nm} \quad \text{and} \quad b_{ij}^{*mn} = b_{ji}^{*mn} = b_{ij}^{*nm} \qquad (19.75)$$

where the well-known symmetry properties of the elastic coefficients c_{ijmn} also have been taken into account.

We wish now to prove the following symmetry properties:

$$\langle b_{ij}^{mn} \rangle = \langle b_{mn}^{ij} \rangle$$

$$\langle z b_{ij}^{mn} \rangle = \langle b_{mn}^{*ij} \rangle \qquad (19.76)$$

$$\langle z b_{ij}^{*mn} \rangle = \langle z b_{mn}^{*ij} \rangle$$

To this end, define

$$\zeta_1 = h_1 \xi_1 \qquad \zeta_2 = h_2 \xi_2 \qquad \zeta_3 = z$$

Equations 19.33 then reduce to

$$L_{ijl} = c_{ijlr} \frac{\partial}{\partial \zeta_r}$$

$$D_{il} = \frac{\partial}{\partial \zeta_p} c_{iplr} \frac{\partial}{\partial \zeta_r}$$

$$C_{imn} = \frac{\partial c_{iqmn}}{\partial \zeta_q}$$

Equations 19.41 and 19.53 become

$$b_{ij}^{mn} = c_{ijlr} \left(\frac{\partial U_l^{mn}}{\partial \zeta_r} + \delta_{lm} \delta_{rn} \right) \tag{19.77}$$

$$b_{ij}^{*mn} = c_{ijlr} \left(\frac{\partial V_l^{mn}}{\partial \zeta_r} + z \delta_{lm} \delta_{rn} \right) \tag{19.78}$$

and the local problems given by Equations 19.55 and 19.56 can be rewritten as

$$\frac{\partial b_{ij}^{mn}}{\partial \zeta_j} = 0 \qquad N_j^{\pm} b_{ij}^{mn} = 0 \qquad (z = z^{\pm}) \tag{19.79}$$

$$\frac{\partial b_{ij}^{*mn}}{\partial \zeta_j} = 0 \qquad N_j^{\pm} b_{ij}^{*mn} = 0 \qquad (z = z^{\pm}) \tag{19.80}$$

Now if the function $\varphi(\zeta_1, \zeta_2, \zeta_3)$ is periodic in ζ_1 and ζ_2 with unit cell Ω, then the equations

$$\left\langle b_{qp}^{mn} \frac{\partial \varphi}{\partial \zeta_p} \right\rangle = 0 \quad \text{and} \quad \left\langle b_{qp}^{*mn} \frac{\partial \varphi}{\partial \zeta_p} \right\rangle = 0 \tag{19.81}$$

hold. To show this, employ Equations 19.79 to obtain

$$\frac{\partial}{\partial \zeta_p} (b_{qp}^{mn} \varphi) = \varphi \frac{\partial b_{qp}^{mn}}{\partial \zeta_p} + b_{qp}^{mn} \frac{\partial \varphi}{\partial \zeta_p} = b_{qp}^{mn} \frac{\partial \varphi}{\partial \zeta_p}$$

Applying Equation 19.28 and making use of the conditions at the $z = z^{\pm}$ surfaces, we then find

$$\left\langle \frac{\partial}{\partial \zeta_p}(b_{qp}^{mn}\varphi) \right\rangle = \int_{-1/2}^{1/2}\int_{-1/2}^{1/2}\left[(\varphi b_{qp}^{mn})^{+} N_p^{+} - (\varphi b_{qp}^{mn})^{-} N_p^{-}\right] dy_1\, dy_2 = 0$$

thereby proving the first of Equations 19.81. The second is proved in exactly the same way by taking Equations 19.80 as the starting point.

Next, we observe that adding the identity

$$\langle b_{ij}^{mn} \rangle = \langle b_{qp}^{mn}\delta_{qi}\delta_{pi} \rangle$$

and the equation

$$\left\langle b_{qp}^{mn}\frac{\partial U_q^{ij}}{\partial \zeta_p} \right\rangle = 0$$

(which is simply the first of Equations 19.81, for $\varphi = U_q^{ij}$, results in

$$\langle b_{ij}^{mn} \rangle = \left\langle b_{qp}^{mn}\left(\frac{\partial U_q^{ij}}{\partial \zeta_p} + \delta_{qi}\delta_{pj}\right) \right\rangle$$

$$= \left\langle c_{qplr}\left(\frac{\partial U_l^{mn}}{\partial \zeta_r} + \delta_{lm}\delta_{rn}\right)\left(\frac{\partial U_q^{ij}}{\partial \zeta_p} + \delta_{qi}\delta_{pj}\right) \right\rangle \quad (19.82)$$

which actually proves the first of Equations 19.76 if the symmetry properties of the elastic coefficients are properly taken into account.

Likewise, setting φ equal successively to U_q^{ij} and V_q^{ij}, it follows from Equations 19.77, 19.78, and 19.81 that

$$\langle zb_{ij}^{*mn} \rangle = \left\langle c_{qplr}\left(\frac{\partial V_l^{mn}}{\partial \zeta_r} + z\delta_{lm}\delta_{rn}\right)\left(\frac{\partial V_q^{ij}}{\partial \zeta_p} + z\delta_{qi}\delta_{pj}\right) \right\rangle \quad (19.83)$$

$$\langle b_{ij}^{*mn} \rangle = \left\langle c_{qplr}\left(\frac{\partial V_l^{mn}}{\partial \zeta_r} + z\delta_{lm}\delta_{rn}\right)\left(\frac{\partial U_q^{ij}}{\partial \zeta_p} + \delta_{qi}\delta_{pj}\right) \right\rangle$$

$$\qquad\qquad\qquad\qquad\qquad\qquad\qquad\qquad (19.84)$$

$$\langle zb_{ij}^{mn} \rangle = \left\langle c_{qplr}\left(\frac{\partial U_l^{mn}}{\partial \zeta_r} + \delta_{lm}\delta_{rn}\right)\left(\frac{\partial V_q^{ij}}{\partial \zeta_p} + z\delta_{qi}\delta_{pj}\right) \right\rangle$$

Using again the symmetry of the elastic coefficients, the third of Equations 19.76 is derived from Equation 19.83, and the second from Equations 19.84.

We note that if one or more of the indices is 3, Equations 19.76 produce an identical zero by virtue of Equations 19.42 and 19.62; otherwise, these formulas are nontrivial and secure the symmetry of the coefficient matrix of the elastic relations, Equations 19.71 and 19.72.

19.5. Homogenized-shell model versus thin-shell theory

To carry out this comparison, the basic results we have obtained by homogenizing shell equations will be rewritten here in terms of the standard theory of thin shells as described, for example, in Novozhilov (1962) or Ambartsumyan (1974).

Using Equations 19.16, 19.46, and 19.63, the leading-order expressions for the stress resultants N_1, N_2, and N_{12}, the moment resultants M_1, M_2, and M_{12}, and the shearing forces Q_1 and Q_2 are found to be given by

$$N_1 = \delta^2 \langle \sigma_{11}^{(1)} \rangle \qquad M_1 = \delta^3 \langle z\sigma_{11}^{(1)} \rangle \qquad (1 \leftrightarrow 2)$$

$$N_{12} = \delta^2 \langle \sigma_{12}^{(1)} \rangle \qquad M_{12} = \delta^3 \langle z\sigma_{12}^{(1)} \rangle \qquad (19.85)$$

$$Q_1 = \delta^3 \langle \sigma_{31}^{(2)} \rangle \qquad Q_2 = \delta^3 \langle \sigma_{32}^{(2)} \rangle$$

For determining the components of the displacement vector, again at leading order, Formulas 19.73 must be used. Denoting

$$v_1(\alpha) = \delta v_1^{(1)}(\alpha) \qquad v_2(\alpha) = \delta v_2^{(1)}(\alpha) \qquad \varepsilon_{\mu\nu} = \delta \varepsilon_{\mu\nu}^{(1)}$$

and using Equation 19.9, Equations 19.48 take the form

$$\varepsilon_{11} = \frac{1}{A_1} \frac{\partial v_1}{\partial \alpha_1} + \frac{1}{A_1 A_2} \frac{\partial A_1}{\partial \alpha_2} v_2 + K_1 w$$

$$\varepsilon_{22} = \frac{1}{A_2} \frac{\partial v_2}{\partial \alpha_2} + \frac{1}{A_1 A_2} \frac{\partial A_2}{\partial \alpha_1} v_1 + K_2 w \qquad (19.86)$$

$$\varepsilon_{12} = \varepsilon_{21} = \frac{1}{2} \left[\frac{A_1}{A_2} \frac{\partial}{\partial \alpha_2} \left(\frac{v_1}{A_1} \right) + \frac{A_2}{A_1} \frac{\partial}{\partial \alpha_1} \left(\frac{v_2}{A_2} \right) \right]$$

and Equations 19.73 may be rewritten as

$$u_1 = v_1(\alpha) - \frac{\gamma}{A_1} \frac{\partial w}{\partial \alpha_1} + \delta U_1^{\mu\nu} \varepsilon_{\mu\nu} + \delta^2 V_1^{\mu\nu} \tau_{\mu\nu} + \cdots$$

$$u_2 = v_2(\alpha) - \frac{\gamma}{A_2} \frac{\partial w}{\partial \alpha_2} + \delta U_2^{\mu\nu} \varepsilon_{\mu\nu} + \delta^2 V_2^{\mu\nu} \tau_{\mu\nu} + \cdots \qquad (19.87)$$

$$u_3 = w(\alpha) + \delta U_3^{\mu\nu} \varepsilon_{\mu\nu} + \delta^2 V_3^{\mu\nu} \tau_{\mu\nu} + \cdots$$

A comparison of Relations 19.86, 19.49, and 19.87 with the corresponding thin-shell theory results (Novozhilov 1962, Ambartsumyan 1974) shows that the functions $v_1(\alpha)$, $v_2(\alpha)$, and $w(\alpha)$ are the midsurface displacements of the shell; the quantities $\varepsilon_{11} = \varepsilon_1$, $\varepsilon_{22} = \varepsilon_2$, and $\varepsilon_{12} = \varepsilon_{21} = \omega/2$ are the elongations and shears; and, finally, $\tau_{11} = \kappa_1$, $\tau_{22} = \kappa_2$, and $\tau_{12} = \tau_{21} = \tau$ are the torsional and flexural midsurface strains as calculated in the framework of the Donnell–Mushtari–Vlasov model (see Novozhilov 1962). It will become clear that the correspondence between the two sets of results is due to the asymptotic form adopted previously for the main curvatures of the shell midsurfaces, Equations 19.9.

The elastic relations of the homogenized shell, that is, those between the stress and moment resultants on the one hand, and the midsurface strains on the other, are found from Equations 19.71, 19.72, and 19.85. We have

$$N_\beta = \delta \langle b_{\beta\beta}^{\mu\nu} \rangle \varepsilon_{\mu\nu} + \delta^2 \langle b_{\beta\beta}^{*\mu\nu} \rangle \tau_{\mu\nu}$$

$$N_{12} = \delta \langle b_{12}^{\mu\nu} \rangle \varepsilon_{\mu\nu} + \delta^2 \langle b_{12}^{*\mu\nu} \rangle \tau_{\mu\nu}$$

$$M_\beta = \delta^2 \langle z b_{\beta\beta}^{\mu\nu} \rangle \varepsilon_{\mu\nu} + \delta^3 \langle z b_{\beta\beta}^{*\mu\nu} \rangle \tau_{\mu\nu} \qquad (19.88)$$

$$M_{12} = \delta^2 \langle z b_{12}^{\mu\nu} \rangle \varepsilon_{\mu\nu} + \delta^3 \langle z b_{12}^{*\mu\nu} \rangle \tau_{\mu\nu}$$

where β takes the values 1 and 2 and is not summed; we should remark here that the symmetry of the 6×6 coefficient matrix involved in this equation is ensured by the symmetry properties of the effective elastic moduli (see Equations 19.76).

Using Equations 19.67, 19.69, 19.10, 19.11, and 19.85 and following the notation of Equations 19.9 and 19.12, Equations 19.64 through 19.66 and

19.68 can be written as

$$\frac{\partial(A_2 N_1)}{\partial \alpha_1} - \frac{\partial A_2}{\partial \alpha_1} N_2 + \frac{\partial(A_1 N_{12})}{\partial \alpha_2} + \frac{\partial A_1}{\partial \alpha_2} N_{12} = -A_1 A_2 G_1$$

$$\frac{\partial(A_1 N_2)}{\partial \alpha_2} - \frac{\partial A_1}{\partial \alpha_2} N_1 + \frac{\partial(A_2 N_{12})}{\partial \alpha_1} + \frac{\partial A_2}{\partial \alpha_1} N_{12} = -A_1 A_2 G_2$$

$$k_1 N_1 + k_2 N_2 - \frac{1}{A_1 A_2}\left[\frac{\partial(A_2 Q_1)}{\partial \alpha_1} + \frac{\partial(A_1 Q_2)}{\partial \alpha_2}\right] = G_3 \qquad (19.89)$$

$$Q_1 = \frac{1}{A_1 A_2}\left[\frac{\partial(A_2 M_1)}{\partial \alpha_1} - \frac{\partial A_2}{\partial \alpha_1} M_2 + \frac{\partial(A_1 M_{12})}{\partial \alpha_2} + \frac{\partial A_1}{\partial \alpha_2} M_{12}\right] + m_1$$

$$Q_2 = \frac{1}{A_1 A_2}\left[\frac{\partial(A_1 M_2)}{\partial \alpha_2} - \frac{\partial A_1}{\partial \alpha_2} M_1 + \frac{\partial(A_2 M_{12})}{\partial \alpha_1} + \frac{\partial A_2}{\partial \alpha_1} M_{12}\right] + m_2$$

In Equations 19.89, the external loads are given by

$$G_i = \int_{-1/2}^{1/2}\int_{-1/2}^{1/2}(\omega^+ p_i^+ + \omega^- p_i^-)\, dy_1\, dy_2 + \delta\langle P_i\rangle$$

$$(19.90)$$

$$m_\beta = \int_{-1/2}^{1/2}\int_{-1/2}^{1/2}(\gamma^+ \omega^+ p_\beta^+ + \gamma^- \omega^- p_\beta^-)\, dy_1\, dy_2 + \delta\langle \gamma P_\beta\rangle$$

where the functions $\omega^\pm$, defined by Equation 19.26, are determined by the profiles of the top and bottom shell surfaces, $\gamma = \gamma^\pm$, and where the presence of the coefficient δ is due to the fact that the averaging operation Equation 19.27 involves integration over the coordinate z rather than over $\gamma = \delta z$. We thus see that Equations 19.89 are identical with the equilibrium equations of the engineering formulation of thin-shell theory (Ambartsumyan 1974).

If we take the limiting case of a homogeneous isotropic shell of constant thickness as an example, we have the equations $F^\pm \equiv 0$, $N_i^\pm = \{0, 0, 1\}$, $\omega^\pm \equiv 1$, and

$$c_{ijmn} = \text{constant}$$

$$= \frac{E}{2(1 + \nu)}\left(\frac{2\nu}{1 - 2\nu}\delta_{ij}\delta_{mn} + \delta_{im}\delta_{jn} + \delta_{in}\delta_{jm}\right) \qquad (19.91)$$

as the initial formulation of the problem, with E and ν denoting, as usual, Young's modulus and Poisson's ratio of the material. Because in this case none of the quantities of interest depends on y_1 or y_2, the problem is solvable exactly, giving

$$U_3^{11} = U_3^{22} = -\frac{\nu z}{1 - \nu}$$

$$V_3^{11} = V_3^{22} = -\frac{\nu z^2}{2(1 - \nu)}$$

(19.92)

for the nonzero solutions of the local problems given by Equations 19.55 through 19.57. The set of nonzero elastic moduli then follows from Equations 19.41 and 19.53 as

$$\langle b_{11}^{11} \rangle = \langle b_{22}^{22} \rangle = \frac{E}{1 - \nu^2} \qquad \langle b_{11}^{22} \rangle = \langle b_{22}^{11} \rangle = \frac{E\nu}{1 - \nu^2}$$

$$\langle b_{12}^{12} \rangle = \frac{E}{2(1 + \nu)} \qquad \langle zb_{\mu\theta}^{\beta\lambda} \rangle = 0 \qquad \langle b_{\mu\theta}^{*\beta\lambda} \rangle = 0$$

(19.93)

$$\langle zb_{\mu\theta}^{*\beta\lambda} \rangle = \frac{1}{12} \langle b_{\mu\theta}^{\beta\lambda} \rangle$$

which when substituted into Equations 19.88 leads—the point worth emphasizing—to the elastic relations of the theory of thin isotropic shells (Novozhilov 1962, Ambartsumyan 1974).

An important conclusion to be drawn from the preceding comparison is that we may utilize the well-studied formalism of the theory of anisotropic shells for the solution of the homogenized problem, taking the elastic relations Equations 19.88 to describe the particular type of anisotropy of the problem under consideration.

As regards the edge conditions necessary for the solution of the homogenized boundary value problem, we may take these in the form known from ordinary shell theory, but, obviously enough, edge effects remain unaccounted for in this approach. The remedy is to solve the problem in its exact three-dimensional formulation or, alternatively, to employ the method of boundary layer solutions that has been developed in the framework of the asymptotic homogenization method by a number of investigators (Panasenko 1979, Bakhvalov and Panasenko 1984, Sanchez-Palencia 1987).

19.6. Cylindrical shell

In this special case, of well-known importance for engineering and other purposes, the midsurface of the shell is a cylindrical surface, as shown in Figure 19.3. If we introduce a coordinate system (α_1, α_2) such that α_1 is

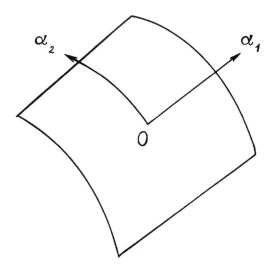

FIGURE 19.3. Cylindrical surface α_1, α_2.

measured along the generator and α_2 along the directrix of the cylinder, we have (Novozhilov 1962, Ambartsumyan 1974)

$$A_1 \equiv A_2 \equiv 1 \qquad k_1 = 0 \qquad k_2 = \frac{1}{r(\alpha_2)}$$

where $r(\alpha_2)$ is constant in the special case of a circular cylinder. Clearly, α_1 and α_2 can be made dimensionless by dividing by a certain characteristic dimension L, which we take to be equal to unity for the sake of simplicity.

From Equations 19.49 and 19.86,

$$\varepsilon_{11} = \frac{\partial v_1}{\partial \alpha_1}$$

$$\varepsilon_{22} = \frac{\partial v_2}{\partial \alpha_2} + \frac{w}{r}$$

$$\varepsilon_{12} = \frac{1}{2}\left(\frac{\partial v_1}{\partial \alpha_2} + \frac{\partial v_2}{\partial \alpha_1} \right)$$

$$\tau_{\lambda\mu} = -\frac{\partial^2 w}{\partial \alpha_\lambda \, \partial \alpha_\mu}$$

$\qquad\qquad(19.94)$

and Equations 19.89 become

$$\frac{\partial N_1}{\partial \alpha_1} + \frac{\partial N_{12}}{\partial \alpha_2} = -G_1$$

$$\frac{\partial N_2}{\partial \alpha_2} + \frac{\partial N_{12}}{\partial \alpha_1} = -G_2 \tag{19.95}$$

$$\frac{\partial^2 M_\beta}{\partial \alpha_\beta^2} + 2\frac{\partial^2 M_{12}}{\partial \alpha_1 \, \partial \alpha_2} - \frac{N_2}{r} = -G_3 - \frac{\partial m_\mu}{\partial \alpha_\mu}$$

The stress and moment resultants are expressed in terms of the midsurface strains, Equations 19.94, by means of the elastic relations, Equations 19.88. Because the quantities A_1 and A_2 are constant in this case, so too are the effective stiffness moduli, because the functions b and b^* are independent of the coordinates α_1 and α_2. The values of the moduli are determined by the functions $c_{ijmn}(y, z)$ and $F^\pm(y)$ and are found from the local problems of the type we have discussed previously.

19.7. Plane shell

This is also a geometry of great practical interest. Because the midsurface is obviously a plane in this case, the coordinate system $\{\alpha_1, \alpha_2, \gamma\}$ is naturally chosen to be Cartesian, so that $A_1 = A_2 \equiv 1$, $k_1 = k_2 = 0$, and the governing relations of the problem follow from Equations 19.94 and 19.95 in the limit as $r \to \infty$. The effective moduli are again constants, and we obtain the following system of three equations for determining the functions $v_1(\alpha)$, $v_2(\alpha)$, and $w(\alpha)$:

$$\delta \langle b_{\theta\mu}^{\beta\lambda} \rangle \frac{\partial^2 v_\beta}{\partial \alpha_\lambda \, \partial \alpha_\mu} - \delta^2 \langle b_{\theta\mu}^{*\beta\lambda} \rangle \frac{\partial^3 w}{\partial \alpha_\beta \, \partial \alpha_\lambda \, \partial \alpha_\mu} = -G_\theta \qquad (\theta = 1, 2)$$

$$\delta^2 \langle z b_{\mu\nu}^{\beta\lambda} \rangle \frac{\partial^3 v_\beta}{\partial \alpha_\lambda \, \partial \alpha_\mu \, \partial \alpha_\nu} - \delta^3 \langle z b_{\mu\nu}^{*\beta\lambda} \rangle \frac{\partial^4 w}{\partial \alpha_\beta \, \partial \alpha_\lambda \, \partial \alpha_\mu \, \partial \alpha_\nu} = -G_3 \tag{19.96}$$

$$- \frac{\partial m_\mu}{\partial \alpha_\mu}$$

where the elastic relations (Equations 19.88) and the properties given in Equations 19.75 of the stiffness moduli have been used.

Probably the primary thing to note in this system is that, unless in some special cases, it does not separate into an equation for the deflection $w(\alpha)$ and equations for the plate surface displacements $v_1(\alpha)$ and $v_2(\alpha)$. It is obvious, however, that the separation does occur when either $\langle z b_{\mu\nu}^{\beta\lambda} \rangle = 0$

or $\langle b_{\theta\mu}^{*\beta\lambda} \rangle = 0$, which two conditions are equivalent in view of Equations 19.76 and may be satisfied if certain types of "good symmetry" happen to occur in material properties and/or unit cell geometry. This is illustrated by means of examples below.

In the limiting case of a homogeneous, isotropic plane plate of constant thickness, the effective elastic moduli are determined by Equations 19.93, the system of Equations 19.96 separates, and well-known equations of ordinary plate theory (cf. Ambartsumyan 1987) are obtained.

19.8. A laminated shell composed of homogeneous isotropic layers

The mathematical apparatus we developed previously can be successfully applied to the calculation of the effective moduli of a laminated shell composed of homogeneous isotropic layers. If we assume that the layers are perfectly bonded and are all parallel to the midsurface of the shell, all the variables involved will be independent of the coordinates y_1 and y_2 (or, equivalently, of ξ_1 and ξ_2) and the local problems of relevance will be solved in an elementary fashion.

The unit cell of the problem, as shown in Figure 19.4, is referred to the coordinate system (ξ_1, ξ_2, z) and is completely determined by the set of parameters $\delta_1, \delta_2, \ldots, \delta_M$, where M is the number of the layers. The thickness of the mth layer is, in these coordinates, $\delta_m - \delta_{m-1}$, with δ_0 and δ_M being zero and unity, respectively. The real thickness of the mth layer (referred to the original coordinate system $\alpha_1, \alpha_2, \gamma$) is $\delta(\delta_m - \delta_{m-1})$, and the total thickness of the packet is δ.

Because

$$z^{\pm} = \frac{1}{2} \qquad N_i^{\pm} = n_i^{(k)} = \{0, 0, 1\} \qquad L_{ijn} = c_{ijn3}\left(\frac{d}{dz}\right)$$

in this case, the first local problem (Equations 19.41, 19.55, 19.57, and 19.61 is transformed to

$$b_{ij}^{\lambda\mu} = c_{ijn3}\frac{dU_n^{\lambda\mu}}{dz} + c_{ij\lambda\mu}$$

$$\frac{db_{i3}^{\lambda\mu}}{dz} = 0 \qquad b_{i3}^{\lambda\mu} = 0 \qquad \left(z = \pm\frac{1}{2}\right)$$

(19.97)

$$[U_i^{\lambda\mu}] = 0 \qquad [b_{i3}^{\lambda\mu}] = 0 \qquad \left(z = \delta_m - \frac{1}{2};\right.$$

$$\left. m = 1, 2, \ldots, M - 1\right)$$

(19.98)

$$U_i^{\lambda\mu} = 0 \qquad\qquad (z = 0)$$

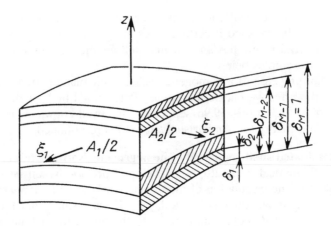

FIGURE 19.4. Laminated shell composed of M homogeneous layers.

and the corresponding form of the second local problem (Equations 19.53, 19.55, 19.57, and 19.61 differs from Equations 19.97 and 19.98 by having

$$b_{ij}^{*\lambda\mu} = c_{ijn3}\frac{dV_n^{\lambda\mu}}{dz} + zc_{ij\lambda\mu} \tag{19.99}$$

instead of Equations 19.97 and by the replacements

$$b_{i3}^{\lambda\mu} \rightarrow b_{i3}^{*\lambda\mu} \quad \text{and} \quad U_i^{\lambda\mu} \rightarrow V_i^{\lambda\mu}$$

everywhere in Equations 19.98. Because of the assumed isotropy, the elastic coefficients in the preceding equations are determined by means of Equation 19.91 in which the mth layer of the composite shell will be referenced by employing an index m on the quantities c, E, and ν.

From Equations 19.97 and 19.98, it follows that $b_{i3} = 0$ so that, under the isotropy assumption,

$$\frac{dU_1^{\lambda\mu(m)}}{dz} = \frac{dU_2^{\lambda\mu(m)}}{dz} = \frac{dU_3^{12(m)}}{dz} = 0$$

$$\frac{dU_3^{11(m)}}{dz} = \frac{dU_3^{22(m)}}{dz} = -\frac{\nu_m}{1-\nu_m} \quad (m = 1, 2, \ldots, M) \tag{19.100}$$

It will be understood that the condition at $z = 0$ and the interface conditions (Equations 19.98) with respect to the functions $U_i^{\lambda\mu}$ are satisfied by properly choosing the values of the constants arising from the integration of the Relations 19.100. According to the total number of layers, there are M such constants for each $\{i, \lambda, \mu\}$ set, and there are precisely M algebraic equations for their determination, of which $M - 1$ express interface conditions, and one expresses the condition at $z = 0$. Note, however, that the evaluation of the elastic moduli does not require the knowledge of these constants, because the quantities b, as given by Equations 19.97, only contain the derivatives of the functions $U_i^{\lambda\mu}$ and may therefore be determined by the use of Equations 19.100.

Substituting Equations 19.100 into Equations 19.97, we have, after averaging over the thickness,

$$\langle b_{11}^{11} \rangle = \langle b_{22}^{22} \rangle = \sum_{m=1}^{M} \frac{(\delta_m - \delta_{m-1})E_m}{1 - \nu_m^2}$$

$$\langle b_{11}^{22} \rangle = \sum_{m=1}^{M} \frac{(\delta_m - \delta_{m-1})\nu_m E_m}{1 - \nu_m^2} \tag{19.101}$$

$$\langle b_{12}^{12} \rangle = \sum_{m=1}^{M} \frac{(\delta_m - \delta_{m-1})E_m}{2(1 + \nu_m)}$$

$$\langle z b_{11}^{11} \rangle = \langle z b_{22}^{22} \rangle$$

$$= \frac{1}{2} \sum_{m=1}^{M} \frac{E_m}{1 - \nu_m^2} \left[\delta_m^2 - \delta_{m-1}^2 - (\delta_m - \delta_{m-1}) \right]$$

$$\langle z b_{11}^{22} \rangle = \langle z b_{22}^{11} \rangle \tag{19.102}$$

$$= \frac{1}{2} \sum_{m=1}^{M} \frac{\nu_m E_m}{1 - \nu_m^2} \left[\delta_m^2 - \delta_{m-1}^2 - (\delta_m - \delta_{m-1}) \right]$$

$$\langle z b_{12}^{12} \rangle = \frac{1}{2} \sum_{m=1}^{M} \frac{E_m}{2(1 + \nu_m)} \left[\delta_m^2 - \delta_{m-1}^2 - (\delta_m - \delta_{m-1}) \right]$$

For the second local problem, we find in an identical manner that

$$\frac{dV_1^{\lambda\mu(m)}}{dz} = \frac{dV_2^{\lambda\mu(m)}}{dz} = \frac{dV_3^{\lambda\mu(m)}}{dz} = 0$$

$$\frac{dV_3^{11(m)}}{dz} = \frac{dV_3^{22(m)}}{dz} = -\frac{\nu_m z}{1 - \nu_m} \qquad (m = 1, 2, \ldots, M) \tag{19.103}$$

which when substituted into Equation 19.99 gives

$$
\langle zb_{11}^{*11} \rangle = \langle zb_{22}^{*22} \rangle = \frac{1}{3} \sum_{m=1}^{M} \frac{E_m}{1 - \nu_m^2} \Bigg[\delta_m^3 - \delta_{m-1}^3 - \frac{3}{2}(\delta_m^2 - \delta_{m-1}^2)
$$

$$
+ \frac{3}{4}(\delta_m - \delta_{m-1}) \Bigg]
$$

$$
\langle zb_{11}^{*22} \rangle = \frac{1}{3} \sum_{m=1}^{M} \frac{\nu_m E_m}{1 - \nu_m^2} \Bigg[\delta_m^3 - \delta_{m-1}^3 - \frac{3}{2}(\delta_m^2 - \delta_{m-1}^2)
$$

$$(19.104)$$

$$
+ \frac{3}{4}(\delta_m - \delta_{m-1}) \Bigg]
$$

$$
\langle zb_{12}^{*12} \rangle = \frac{1}{3} \sum_{m=1}^{M} \frac{E_m}{2(1 + \nu_m)} \Bigg[\delta_m^3 - \delta_{m-1}^3 - \frac{3}{2}(\delta_m^2 - \delta_{m-1}^2)
$$

$$
+ \frac{3}{4}(\delta_m - \delta_{m-1}) \Bigg]
$$

after averaging over the thickness. Equations 19.101, 19.102, and 19.104, together with the very general Relations 19.75 and 19.76 proven earlier in this section, enable one to determine all the effective elastic moduli of the laminated shell under study.

Before concluding, we admit that the preceding equations do not actually contain any new information in them because they are totally identical in form to the corresponding laminated shell results based on the traditional nondeformable normal hypothesis (see Ambartsumyan 1974). In Section 20, application of the formalism to composite and stiffened shells of various kinds will lead us to results unachievable by traditional methods.

20. THERMAL CONDUCTIVITY OF A CURVED THIN SHELL OF A REGULARLY NONHOMOGENEOUS MATERIAL WITH CORRUGATED SURFACES

20.1. Orthogonal coordinate formulation

We consider here the problem of heat conductance for a thin, curved, regularly nonhomogeneous (i.e., composite) shell described in Section 19.

We assume that the physical components of the heat flow vector q_i and the temperature θ are related by Fourier's law (see, e.g., Sedov 1972,

Podstrigach et al. 1984), which in terms of the orthogonal coordinates α_1, α_2, and γ, takes the form

$$q_i = -\lambda_{i\mu} \frac{1}{H_\mu} \frac{\partial\theta}{\partial\alpha_\mu} - \lambda_{i3} \frac{\partial\theta}{\partial\gamma} \tag{20.1}$$

where λ_{ij} are the heat conductivity coefficients, and where Latin indices range from 1 to 3, and greek indices from 1 to 2.

In the approximation where the strain-related energy dissipation may be neglected, the heat balance equation can be written as

$$-f + c_v \frac{\partial\theta}{\partial t} = -\frac{1}{H_1 H_2} \left[\frac{\partial(H_2 q_1)}{\partial\alpha_1} + \frac{\partial(H_1 q_2)}{\partial\alpha_2} + \frac{\partial(H_1 H_2 q_3)}{\partial\gamma} \right] \tag{20.2}$$

f denoting the density of internal heat sources and c_v the bulk heat capacity.

The heat transfer conditions to be satisfied on the top and bottom faces of the shell, $S^\pm$, are taken in the form

$$n_\mu^\pm \frac{1}{H_\mu} q_\mu + n_3^\pm q_3 = \pm \alpha_S^\pm \theta \mp g_S^\pm \qquad (\gamma = \gamma^\pm) \tag{20.3}$$

where $\alpha^\pm$ and $g_S^\pm$ are, respectively, the heat transfer coefficients and external heat flows on the surfaces $S^\pm$, and the $n_i^\pm$ are the components of the unit vectors normal to these surfaces and are given by Equations 19.6. In the special case of convective heat transfer on the surfaces $S^\pm$ (heat transfer conditions of the third kind),

$$g_S^\pm = \alpha_S \theta_S^\pm \tag{20.4}$$

where $\theta_S^\pm$ is the value the ambient temperature takes on shell faces. If $\alpha_S^\pm = 0$, Equation 20.3 expresses the heat transfer conditions of the second kind.

The edge surface of the shell, which we shall denote by Σ, is a ruled surface whose directrix is the contour Γ bounding the shell midsurface and whose generatrix is the normal to the midsurface. The heat transfer conditions on this surface may be taken in the form

$$q_\mu n_\mu^\Gamma = \alpha_\Sigma \theta - g_\Sigma(\alpha, \gamma, t) \qquad (\alpha \in \Gamma) \tag{20.5}$$

which embraces the conditions of the second and third kinds and in which α_Σ is the heat transfer coefficient, g_Σ is the external heat flow, and n_μ^Γ is the component of the unit vector normal to Σ. Alternatively, a condition of the first kind may be imposed:

$$\theta|_\Sigma = \theta_\Sigma(\alpha, \gamma, t) \qquad (\alpha \in \Gamma) \tag{20.6}$$

Specification of the initial temperature distribution,

$$\theta|_{t=0} = \theta^* \tag{20.7}$$

completes the formulation of the heat conduction problem for the regularly nonhomogeneous (composite) shell with a rapidly oscillating thickness, it being assumed that all the thermal properties of the material, that is, the quantities $\lambda_{ij}(y_1, y_2, z)$, $c_v(y_1, y_2, z)$, and $\alpha_S(y_1, y_2)$, are periodic functions of y_1 and y_2 with a unit cell Ω. The external factors $f(\alpha, y, z, t)$, $g_S^\pm(\alpha, y, t)$, and $\theta^*(\alpha, y, z)$ may generally depend both on the "slow" coordinates α_1, α_2 and on $y = (y_1, y_2)$, with the same periodicity cell.

20.2. Asymptotic analysis of the problem

We solve the heat conduction problem given by Equation 20.1 through 20.7 by expressing the function θ in the form of a two-scale asymptotic expansion in powers of the small parameter δ:

$$\theta = \theta_1 + z\theta_2$$
$$\theta_\nu = \theta_\nu^{(0)}(\alpha, t) + \delta\theta_\nu^{(1)}(\alpha, t, y, z) + \delta^2\theta_\nu^{(2)}(\alpha, t, y, z) + \cdots \tag{20.8}$$

where the functions $\theta_\nu^{(l)}(\alpha, t, y, z)$ $(\nu = 1, 2, l = 1, 2)$ are periodic in y_1, y_2 with unit cell Ω. The leading term of this expansion,

$$\theta^{(0)} = \theta_1^{(0)}(\alpha, t) + z\theta_1^{(0)}(\alpha, t)$$

corresponds to the linear thickness coordinate dependence of the temperature, a commonly adopted assumption in the treatment of the heat conduction of plate and shell structures (Podstrigach and Shvets 1978, Podstrigach et al. 1984).

The asymptotic behavior of the midsurface curvatures of the shell (see Equations 19.9) may be conveniently rewritten in the form

$$k_\nu = \delta k_\nu'(\alpha) \qquad (\nu = 1, 2) \tag{20.9}$$

for the purposes of the present section, and in Conditions 20.3 we put

$$\alpha_S^\pm = \delta\alpha^\pm(y) \quad \text{and} \quad g_S^\pm = \delta g^\pm(\alpha, y, t) \tag{20.10}$$

It is useful to note—and it will be demonstrated later on in our discussion—that the asymptotic forms given by Equations 20.9 and 20.10

are equivalent to neglecting terms in $k_1\gamma$, $k_2\gamma$ in the derivation of thin-shell heat conduction equations and in the determination of reduced thermal characteristics of a thin-walled structure (see again Podstrigach and Shvets 1978, or Podstrigach et al. 1984).

In a manner similar to that discussed in Section 19, we now separate the rapid and slow coordinates for the purposes of differentiating, and we combine Equations 20.1 and 20.8 to obtain

$$q_i = \delta^{-1}q_i^{(-1)} + q_i^{(0)} + \delta q_i^{(1)} + \cdots \tag{20.11}$$

where

$$q_i^{(-1)} = -\lambda_{3i}\theta_2^{(0)}$$

$$q_i^{(l)} = -\lambda_{i\mu}\frac{1}{h_\mu A_\mu}\frac{\partial}{\partial y_\mu}(\theta_1^{(l+1)} + z\theta_2^{(l+1)}) \tag{20.12}$$

$$- \lambda_{i3}\frac{\partial}{\partial z}(\theta_1^{(l+1)} + z\theta_2^{(l+1)}) - \lambda_{i\mu}\frac{1}{A_\mu}\frac{\partial}{\partial\alpha_\mu}(\theta_1^{(l)} + z\theta_2^{(l)}) \quad (l = 0, 1)$$

We define now the differential operators

$$\partial_1\varphi = \frac{1}{A_1 A_2}\frac{\partial(A_2\varphi)}{\partial\alpha_1} \quad \text{and} \quad \partial_2\varphi = \frac{1}{A_1 A_2}\frac{\partial(A_1\varphi)}{\partial\alpha_2} \tag{20.13}$$

and proceed to consistently expanding Equation 20.2 in powers of δ. In doing so, the expressions for the Lamé coefficients H_1 and H_2, the Gauss–Codazzi relations (Equations 19.2), and Equations 20.8, 20.9, and 20.11 through 20.13 will be employed. For the right-hand side of Equation 20.2, the expansion in terms of δ^m $(m \leq 0)$ can be written as

$$\frac{1}{H_1 H_2}\left[\frac{\partial(H_2 q_1)}{\partial\alpha_1} + \frac{\partial(H_1 q_2)}{\partial\alpha_2} + \frac{\partial(H_1 H_2 q_3)}{\partial\gamma}\right]$$

$$= -\delta^{-1}\lambda_{3\mu}\partial_\mu\theta_2^{(0)} + \partial_\mu q_\mu^{(0)} + \delta^{-2}\frac{1}{h_\mu}\theta_2^{(0)}\frac{\partial\lambda_{3\mu}}{\partial\xi_\mu}$$

$$+ \delta^{-1}\frac{1}{h_\mu}\frac{\partial q_\mu^{(0)}}{\partial\xi_\mu} + \frac{1}{h_\mu}\frac{\partial q_\mu^{(1)}}{\partial\xi_\mu} + z\frac{k_\mu'\theta_2^{(0)}}{h_\mu}\frac{\partial\lambda_{3\mu}}{\partial\xi_\mu}$$

$$- \delta^{-2}\theta_2^{(0)}\frac{\partial\lambda_{33}}{\partial z} + \delta^{-1}\frac{\partial q_3^{(0)}}{\partial z} + \frac{\partial q_3^{(1)}}{\partial z} - (k_1' + k_2')\lambda_{33}\theta_2^{(0)}$$

Substituting this into Equation 20.2 and using Equations 20.8, we obtain the following expansion of Equation 20.2 in powers of δ^m for $m \leq 0$:

$$-f + c_v \left(\frac{\partial \theta_1^{(0)}}{\partial t} + z \frac{\partial \theta_2^{(0)}}{\partial t} \right) = \delta^{-1} \lambda_{3\mu} \partial_\mu \theta_2^{(0)} - \partial_\mu q_\mu^{(0)}$$

$$+ (k_1' + k_2') \lambda_{33} \theta_2^{(0)} - \frac{1}{h_\mu} \frac{\partial}{\partial \xi_\mu}$$

$$\times \left(\delta^{-1} q_\mu^{(0)} + q_\mu^{(1)} + \lambda_{3\mu} k_\mu' z \theta_2^{(0)} - \delta^{-2} \lambda_{3\mu} \theta_2^{(0)} \right)$$

$$- \frac{\partial}{\partial z} \left(\delta^{-1} q_3^{(0)} + q_3^{(1)} - \delta^{-2} \lambda_{33} \theta_2^{(0)} \right) \qquad (20.14)$$

To expand Conditions 20.3, we use Equations 19.21, 19.25, and 19.26 and Relations 20.8 through 20.12. Neglecting terms of the second and higher orders in δ, we find

$$n_\mu^\pm \frac{1}{H_\mu} q_\mu + n_3^\pm q_3 = \frac{N_i^\pm}{\omega^\pm} \left(-\delta^{-1} \lambda_{3i} \theta_2^{(0)} + q_i^{(0)} + \delta q_i^{(1)} \right)$$

$$+ \frac{N_\mu^\pm}{\omega^\pm} \delta \lambda_{3\mu} k_\mu' z^\pm \theta_2^{(0)} - \frac{N_i^\pm}{\omega^\pm} \delta \zeta^\pm \lambda_{3i} \theta_2^{(0)}$$

If one substitutes this into Conditions 20.3 and uses Equations 20.8 and 20.10, the expansion of the conditions at $z = z^\pm$ in powers of δ^m takes the form, for $m \leq 1$,

$$N_\mu^\pm \left(q_\mu^{(1)} + \delta^{-1} q_\mu^{(0)} + \lambda_{3\mu} k_\mu' z^\pm \theta_2^{(0)} - \delta^{-2} \lambda_{3\mu} \theta_2^{(0)} \right)$$

$$+ N_3^\pm \left(q_3^{(1)} + \delta^{-1} q_3^{(0)} - \delta^{-2} \lambda_{33} \theta_2^{(0)} \right)$$

$$= \pm \omega^\pm \left[\alpha^\pm (\theta_1^{(0)} + z^\pm \theta_2^{(0)}) - g^\pm \right] + N_i^\pm \zeta^\pm \lambda_{3i} \theta_2^{(0)} \qquad (z = z^\pm)$$

$$(20.15)$$

where the functions $\zeta^\pm$, $N_i^\pm$, $\omega^\pm$, are defined by Formulas 19.21, 19.25, and 19.26, respectively.

Now Equations 20.14 and 20.15 may be reduced to the problems

$$\frac{1}{h_\mu}\frac{\partial q_\mu^{(l)}}{\partial \xi_\mu} + \frac{\partial q_3^{(l)}}{\partial z} = 0$$

$$N_i^\pm q_i^{(l)} = 0 \qquad (z = z^\pm, l = 0,1) \qquad (20.16)$$

$$-f + c_v\left(\frac{\partial \theta_1^{(0)}}{\partial t} + z\frac{\partial \theta_2^{(0)}}{\partial t}\right)$$

$$= \delta^{-1}\lambda_{3\mu}\partial_\mu\theta_2^{(0)} - \partial_\mu q_\mu^{(0)} + (k_1' + k_2')\lambda_{33}\theta_2^{(0)}$$

$$+ \frac{1}{h_\mu}\frac{\partial}{\partial \xi_\mu}\left(\delta^{-2}\lambda_{3\mu}\theta_2^{(0)} - \lambda_{3\mu}k_\mu' z\theta_2^{(0)}\right) + \frac{\partial}{\partial z}(\delta^{-2}\lambda_{33}\theta_2^{(0)})$$

$$N_\mu^\pm\left(\delta^{-2}\lambda_{3\mu}\theta_2^{(0)} - \lambda_{3\mu}k_\mu' z^\pm\theta_2^{(0)}\right) + N_3^\pm\delta^{-2}\lambda_{33}\theta_2^{(0)}$$

$$= \mp\omega^\pm\left[\alpha^\pm(\theta_1^{(0)} + z^\pm\theta_2^{(0)}) - g^\pm\right] - N_i^\pm\lambda_{3i}\zeta^\pm\theta_2^{(0)} \qquad (z = z^\pm)$$

$$(20.17)$$

Let us introduce the notation

$$L_i = \frac{\lambda_{i\nu}}{h_\nu}\frac{\partial}{\partial \xi_\nu} + \lambda_{i3}\frac{\partial}{\partial z}$$

$$D = \frac{1}{h_\mu}\frac{\partial}{\partial \xi_\mu}L_\mu + \frac{\partial}{\partial z}L_3 \qquad (20.18)$$

and

$$\Lambda_i = \frac{1}{h_\mu}\frac{\partial\lambda_{i\mu}}{\partial \xi_\mu} + \frac{\partial\lambda_{i3}}{\partial z} \qquad (20.19)$$

Taking $l = 0$ in Equations 20.16 and using $q_i^{(0)}$ from Equations 20.12, we then obtain

$$D(\theta_1^{(1)} + z\theta_2^{(1)}) = -\Lambda_\mu\frac{1}{A_\mu}\frac{\partial\theta_1^{(0)}}{\partial\alpha_\mu} - (\lambda_{3\mu} + z\Lambda_\mu)\frac{1}{A_\mu}\frac{\partial\theta_2^{(0)}}{\partial\alpha_\mu}$$

$$N_i^\pm\left[L_i(\theta_1^{(1)} + z^\pm\theta_2^{(1)}) + \frac{\lambda_{i\mu}}{A_\mu}\left(\frac{\partial\theta_1^{(0)}}{\partial\alpha_\mu} + z^\pm\frac{\partial\theta_2^{(0)}}{\partial\alpha_\mu}\right)\right] = 0$$

$$(z = z^\pm) \quad (20.20)$$

We may take a solution of Equations 20.20 in the form

$$\theta_1^{(1)} = W_\mu(\xi, z)\frac{1}{A_\mu}\frac{\partial \theta_1^{(0)}}{\partial \alpha_\mu}$$

$$z\theta_2^{(1)} = W_\mu^*(\xi, z)\frac{1}{A_\mu}\frac{\partial \theta_2^{(0)}}{\partial \alpha_\mu} \tag{20.21}$$

provided the functions $W_\mu(\xi, z)$ and $W_\mu^*(\xi, z)$ are periodic in ξ_1 and ξ_2 (with respective periods A_1 and A_2) and solve the problems

$$DW_\mu = -\Lambda_\mu$$

$$N_i^\pm(L_iW_\mu + \lambda_{i\mu}) = 0 \qquad (z = z^\pm) \tag{20.22}$$

and

$$DW_\mu^* = -(\lambda_{3\mu} + z\Lambda_\mu)$$

$$N_i^\pm\left(L_iW_\mu^* + z^\pm\lambda_{i\mu}\right) = 0 \qquad (z = z^\pm) \tag{20.23}$$

Let

$$l_{i\mu}(\xi, z) = L_iW_\mu + \lambda_{i\mu} \quad \text{and} \quad l_{i\mu}^*(\xi, z) = L_iW_\mu^* + z\lambda_{i\mu} \tag{20.24}$$

Then, from Equations 20.12 with $l = 0$, using Equations 20.21, we find that

$$q_i^{(0)} = -l_{i\mu}\frac{1}{A_\mu}\frac{\partial \theta_1^{(0)}}{\partial \alpha_\mu} - l_{i\mu}^*\frac{1}{A_\mu}\frac{\partial \theta_2^{(0)}}{\partial \alpha_\mu} \tag{20.25}$$

Using the notation introduced in Equations 20.18, 20.19, and 20.24, the problems given by Equations 20.22 and 20.23 may be rewritten as

$$\frac{1}{h_\beta}\frac{\partial}{\partial \xi_\beta}l_{\beta\mu} + \frac{\partial}{\partial z}l_{3\mu} = 0$$

$$N_i^\pm l_{i\mu} = 0 \qquad (z = z^\pm) \tag{20.26}$$

$$\frac{1}{h_\beta}\frac{\partial}{\partial \xi_\beta}l_{\beta\mu}^* + \frac{\partial}{\partial z}l_{3\mu}^* = 0$$

$$N_i^\pm l_{i\mu}^* = 0 \qquad (z = z^\pm) \tag{20.27}$$

Equations 20.26 and 20.27 are the local problems of heat conduction theory. They are very similar in form to the local problems of the theory of elasticity, Equations 19.55 and 19.56, and their solutions are unique up to a constant term, an ambiguity that is removed by imposing the conditions

$$\langle W_\mu \rangle_\xi = 0 \qquad \langle W_\mu^* \rangle_\xi = 0 \quad \text{for } z = 0 \qquad (\mu = 1, 2) \qquad (20.28)$$

analogous to those given by (19.57). Together with Equations 20.8 and 20.21, these conditions elucidate the meaning of the function $\theta_1^{(0)}$, which characterizes the distribution of temperature over the midsurface of the shell, $z = 0$.

The conditions for the faces $z = z^\pm$ involved in the local problems given by Equations 20.26 and 20.27 may be represented in the same form as Equations 19.58 and 19.59, namely,

$$\frac{1}{h_\beta} n_\beta^{\pm(\xi)} l_{\beta\mu} + n_3^{\pm(\xi)} l_{3\mu} = 0 \qquad z = z^\pm \qquad (l_{i\mu} \leftrightarrow l_{i\mu}^*) \quad (20.29)$$

As was the case in Section 19, the local heat conduction problems may also be generalized to the case of piecewise-smooth heat conduction coefficients $\lambda_{ij}(y, z)$, with discontinuities of the first kind on the (noninter-secting) interfaces between dissimilar phases of the composite. The formulations of the local problems are then augmented by the conditions of continuity on the interfaces,

$$[W_\mu] = 0 \qquad \left[\frac{1}{h_\beta} n_\beta^{(k)} l_{\beta\mu} + n_3^{(k)} l_{3\mu} \right] = 0 \qquad (20.30)$$

$$[W_\mu^*] = 0 \qquad \left[\frac{1}{h_\beta} n_\beta^{(k)} l_{\beta\mu}^* + n_3^{(k)} l_{3\mu}^* \right] = 0 \qquad (20.31)$$

which are analogous to Conditions 19.60 and 19.61 and correspond to an ideal thermal contact in the sense that both the temperature and the heat flow vector have no jumps at the surface of a contact.

20.3. Governing equations for the conduction of the homogenized shell

To derive homogenized heat conduction equations use will be made of Equations 19.28, which we proved earlier. We begin by averaging Equation 20.17, using Conditions 20.17 at $z = z^\pm$ in doing so. Multiplying by δ and employing the notation described in Equations 20.9 and 20.10, we find the first of the governing equations of heat conduction for the homogenized shell,

$$-\delta\langle f\rangle + \delta\langle c_v\rangle\frac{\partial\theta_1^{(0)}}{\partial t} + \delta\langle zc_v\rangle\frac{\partial\theta_2^{(0)}}{\partial t}$$

$$= \langle\lambda_{3\mu}\rangle\partial_\mu\theta_2^{(0)} - \delta\partial_\mu\langle q_\mu^{(0)}\rangle - \mathscr{I}_0\theta_1^{(0)}$$

$$-[\mathscr{I}_1 - (k_1 + k_2)\langle\lambda_{33}\rangle + \delta Z_0]\theta_2^{(0)} + G_0 \qquad (20.32)$$

where

$$\mathscr{I}_m = \int_{-1/2}^{1/2}\int_{-1/2}^{1/2}\left[(z^+)^m\omega^+\alpha_S^+ + (z^-)^m\omega^-\alpha_S^-\right]dy_1\,dy_2$$

$$(m = 0, 1, 2)$$

$$G_m = \int_{-1/2}^{1/2}\int_{-1/2}^{1/2}\left[(z^+)^m\omega^+g_S^+ + (z^-)^m\omega^-g_S^-\right]dy_1\,dy_2$$

$$(m = 0, 1)$$

$$Z_m = \int_{-1/2}^{1/2}\int_{-1/2}^{1/2}\left[(z^+)^m N_i^+\lambda_{3i}(y, z^+)\zeta^+\right.$$

$$\left.-(z^-)^m N_i^-\lambda_{3i}(y, z^-)\zeta^-\right]dy_1\,dy_2 \qquad (20.33)$$

The second equation, one for the functions $\theta_1^{(0)}(\alpha, t)$ and $\theta_2^{(0)}(\alpha, t)$, may be derived by averaging Conditions 20.17 after first multiplying by z. Multiplying by z Conditions 20.17 at $z = z^\pm$, one obtains, using Equation 19.28,

$$-\delta\langle zf\rangle + \delta\langle zc_v\rangle\frac{\partial\theta_1^{(0)}}{\partial t} + \delta\langle z^2c_v\rangle\frac{\partial\theta_2^{(0)}}{\partial t}$$

$$= \langle z\lambda_{3\mu}\rangle\partial_\mu\theta_2^{(0)} - \delta\partial_\mu\langle zq_\mu^{(0)}\rangle - \mathscr{I}_1\theta_1^{(0)}$$

$$-[\mathscr{I}_2 - (k_1 + k_2)\langle z\lambda_{33}\rangle + \delta^{-1}\langle\lambda_{33}\rangle + \delta Z_1]\theta_2^{(0)} + G_1 \qquad (20.34)$$

The components $\langle q_\mu^{(0)} \rangle$ and $\langle z q_\mu^{(0)} \rangle$ involved in Equations 20.32 and 20.34 can be evaluated by averaging Relations 20.35 to give

$$\langle q_\mu^{(0)} \rangle = -\langle l_{\mu\nu} \rangle \frac{1}{A_\nu} \frac{\partial \theta_1^{(0)}}{\partial \alpha_\nu} - \langle l_{\mu\nu}^* \rangle \frac{1}{A_\nu} \frac{\partial \theta_2^{(0)}}{\partial \alpha_\nu}$$

$$ \tag{20.35}$$

$$\langle z q_\mu^{(0)} \rangle = -\langle z l_{\mu\nu} \rangle \frac{1}{A_\nu} \frac{\partial \theta_1^{(0)}}{\partial \alpha_\nu} - \langle z l_{\mu\nu}^* \rangle \frac{1}{A_\nu} \frac{\partial \theta_2^{(0)}}{\partial \alpha_\nu}$$

Using Equations 20.26 and 20.27, it is readily proved that

$$\langle l_{3\mu} \rangle = \langle z l_{3\mu} \rangle = \langle l_{3\mu}^* \rangle = \langle z l_{3\mu}^* \rangle = 0 \tag{20.36}$$

and therefore by Equation 20.25

$$\langle q_3^{(0)} \rangle = \langle z q_3^{(0)} \rangle = 0$$

Substituting Equations 20.35 into Equations 20.32 and 20.34 now gives a system of two governing equations for $\theta_1^{(0)}(\alpha, t)$ and $\theta_2^{(0)}(\alpha, t)$, which functions, together with the solutions of the local problems given by Equations 20.26 through 20.31, enable one to estimate quite closely the three-dimensional local structure of the temperature and heat flow fields of the problem. The necessary formulas follow from Equations 20.8 and 20.11, 20.12, 20.21, and 20.25 as

$$\theta = \theta_1^{(0)} + z \theta_2^{(0)} + \delta \left(W_\mu \frac{1}{A_\mu} \frac{\partial \theta_1^{(0)}}{\partial \alpha_\mu} + W_\mu^* \frac{1}{A_\mu} \frac{\partial \theta_2^{(0)}}{\partial \alpha_\mu} \right) + \cdots$$

$$ \tag{20.37}$$

$$q_i = -\delta^{-1} \lambda_{i3} \theta_2^{(0)} - l_{i\mu} \frac{1}{A_\mu} \frac{\partial \theta_1^{(0)}}{\partial \alpha_\mu} - l_{i\mu}^* \frac{1}{A_\mu} \frac{\partial \theta_2^{(0)}}{\partial \alpha_\mu} + \cdots$$

The functions $\theta_1^{(2)}$ and $\theta_2^{(2)}$ are determined from the higher-order local problems that result from combining the problem given by Equations 20.16 for $l = 1$ with the expression for $q_i^{(1)}$ from Equations 20.12. These problems will not be considered here, however, because the accuracy provided by Equations 20.37 is quite enough for our purposes.

We next turn our attention to the initial and boundary conditions at the contour Γ. To derive them, we average Equations 20.5 through 20.7 and retain only the leading terms in the expansions for the temperature and the heat flow components, Equations 20.37. In the case of boundary equations of the second or third kind, we see from Equation 20.5 that

$$\left(-\delta^{-1}\langle z^m \lambda_{\mu 3}\rangle \theta_2^{(0)} + \langle z^m q_\mu^{(0)}\rangle\right) n_\mu^\Gamma$$

$$= \langle z^m \alpha_\Sigma \rangle \theta_1^{(0)} + \langle z^{m+1} \alpha_\Sigma \rangle \theta_2^{(0)} - \langle z^m g_\Sigma \rangle \qquad (\alpha \in \Gamma, m = 0, 1)$$

$$(20.38)$$

and for boundary conditions of the first kind we find

$$\langle z^m \rangle \theta_1^{(0)} + \langle z^{m+1}\rangle \theta_2^{(0)} = \langle z^m \theta_\Sigma \rangle \qquad \alpha \in \Gamma, (m = 0, 1) \quad (20.39)$$

using Equation 20.6. The initial conditions for the functions $\theta_1^{(0)}$ and $\theta_2^{(0)}$ follow from Equation 20.7 as

$$(\langle z^m \rangle \theta_1^{(0)} + \langle z^{m+1}\rangle \theta_2^{(0)})|_{t=0} = \langle z^m \theta * \rangle \qquad (m = 0, 1) \quad (20.40)$$

20.4. Symmetry of the effective heat conduction properties

By the effective heat conduction properties we mean here the quantities $\langle l_{\mu\nu}\rangle$, $\langle zl_{\mu\nu}\rangle$, $\langle l_{\mu\nu}^*\rangle$, and $\langle zl_{\mu\nu}^*\rangle$ ($\mu, \nu = 1, 2$), which enter as coefficients in Equations 20.35 and are determined from two local problems—one of which is posed by Equations 20.24, 20.26, 20.28, and 20.30, and the other by 20.24, 20.27, 20.28, and 20.31. Note that, quite analogously to Section 19, these problems involve the coefficients $A_1(\alpha)$ and $A_2(\alpha)$ of the first quadratic form of the shell midsurface, and the effective properties may therefore depend on the slow coordinates α_1 and α_2 even in the case of an originally homogeneous material. Our object here is to prove that

$$\langle l_{\mu\nu}\rangle = \langle l_{\nu\mu}\rangle \qquad \langle zl_{\mu\nu}\rangle = \langle l_{\nu\mu}^*\rangle \qquad \langle zl_{\mu\nu}^*\rangle = \langle zl_{\nu\mu}\rangle \quad (20.41)$$

From Equations 20.18 and 20.19, using the auxiliary coordinates ζ_i defined in Section 19, we find

$$L_i = \lambda_{ir}\frac{\partial}{\partial \zeta_r} \qquad D = \frac{\partial}{\partial \zeta_p}\lambda_{pr}\frac{\partial}{\partial \zeta_r} \qquad \Lambda_i = \frac{\partial \lambda_{ir}}{\partial \zeta_r}$$

Now if we redefine $l_{i\mu}$ and $l_{i\mu}^*$ for the entire range of indices, that is, for 1, 2, 3, we obtain

$$l_{ij} = \lambda_{ir}\left(\frac{\partial W_j}{\partial \zeta_r} + \delta_{rj}\right)$$

$$(20.42)$$

$$l_{ij}^* = \lambda_{ir}\left(\frac{\partial W_j}{\partial \zeta_r} + z\delta_{rj}\right)$$

from Equations 20.24, and the local problems given by Equations 20.26 and 20.27 can be rewritten as

$$\frac{\partial l_{pj}}{\partial \zeta_p} = 0 \qquad N_p^{\pm} l_{pj} = 0 \qquad (z = z^{\pm}) \qquad l_{pj} \leftrightarrow l_{pj}^* \qquad (20.43)$$

which, combined with Equation 19.28, leads to the results

$$\left\langle l_{pj} \frac{\partial \varphi}{\partial \zeta_p} \right\rangle = 0 \quad \text{and} \quad \left\langle l_{pj}^* \frac{\partial \varphi}{\partial \zeta_p} \right\rangle = 0 \qquad (20.44)$$

which hold for any function $\varphi(\alpha_1, \alpha_2, \alpha_3)$ periodic in α_1 and α_2 with unit cell Ω.

Setting $\varphi = W_i$ and then $\varphi = W_i^*$ in Equations 20.44 and using Equations 20.43, it is readily verified that

$$\langle l_{ij} \rangle = \left\langle \lambda_{pr} \left(\frac{\partial W_i}{\partial \zeta_p} + \delta_{pi} \right) \left(\frac{\partial W_j}{\partial \zeta_r} + \delta_{rj} \right) \right\rangle$$

$$\langle z l_{ij}^* \rangle = \left\langle \lambda_{pr} \left(\frac{\partial W_i^*}{\partial \zeta_p} + z \delta_{pi} \right) \left(\frac{\partial W_j^*}{\partial \zeta_r} + z \delta_{rj} \right) \right\rangle$$

$$(20.45)$$

$$\langle z l_{ij} \rangle = \left\langle \lambda_{pr} \left(\frac{\partial W_j}{\partial \zeta_r} + \delta_{rj} \right) \left(\frac{\partial W_i^*}{\partial \zeta_p} + z \delta_{pi} \right) \right\rangle$$

$$\langle l_{ij}' \rangle = \left\langle \lambda_{pr} \left(\frac{\partial W_i}{\partial \zeta_p} + \delta_{pi} \right) \left(\frac{\partial W_j^*}{\partial \zeta_r} + z \delta_{rj} \right) \right\rangle$$

and the symmetry properties given by Equations 20.41 now immediately follow from Equations 20.45 and the symmetry of the heat conductivity tensor λ_{pr}. If one or more of the indices are equal to 3, it is seen that both sides of Equations 20.45 turn to zero in view of Equation 20.36.

20.5. Homogenized heat conduction problem versus thin-walled model results

In this section we apply the preceding analysis to the limiting case of a homogeneous shell of constant thickness, which implies that the quantities λ_{ij}, c_v, and $\alpha^{\pm}$ are constant and $F^{\pm}$ is identically zero. Because there is no dependence on ξ_1 and ξ_2 in this case, the local problems are solved without difficulty, giving

$$W_\mu = -z\frac{\lambda_{3\mu}}{\lambda_{33}} \quad \text{and} \quad W_\mu^* = -\frac{z^2}{2}\frac{\lambda_{3\mu}}{\lambda_{33}} \tag{20.46}$$

and substitution into Equation 20.24 results in

$$l_{\mu\nu} = \lambda_{\mu\nu} - \frac{\lambda_{3\mu}\lambda_{3\nu}}{\lambda_{33}} \quad \text{and} \quad l_{\mu\nu}^* = zl_{\mu\nu} \tag{20.47}$$

Now because $z^\pm = \pm(\tfrac{1}{2})$ in this particular case, it follows from Equations 20.35 that

$$\langle q_\mu^{(0)} \rangle = -l_{\mu\nu}\frac{1}{A_\nu}\frac{\partial\theta_1^{(0)}}{\partial\alpha_\nu}$$

$$\langle zq_\mu^{(0)} \rangle = -\frac{l_{\mu\nu}}{12}\frac{1}{A_\nu}\frac{\partial\theta_2^{(0)}}{\partial\alpha_\nu} \tag{20.48}$$

Using Equations 20.33, 20.47, and 20.48, Equations 20.32 and 20.34 can be written in the form

$$-\delta\langle f \rangle + \delta c_\nu\frac{\partial\theta_1^{(0)}}{\partial t} = \lambda_{3\mu}\partial_\mu\theta_2^{(0)} + \delta\left(\lambda_{\mu\nu} - \frac{\lambda_{3\mu}\lambda_{3\nu}}{\lambda_{33}}\right)\partial_\mu$$

$$\times \left(\frac{1}{A_\nu}\frac{\partial\theta_1^{(0)}}{\partial\alpha_\nu}\right) - (\alpha_S^+ + \alpha_S^-)\theta_1^{(0)}$$

$$-\left[\frac{\alpha_S^+ - \alpha_S^-}{2} - (k_1 + k_2)\lambda_{33}\right]\theta_2^{(0)}$$

$$+ g_S^+ + g_S^-$$

$$-12\delta\langle zf \rangle + \delta c_\nu\frac{\partial\theta_2^{(0)}}{\partial t} = \delta\left(\lambda_{\mu\nu} - \frac{\lambda_{3\mu}\lambda_{3\nu}}{\lambda_{33}}\right)\partial_\mu\left(\frac{1}{A_\nu}\frac{\partial\theta_2^{(0)}}{\partial\alpha_\nu}\right)$$

$$- 6(\alpha_S^+ - \alpha_S^-)\theta_1^{(0)} - 3(\alpha_S^+ + \alpha_S^- + 48^{-1}\lambda_{33})\theta_2^{(0)}$$

$$+ \frac{g_S^+ - g_S^-}{2}.7 \tag{20.49}$$

which is the system of equations for the heat conduction problem of an anisotropic homogeneous shell (or plate). The quantities $g_s^{\pm}$ involved in Equations 20.49 are determined by Equation 20.4 if heat transfer proceeds by convection across the faces of the shell.

Returning to Equation 20.37, we see that the leading term in the temperature equation,

$$\theta \approx \theta_1^{(0)}(\alpha, t) + \frac{\gamma}{\delta} \theta_2^{(0)}(\alpha, t) \tag{20.50}$$

describes a linear dependence of temperature on the thickness coordinate, and we may use this dependence to calculate the integral temperature characteristics usually employed in the analysis of heat conduction in thin plates and shells. From Podstrigach and Shvets (1978) (see also Podstrigach et al. 1984), these characteristics are

$$T = \frac{1}{\delta} \int_{-\delta/2}^{\delta/2} \theta \, d\gamma = \theta_1^{(0)} \quad \text{and} \quad T^* = \frac{6}{\delta^2} \int_{-\delta/2}^{\delta/2} \gamma \theta \, d\gamma = \frac{\theta_2^{(0)}}{2} \tag{20.51}$$

and their substitution into Equation 20.50 gives the relation between the temperature and its integral characteristics,

$$\theta \approx T + \frac{2\gamma}{\delta} T^*$$

which is well known from the heat conduction theory of thin shells. Referring to the same authors, similar integral expressions for the density of heat sources and for its first moment may be constructed,

$$\Phi = \int_{-\delta/2}^{\delta/2} f \, d\gamma = \delta \langle f \rangle$$

$$\Phi^* = \frac{6}{\delta} \int_{-\delta/2}^{\delta/2} \gamma f \, d\gamma = 6\delta \langle zf \rangle \tag{20.52}$$

Equations 20.50 through 20.52 thus show that the system of Equations 20.49 coincides with the analogous heat conduction equations known for the cases when (a) the shell is homogeneous and isotropic ($\lambda_{ij} = \lambda \delta_{ij}$, see Podstrigach and Shvets 1978) or when (b) the shell is anisotropic, but there exists a thermal symmetry plane at each point of it that is parallel to the plane of the shell ($\lambda_{31} = \lambda_{32} = 0$, $k_1 = k_2 = 0$, $A_1 = A_2 = 1$, see Podstrigachet al. 1984).

20.6. Laminated shell composed of homogeneous anisotropic layers

In this section the effective heat conduction coefficients of the laminated shell of Figure 19.4 will be evaluated under the assumption that the layers (or laminae) of the shell are anisotropic while being homogeneous.

Of the two local problems that must be considered, the first—expressed by Equations 20.24, 20.26, 20.28, and 20.30—takes the form

$$l_{i\mu} = \lambda_{i3}\frac{dW_\mu}{dz} + \lambda_{i\mu} \tag{20.53}$$

$$\frac{dl_{3\mu}}{dz} = 0 \qquad l_{3\mu} = 0 \qquad \left(z = \pm\frac{1}{2}\right)$$

$$[W_\mu] = 0 \qquad [l_{3\mu}] = 0 \qquad \left(z = \delta_m - \frac{1}{2}\right),$$

$$m = 1, 2, \ldots, M-1 \Big) \tag{20.54}$$

$$W_\mu = 0 \qquad\qquad\qquad (z = 0)$$

in this case. The relevant form of the second local problem—Equations 20.24, 20.27, 20.28, and 20.31—is obtained from Equations 20.53 and 20.54 by writing

$$l^*_{i\mu} = \lambda_{i3}\frac{dW^*_\mu}{dz} + z\lambda_{i\mu} \tag{20.55}$$

in place of Equations 20.53 and making the replacements

$$l_{3\mu} \rightarrow l^*_{3\mu} \quad \text{and} \quad W_\mu \rightarrow W^*_\mu$$

in Equations 20.54.

Because $l_{3\mu} = 0$ and $l^*_{3\mu} = 0$ from the solution of the local problems, it follows that

$$\frac{dW^{(m)}_\mu}{dz} = -\frac{\lambda^{(m)}_{3\mu}}{\lambda^{(m)}_{33}} \quad \text{and} \quad \frac{dW^{*(m)}_\mu}{dz} = -z\frac{\lambda^{(m)}_{3\mu}}{\lambda^{(m)}_{33}} \tag{20.56}$$

with superscript m indicating the layer. Analogously to Section 19, here again the interface conditions and $z = 0$ conditions are satisfied by the appropriate choice of the constants of integration in Equations 20.56.

The effective heat conduction coefficients of the laminated shell are obtained by substituting Equations 20.56 into Equations 20.53 and 20.55 and then averaging over the shell thickness to give

$$\langle l_{\beta\mu} \rangle = \sum_{m=1}^{M} \left(\lambda_{\beta\mu}^{(m)} - \frac{\lambda_{3\beta}^{(m)}\lambda_{3\mu}^{(m)}}{\lambda_{33}^{(m)}} \right) (\delta_m - \delta_{m-1})$$

$$\langle z l_{\beta\mu} \rangle = \langle l_{\beta\mu}^* \rangle$$

$$= \frac{1}{2} \sum_{m=1}^{M} \left(\lambda_{\beta\mu}^{(m)} - \frac{\lambda_{3\beta}^{(m)}\lambda_{3\mu}^{(m)}}{\lambda_{33}^{(m)}} \right) \left[\delta_m^2 - \delta_{m-1}^2 - (\delta_m - \delta_{m-1}) \right] \quad (20.57)$$

$$\langle z l_{\beta\mu}^* \rangle = \frac{1}{3} \sum_{m=1}^{M} \left(\lambda_{\beta\mu}^{(m)} - \frac{\lambda_{3\beta}^{(m)}\lambda_{3\mu}^{(m)}}{\lambda_{33}^{(m)}} \right)$$

$$\times \left[\delta_m^3 - \delta_{m-1}^3 - \frac{3}{2}(\delta_m^2 - \delta_{m-1}^2) + \frac{3}{4}(\delta_m - \delta_{m-1}) \right]$$

If the layer material is isotropic, then $\lambda_{ij}^{(m)} = \lambda^{(m)}\delta_{ij}$, with resulting simplifications in Equations 20.57 (in particular, $\lambda_3^{(m)} = 0$).

21. THERMOELASTICITY OF A CURVED SHELL OF REGULARLY NONHOMOGENEOUS MATERIAL WITH CORRUGATED SURFACES

21.1. Governing thermoelastic equations for the homogenized shell

In Sections 19 and 20, we derived an asymptotic formalism necessary for treating the elasticity and heat conduction problems associated with a thin, curved, regularly nonhomogeneous (or composite) shell with rapidly oscillating thickness. With the results obtained, we are now in a position to consider the problem of thermoelasticity for such a structure.

As is customary in the theory of thermoelasticity, the stresses and temperature changes will be taken to be related by the usual Duhamel–Neumann law,

$$\sigma_{ij} = c_{ijmn}(e_{mn} - \alpha_{mn}^T \theta) \quad (21.1)$$

where α_{mn}^T are the thermal elongation and shear coefficients. We assume that the thermal relaxation time is much greater than the attenuation time associated with mechanical oscillations, and we accordingly limit our consideration to the quasistatic formulation of the problem. This means that inertial effects may be neglected in the equations of motion, and these reduce therefore to the equilibrium equations (Equations 19.3), with time as a parameter.

If the thermoelastic problem is an uncoupled one (which is the case we consider), the heat conduction problem set by Equations 20.1 through 20.7 is solved independently. Its solution, the temperature θ, then enters into Equations 19.3 by means of Equation 21.1 and determines the additional external forces caused by thermal deformation effects.

Because there are rapidly oscillating components in the coefficients in Equation 21.1, clearly a homogenization technique must be applied here, along lines similar to those discussed in Section 19, but with Equation 21.1 instead of Hooke's law (Equation 19.4).

As will be seen in the following discussion, the stress equations, Equations 19.64 and 19.70, are not affected by the thermal term in Equation 21.1, but in the elastic relations, Equations 19.71 and 19.72, additional terms associated with thermal deformations will appear. These are expressible in terms of $\theta_1^{(0)}(\alpha, t)$ and $\theta_2^{(0)}(\alpha, t)$, for which functions the homogenized problem given by Equations 20.32 through 20.35 and Equations 20.38 through 20.40 was formulated in the previous section.

Assuming the asymptotic form

$$\alpha_{mn}^T = \delta\alpha_{mn}(y, z) \tag{21.2}$$

where the functions $\alpha_{mn}(y, z)$ are periodic in y_1, y_2 with unit cell Ω, and using Equations 19.13, 19.14, and 20.8 in combination with Equation 21.1 we obtain for $\sigma_{ij}^{(0)}$ an expression identical in form with Equation 19.31. Proceeding in a manner parallel to that described in Section 19, we also retrieve Equations 19.45 for the leading-order terms of displacement components, and Relations 19.46 for the stresses $\sigma_{ij}^{(0)}$.

From Equations 21.1, 21.2, 19.15, and 20.8, it is found that

$$\sigma_{ij}^{(1)} = L_{ijm}u_m^{(2)} + c_{ijm\nu}\varepsilon_{m\nu}^{(1)} + zc_{ij\mu\nu}\tau_{\mu\nu} - c_{ijmn}\alpha_{mn}(\theta_1^{(0)} + z\theta_2^{(0)}) \tag{21.3}$$

which differs from Equation 19.47 in having terms corresponding to thermal stresses. As before, the functions $\varepsilon_m^{(1)}$ and $\tau_{\mu\nu}$ are determined by Equations 19.48 and 19.49.

Now substitute Equation 21.3 into Equation 19.22 and into Conditions 19.22 with $l = 1$. Using Equation 19.46 and following the notation introduced in Equations 19.33 through 19.35, we arrive at

$$D_{im}u_m^{(2)} = -C_{im\nu}\varepsilon_{m\nu}^{(1)} - (c_{i3\mu\nu} + zC_{i\mu\nu})\tau_{\mu\nu}$$

$$+ B_i\theta_1^{(0)} + (\beta_{i3} + zB_i)\theta_2^{(0)}$$

$$N_j^\pm\Big[L_{ijm}u_m^{(2)} + c_{ijm\nu}\varepsilon_{m\nu}^{(1)} + z^\pm c_{ij\mu\nu}\tau_{\mu\nu}$$

$$- \beta_{ij}(\theta_1^{(0)} + z^\pm\theta_2^{(0)})\Big] = 0 \quad (z = z^\pm)$$

(21.4)

defining

$$\beta_{ij} = c_{ijmn}\alpha_{mn} \quad \text{and} \quad B_i = \frac{1}{h_\nu}\frac{\partial\beta_{i\nu}}{\partial\xi_\nu} + \frac{\partial\beta_{i3}}{\partial z}$$

(21.5)

We shall have satisfied Equations 21.4 and secured periodicity in ξ_1 and ξ_2 (with respective periods A_1 and A_2) by writing

$$u_m^{(2)} = U_m^{l\nu}\varepsilon_{l\nu}^{(1)} + V_m^{\mu\nu}\tau_{\mu\nu} + S_m\theta_1^{(0)} + S_m^*\theta_2^{(0)}$$

(21.6)

for the solution, where $U_m^{l\nu}(\xi, z)$ and $V_m^{\mu\nu}(\xi, z)$ are the solutions of the local elastic problems given by Equations 19.39 and 19.52, and the functions $S_m(\xi, z)$ and $S_m^*(\xi, z)$ are periodic in ξ_1 and ξ_2 (with periods indicated) and solve the problems

$$D_{im}S_m = B_i$$

$$(L_{ijm}S_m - \beta_{ij})N_j^\pm = 0 \quad (z = z^\pm)$$

(21.7)

and

$$D_{im}S_m^* = \beta_{i3} + zB_i$$

$$\Big(L_{ijm}S_m^* - z^\pm\beta_{ij}\Big)N_j^\pm = 0 \quad (z = z^\pm)$$

(21.8)

Now from Equation 21.3, denoting

$$s_{ij} = \beta_{ij} - L_{ijm}S_m \quad \text{and} \quad s_{ij}^* = z\beta_{ij} - L_{ijm}S_m^*$$

(21.9)

and using Equations 21.6 and 21.9 along with Equations 19.41, 19.42, and 19.53, one obtains

$$\sigma_{ij}^{(1)} = b_{ij}^{\mu\nu}\varepsilon_{\mu\nu}^{(1)} + b_{ij}^{*\mu\nu}\tau_{\mu\nu} - s_{ij}\theta_1^{(0)} - s_{ij}^*\theta_2^{(0)}$$

(21.10)

which generalizes Equation 19.54 to include the effects of thermal stresses on material behavior.

Equations 21.7 and 21.8 represent the local problems of thermoelasticity and may be written in the form

$$\frac{1}{h_\mu} \frac{\partial}{\partial \xi_\mu} s_{i\mu} + \frac{\partial}{\partial z} s_{i3} = 0$$

$$\tag{21.11}$$

$$s_{ij} N_j^\pm = 0 \qquad \left(z = z^\pm, \, s_{ij} \leftrightarrow s_{ij}^* \right)$$

using Equations 21.5 and 21.9. From this, multiplying by z and z^2 and averaging by means of Equation 19.28, we find

$$\langle s_{i3} \rangle = \langle z s_{i3} \rangle = \langle s_{i3}^* \rangle = \langle z s_{i3}^* \rangle = 0 \tag{21.12}$$

and using this last equation together with Equation 19.62, it follows from Equation 21.10 that

$$\langle \sigma_{i3}^{(1)} \rangle = \langle z \sigma_{i3}^{(1)} \rangle = 0$$

indicating that Formulas 19.63 retain their truth in this case. Because, further, Relations 19.46 and 19.63 remain unchanged, so do Equations 19.64 and 19.70, because these latter were derived by essentially using the former in Section 19. Instead of the elastic relations (Equations 19.71 and 19.72) for the homogenized Equations, from Equation 21.10 we find

$$\langle \sigma_{\theta\kappa}^{(1)} \rangle = \langle b_{\theta\kappa}^{\mu\nu} \rangle \varepsilon_{\mu\nu}^{(1)} + \langle b_{\theta\kappa}^{*\mu\nu} \rangle \tau_{\mu\nu} - \langle s_{\theta\kappa} \rangle \theta_1^{(0)} - \langle s_{\theta\kappa}^* \rangle \theta_2^{(0)}$$

$$\tag{21.13}$$

$$\langle z \sigma_{\theta\kappa}^{(1)} \rangle = \langle z b_{\theta\kappa}^{\mu\nu} \rangle \varepsilon_{\mu\nu}^{(1)} + \langle z b_{\theta\kappa}^{*\mu\nu} \rangle \tau_{\mu\nu} - \langle z s_{\theta\kappa} \rangle \theta_1^{(0)} - \langle z s_{\theta\kappa}^* \rangle \theta_2^{(0)}$$

Now if this is substituted into Equations 19.64 and 19.70 we obtain, using Equations 19.48 and 19.49, a system of three governing equations for the functions $v_1^{(1)}(\alpha, t)$, $v_2^{(1)}(\alpha, t)$, and $w(\alpha, t)$ determining (at leading order) the components of the displacement vector, Equations 19.45. The functions $\theta_1^{(0)}(\alpha, t)$ and $\theta_2^{(0)}(\alpha, t)$, also involved in these equations, are found from the homogenized heat conduction problem.

21.2. Effective properties

The local thermoelastic problems posed by Equations 21.9 and 21.12 are similar to the problems given by Equations 19.41. Equations 19.55 and 19.56, and their solutions are unique only up to constant terms. This ambiguity is removed by imposing the conditions

$$\langle S_m \rangle_\xi = 0 \qquad \xi \langle S_m^* \rangle_\xi = 0 \quad \text{for } z = 0 \tag{21.14}$$

in accordance with Equation 21.6 and in analogy with Conditions 19.57.

The $z = z^\pm$ conditions relevant to the local problems given by Equations 21.9 and 21.11 may be written in a form analogous to Equations 19.58 and 19.59. That is,

$$\frac{1}{h_\mu} n_\mu^{\pm(\xi)} s_{i\mu} + n_3^{\pm(\xi)} s_{i3} = 0 \qquad (z = z^\pm) \tag{21.15}$$

$$s_{ij} \leftrightarrow s_{ij}^*$$

If the functions $\alpha_{mn}(y, z)$ and $c_{ijmn}(y, z)$ are piecewise-smooth and undergo discontinuities of the first kind at the contact surfaces between dissimilar components of the composite, the interface continuity conditions should be added,

$$[S_m] = 0 \qquad \left[\frac{1}{h_\mu} n_\mu^{(k)} s_{i\mu} + n_3^{(k)} s_{i3} \right] = 0 \tag{21.16}$$

$$[S_m^*] = 0 \qquad \left[\frac{1}{h_\mu} n_\mu^{(k)} s_{i\mu}^* + n_3^{(k)} s_{i3}^* \right] = 0$$

readily recognizable as having the same form as Equations 19.60 and 19.61.

Equations 21.9, 21.11, and 21.14 through 21.16 represent a complete formulation of the local thermoelastic problems, and their solutions enable one to reveal the local structure of the thermal stresses in Equation 21.10 and to calculate the effective properties $\langle s_{\theta\kappa} \rangle$, $\langle s_{\theta\kappa}^* \rangle$, $\langle zs_{\theta\kappa} \rangle$, and $\langle zs_{\theta\kappa}^* \rangle$ occurring in Equation 21.13.

By comparing the local problems of thermoelasticity with those of the theory of elasticity it can be shown that the following equations hold:

$$\delta \langle s_{\theta\kappa} \rangle = \langle \alpha_{ij}^T b_{ij}^{\theta\kappa} \rangle$$

$$\delta \langle zs_{\theta\kappa} \rangle = \langle \alpha_{ij}^T b_{ij}^{*\theta\kappa} \rangle$$

$$\delta \langle s_{\theta\kappa}^* \rangle = \langle z\alpha_{ij}^T b_{ij}^{\theta\kappa} \rangle \tag{21.17}$$

$$\delta \langle zs_{\theta\kappa}^* \rangle = \langle z\alpha_{ij}^T b_{ij}^{*\theta\kappa} \rangle$$

with the implication that it suffices to know the solutions of the local problems of ordinary elasticity in order to derive all the effective thermoelastic properties of the system.

In proving Equations 21.17, the notation developed in Section 19 will be employed. Rewriting Equations 21.11 in the form

$$\frac{\partial s_{ij}}{\partial \zeta_j} = 0 \qquad N_j^{\pm} s_{ij} = 0 \quad \text{for } z = z^{\pm}$$

(21.18)

$$s_{ij} \leftrightarrow s_{ij}^*$$

and following the procedure similar to that outlined in Section 19 we find that for any function $\varphi(\zeta_1, \zeta_2, \zeta_3)$, periodic in ζ_1 and ζ_2 with unit cell Ω,

$$\left\langle s_{qp} \frac{\partial \varphi}{\partial \zeta_p} \right\rangle = 0 \quad \text{and} \quad \left\langle s_{qp}^* \frac{\partial \varphi}{\partial \zeta_p} \right\rangle = 0$$

(21.19)

Putting $\varphi = U_q^{\theta\kappa}$ and using Equations 21.9, it follows from the first of Equations 21.19 that

$$0 = \left\langle s_{qp} \frac{\partial U_q^{\theta\kappa}}{\partial \zeta_p} \right\rangle = \left\langle \beta_{qp} \frac{\partial U_q^{\theta\kappa}}{\partial \zeta_p} \right\rangle - \left\langle c_{qplr} \frac{\partial S_l}{\partial \zeta_r} \frac{\partial U_q^{\theta\kappa}}{\partial \zeta_p} \right\rangle$$

giving

$$\left\langle \beta_{qp} \frac{\partial U_q^{\theta\kappa}}{\partial \zeta_p} \right\rangle = \left\langle c_{qplr} \frac{\partial S_l}{\partial \zeta_r} \frac{\partial U_q^{\theta\kappa}}{\partial \zeta_p} \right\rangle$$

(21.20)

Similarly, by inserting $\varphi = S_q$ into the first of Equations 19.81, we have

$$0 = \left\langle b_{qp}^{\theta\kappa} \frac{\partial S_q}{\partial \zeta_p} \right\rangle = \left\langle c_{qp\theta\kappa} \frac{\partial S_q}{\partial \zeta_p} \right\rangle + \left\langle c_{qplr} \frac{\partial U_l^{\theta\kappa}}{\partial \zeta_r} \frac{\partial S_q}{\partial \zeta_p} \right\rangle$$

which yields

$$\left\langle c_{qp\theta\kappa} \frac{\partial S_q}{\partial \zeta_p} \right\rangle = -\left\langle c_{qplr} \frac{\partial U_l^{\theta\kappa}}{\partial \zeta_r} \frac{\partial S_q}{\partial \zeta_p} \right\rangle$$

(21.21)

However, comparing Equations 21.20 and 21.21 we see that

$$
\left\langle \beta_{qp} \frac{\partial U_q^{\theta\kappa}}{\partial \zeta_p} \right\rangle = -\left\langle c_{qp\theta\kappa} \frac{\partial S_q}{\partial \zeta_p} \right\rangle \tag{21.22}
$$

using the symmetry of the elastic coefficients c_{qplr}, and therefore, by Equation 19.41, the first of Equations 21.5, and Equations 21.9,

$$
\langle s_{\theta\kappa} \rangle = \left\langle \beta_{\theta\kappa} - c_{\theta\kappa qp} \frac{\partial S_q}{\partial \zeta_p} \right\rangle
$$

$$
= \langle \beta_{\theta\kappa} \rangle + \left\langle \beta_{qp} \frac{\partial U_q^{\theta\kappa}}{\partial \zeta_p} \right\rangle
$$

$$
= \left\langle \alpha_{ij} c_{ij\theta\kappa} + \alpha_{ij} c_{ijqp} \frac{\partial U_q^{\theta\kappa}}{\partial \zeta_p} \right\rangle
$$

$$
= \langle \alpha_{ij} b_{ij}^{\theta\kappa} \rangle
$$

which, in view of Equation 21.12, proves the first of Relations 21.17.

Now if we substitute $\varphi = S_q$ into the second of Equations 19.81 and $\varphi = V_q^{\theta\kappa}$ into the first of Equations 21.19, there results

$$
\left\langle \beta_{qp} \frac{\partial V_q^{\theta\kappa}}{\partial \zeta_p} \right\rangle = -\left\langle zc_{qp\theta\kappa} \frac{\partial S_q}{\partial \zeta_p} \right\rangle
$$

leading to the second of Relations 21.17 after use of Equations 19.53, 21.2, 21.5, and 21.9. The third of Equations 21.17 is arrived at by setting $\varphi = S_q^*$ in the first of Equations 19.81 and $\varphi = U_q^{\theta\kappa}$ in the second of Equations 21.19. This gives the result

$$
\left\langle z\beta_{qp} \frac{\partial U_q^{\theta\kappa}}{\partial \zeta_p} \right\rangle = -\left\langle c_{qp\theta\kappa} \frac{\partial S_q^*}{\partial \zeta_p} \right\rangle
$$

which must then be combined with Equations 19.41, 21.2, 21.5, and 21.9. Finally, substitute $\varphi = S_q^*$ into the second of Equations 19.81 and $\varphi = V_q^{\theta\kappa}$ into the second of Equations 21.19, to obtain

$$
\left\langle z\beta_{qp} \frac{\partial V_q^{\theta\kappa}}{\partial \zeta_p} \right\rangle = -\left\langle zc_{qp\theta\kappa} \frac{\partial S_q^*}{\partial \zeta_p} \right\rangle
$$

The last of Relations 21.17 then follows with the help of Equations 19.53, 21.2, 21.5, and 21.9.

21.3. Homogenized thermoelastic problem versus thin-shell thermoelasticity results

Back in Section 19, we were able to rewrite the elastic relations, Equations 19.71 and 19.72, in the form of Equations 19.88 using the stress resultants N and moment resultant M defined in Equations 19.85. Exactly analogous constitutive relations may be derived for the homogenized shell in the context of thermoelasticity theory. Using Equations 19.85 again, Equations 21.13 become

$$N_\beta = \delta \langle b_{\beta\beta}^{\mu\nu} \rangle \varepsilon_{\mu\nu} + \delta^2 \langle b_{\beta\beta}^{*\mu\nu} \rangle \tau_{\mu\nu} - \delta^2 \langle s_{\beta\beta} \rangle \theta_1^{(0)} - \delta^2 \langle s_{\beta\beta}^* \rangle \theta_2^{(0)}$$

$$N_{12} = \delta \langle b_{12}^{\mu\nu} \rangle \varepsilon_{\mu\nu} + \delta^2 \langle b_{12}^{*\mu\nu} \rangle \tau_{\mu\nu} - \delta^2 \langle s_{12} \rangle \theta_1^{(0)} - \delta^2 \langle s_{12}^* \rangle \theta_2^{(0)}$$

$$M_\beta = \delta^2 \langle z b_{\beta\beta}^{\mu\nu} \rangle \varepsilon_{\mu\nu} + \delta^3 \langle z b_{\beta\beta}^{*\mu\nu} \rangle \tau_{\mu\nu} - \delta^3 \langle z s_{\beta\beta} \rangle \theta_1^{(0)} - \delta^3 \langle z s_{\beta\beta}^* \rangle \theta_2^{(0)}$$

$$M_{12} = \delta^2 \langle z b_{12}^{\mu\nu} \rangle \varepsilon_{\mu\nu} + \delta^3 \langle z b_{12}^{*\mu\nu} \rangle \tau_{\mu\nu} - \delta^3 \langle z s_{12} \rangle \theta_1^{(0)} - \delta^3 \langle z s_{12}^* \rangle \theta_2^{(0)}$$

$$(21.23)$$

where β takes the values 1 and 2 and is not summed.

Over the middle surface of the shell ($z = 0$), the strains ($\varepsilon_{\mu\nu}$ and $\tau_{\mu\nu}$) and displacements ($v_1 = \delta v_1^{(1)}$, $v_2 = \delta v_2^{(1)}$, and w) will again be related by Equations 19.86 and 19.49.

For the displacement component, using Equations 19.8 and 19.45, and 21.6, we obtain the relations

$$u_1 = v_1(\alpha, t) - \frac{\gamma}{A_1} \frac{\partial w}{\partial \alpha_1} + \delta U_1^{\mu\nu} \varepsilon_{\mu\nu} + \delta^2 V_1^{\mu\nu} \tau_{\mu\nu} + \delta^2 S_1 \theta_1^{(0)}$$

$$+ \, \delta^2 S_1^* \theta_2^{(0)} + \cdots$$

$$u_2 = v_2(\alpha, t) - \frac{\gamma}{A_2} \frac{\partial w}{\partial \alpha_2} + \delta U_2^{\mu\nu} \varepsilon_{\mu\nu} + \delta^2 V_2^{\mu\nu} \tau_{\mu\nu} + \delta^2 S_2 \theta_1^{(0)} \qquad (21.24)$$

$$+ \, \delta^2 S_2^* \theta_2^{(0)} + \cdots$$

$$u_3 = w(\alpha, t) + \delta U_3^{\mu\nu} \varepsilon_{\mu\nu} + \delta^2 V_3^{\mu\nu} \tau_{\mu\nu} + \delta^2 S_3 \theta_1^{(0)} + \delta^2 S_3^* \theta_2^{(0)} + \cdots$$

which complement Equations 19.87 derived in the framework of the theory of elasticity.

The functions $\theta_1^{(0)}(\alpha, t)$ and $\theta_2^{(0)}(\alpha, t)$ are found from the homogenized heat condition problem; they determine the temperature and heat flow vector by means of Equations 20.37. The integral temperature characteristics needed for the thin-shell–thin-plate theory of thermoelasticity (Podstrigach and Shvets 1978, Podstrigach et al. 1984) follow from Equations 20.37 as

$$T = \langle \theta \rangle \approx \theta_1^{(0)} + \langle z \rangle \theta_2^{(0)}$$

$$T^* = 6\langle z\theta \rangle \approx 6\langle z \rangle \theta_1^{(0)} + 6\langle z^2 \rangle \theta_2^{(0)}$$

(21.25)

at leading order. Note that if the shell is of constant thickness, that is, $z^{\pm} = \pm(\frac{1}{2})$, then $\langle z \rangle = 0$, $\langle z^2 \rangle = \frac{1}{12}$, and Equations 20.51 follow in an obvious way from Equations 21.25.

It should be reemphasized here that Equations 19.64 and 19.70, the homogenized shell equations expressed in terms of stresses, and Equations 19.89 written in terms of forces and moments, remain unchanged in the framework of thermoelasticity theory and are identical to the corresponding thin shell equations.

Let us consider here the limiting case of a homogeneous isotropic shell of constant thickness, when $F^{\pm} \equiv 0$, and $\alpha_{ij}^T = \alpha^T \cdot \delta_{ij}$, α^T being the (linear) thermal expansion coefficient. It was shown in Sections 19 and 20 that the elastic part of Relations 21.23 and the homogenized heat conduction problem reduce in this case to the corresponding thin-shell relations known from the theories of elasticity and heat conduction. We wish to show now that the temperature part of the constitutive equations (Equations 21.23) of such a shell is also reducible to the thin-shell thermoelastic results.

The nonzero solutions of the local thermoelastic problems given by Equations 21.9, 21.11, and 21.14 are found to be given by

$$\delta S_3 = z\alpha^T \frac{1 + \nu}{1 - gn} \quad \text{and} \quad \delta S_3^* = \frac{z^2}{2} \alpha^T \frac{1 + \nu}{1 - \nu}$$

(21.26)

using Equations 19.91 and 21.5. Substituting this into Equations 21.9 and making use of Equations 21.1 and 21.5 again, we find

$$\delta\langle s_{11} \rangle = \delta\langle s_{22} \rangle = \frac{\alpha^T E}{1 - \nu}$$

(21.27)

$$\delta\langle zs_{11}^* \rangle = \delta\langle zs_{22}^* \rangle = \frac{\alpha^T E}{12(1 - \nu)}$$

after averaging the expressions for all nonzero effective thermoelastic properties. (We note that the same equations may be derived from Equations 21.17 using the solutions obtained in Section 19 for the local elastic problems given by Equations 19.92).

Now if we substitute Equations 19.93 and 21.27 into Equations 21.23 and use Equations 20.51, the corresponding relations of the thermoelasticity theory of thin isotropic shells will result (Podstrigach and Shvets 1978).

21.4. Laminated shell composed of homogeneous isotropic layers

Referring once again to Figure 19.4, we proceed to obtain the effective thermoelastic properties of a shell composed of homogeneous isotropic layers. The effective elastic and heat conduction properties of this structure have already been determined (in Sections 19 and 20) from the corresponding local problems.

There are two principal ways of handling the problem of thermoelasticity in this case, one by employing the solutions of the local thermoelastic problems given by Equations 21.9, 21.11, and 21.14 through 21.16; the other by combining Equations 21.17 and Relations 19.97 and 19.99 with the solutions of the laminated shell local elastic problems set by Equations 19.100 and 19.103. Because of its relative simplicity, the second approach is preferred here.

Accordingly, we substitute Equations 19.100 and 19.103 into, respectively, Equations 19.97 and 19.99 to obtain, for the case of isotropic layer materials,

$$b_{11}^{11(m)} = b_{22}^{22(m)} = \frac{E_m}{1 - \nu_m^2}$$

$$b_{11}^{22(m)} = b_{22}^{11(m)} = \frac{\nu_m E_m}{1 - \nu_m^2}$$

$$b_{11}^{*11(m)} = b_{22}^{*22(m)} = \frac{z E_m}{1 - \nu_m^2} \qquad (21.28)$$

$$b_{11}^{*22(m)} = b_{22}^{*11(m)} = \frac{z \nu_m E_m}{1 - \nu_m^2}$$

$$b_{33}^{\theta\kappa(m)} = b_{33}^{*\theta\kappa(m)} = 0$$

with m labeling the layer ($m = 1, 2, \ldots, M$).

Using Equations 21.28 in 21.17, averaging over the thickness, and noting that $\alpha_{ij}^{T(m)} = \alpha_m^T \delta_{ij}$ because of the assumed elastic isotropy, we obtain the following set of nonzero effective thermoelastic properties:

$$\delta\langle s_{11}\rangle = \delta\langle s_{22}\rangle = \sum_{m=1}^{M} \frac{\alpha_m^T E_m}{1 - \nu_m}(\delta_m - \delta_{m-1})$$

$$\delta\langle zs_{11}\rangle = \delta\langle zs_{22}\rangle = \delta\langle s_{11}^*\rangle = \delta\langle s_{22}^*\rangle$$

$$= \frac{1}{2}\sum_{m=1}^{M} \frac{\alpha_m^T E_m}{1 - \nu_m}\left[\delta_m^2 - \delta_{m-1}^2 - (\delta_m - \delta_{m-1})\right] \tag{21.29}$$

$$\delta\langle zs_{11}^*\rangle = \delta\langle zs_{22}^*\rangle$$

$$= \frac{1}{3}\sum_{m=1}^{M} \frac{\alpha_m^T E_m}{1 - \nu_m}\left[\delta_m^3 - \delta_{m-1}^3 - \frac{3}{2}(\delta_m^2 - \delta_{m-1}^2) + \frac{3}{4}(\delta_m - \delta_{m-1})\right]$$

These, together with Equations 19.101, 19.102, 19.104, and 20.57, determine all the (nonzero) effective properties of the laminated shell that are involved in the homogenized equations, Equations 21.23 (for thermoelasticity) and Equations 20.35 (for heat conduction).

6

Structurally Nonhomogeneous Periodic Plates and Shells

There are many applications in a number of technical areas that may and do benefit from the advantages offered by homogeneous thin-walled structural elements with a regular structure. The unit cell of the material system consists in this case of several thin-walled parts made of homogeneous materials (c.f. Figure 19.2), and the inhomogeneity in question is associated with the structural features of the object and is also determined by the specific form of reinforcement used (we remind the reader that voids may also be considered as reinforcement). The structurally nonhomogeneous elements of these types include the rib-and-wafer types of reinforced plates and shells (see Figures 23.1, 23.2, and 26.1), three-layered shells with a honeycomb filler (Figures 24.1 and 24.3), and corrugated surface shells (Figures 27.2, 27.7, and 27.12). All these structures are considered in this chapter on the basis of the general theory developed previously for regularly nonhomogeneous shells with rapidly oscillating thickness.

22. LOCAL PROBLEM FORMULATION FOR STRUCTURALLY NONHOMOGENEOUS PLATES AND SHELLS OF ORTHOTROPIC MATERIAL

We consider here a structurally nonhomogeneous shell (or plate) each element of which is made of a homogeneously orthotropic material. By the definition of orthotropy, such a material possesses elastic symmetry with respect to three mutually orthogonal planes, so that its tensor of elastic moduli, or stiffness tensor, has nine independent components and may be represented as a symmetric 6×6 matrix as follows:

$$[c] = \begin{bmatrix} c_{11} & c_{12} & c_{13} & 0 & 0 & 0 \\ & c_{22} & c_{23} & 0 & 0 & 0 \\ & & c_{33} & 0 & 0 & 0 \\ & & & 2c_{44} & 0 & 0 \\ & & & & 2c_{55} & 0 \\ & & & & & 2c_{66} \end{bmatrix} \qquad (22.1)$$

In terms of the engineering properties of the material, its compliance tensor may also be represented as a six by six matrix, namely,

$$[J] = [c]^{-1} = \begin{bmatrix} \dfrac{1}{E_1} & -\dfrac{\nu_{12}}{E_2} & -\dfrac{\nu_{13}}{E_3} & 0 & 0 & 0 \\[2ex] -\dfrac{\nu_{21}}{E_1} & \dfrac{1}{E_2} & -\dfrac{\nu_{23}}{E_3} & 0 & 0 & 0 \\[2ex] -\dfrac{\nu_{31}}{E_1} & -\dfrac{\nu_{32}}{E_2} & \dfrac{1}{E_3} & 0 & 0 & 0 \\[2ex] & & & \dfrac{1}{2G_{12}} & 0 & 0 \\[2ex] & & & & \dfrac{1}{2G_{13}} & 0 \\[2ex] & & & & & \dfrac{1}{2G_{23}} \end{bmatrix}$$

$$(22.2)$$

where E_1, E_2, and E_3 are the Young's moduli corresponding to the three principal elastic directions, G_{23} and G_{13} are the shear moduli, and ν_{12}, ν_{21}, ν_{13}, ν_{23}, and ν_{32} are the the Poisson ratios. Note that

$$\frac{\nu_{12}}{E_2} = \frac{\nu_{21}}{E_1} \qquad \frac{\nu_{13}}{E_3} = \frac{\nu_{31}}{E_1} \qquad \frac{\nu_{23}}{E_3} = \frac{\nu_{32}}{E_2} \qquad (22.3)$$

in view of the symmetry of the matrix in Equation 22.2.

The relations between the elastic moduli and engineering properties follow from Equations 22.1 through 22.3 as

$$E_1 = c_{11} + \frac{2c_{12}c_{13}c_{23} - c_{12}^2 c_{33} - c_{13}^2 c_{22}}{c_{22}c_{33} - c_{23}^2}$$

$$E_2 = c_{22} + \frac{2c_{12}c_{13}c_{23} - c_{12}^2 c_{33} - c_{23}^3 c_{11}}{c_{11}c_{33} - c_{13}^2}$$

$$E_3 = c_{33} + \frac{2c_{12}c_{13}c_{23} - c_{13}^2 c_{22} - c_{23}^2 c_{11}}{c_{11}c_{22} - c_{12}^2}$$

$$\nu_{12} = \frac{c_{12}c_{33} - c_{13}c_{23}}{c_{11}c_{33} - c_{13}^2} \qquad \nu_{21} = \frac{c_{12}c_{33} - c_{13}c_{23}}{c_{22}c_{33} - c_{23}^2}$$

$$\nu_{13} = \frac{c_{22}c_{13} - c_{12}c_{23}}{c_{11}c_{22} - c_{12}^2} \qquad \nu_{31} = \frac{c_{22}c_{13} - c_{12}c_{23}}{c_{22}c_{33} - c_{23}^2}$$

$$\nu_{23} = \frac{c_{11}c_{23} - c_{13}c_{12}}{c_{11}c_{22} - c_{12}^2} \qquad \nu_{32} = \frac{c_{11}c_{23} - c_{13}c_{12}}{c_{11}c_{33} - c_{13}^2}$$

$$G_{12} = c_{66} = c_{1212} \qquad G_{13} = c_{55} = c_{1313} \qquad G_{23} = c_{44} = c_{2323}$$

(22.4)

If the material is transversely isotropic, that is, if it is elastically symmetric with respect to all directions perpendicular to the third coordinate axis, then equations 22.1 through 22.4 are also true, but

$$c_{11} = c_{22} \qquad c_{13} = c_{23} \qquad c_{55} = c_{44} \qquad c_{66} = \frac{1}{2}(c_{11} - c_{12})$$

$$E_1 = E_2 \qquad \nu_{13} = \nu_{23}$$

(22.5)

$$G_{13} = G_{23} \qquad G_{12} = \frac{E_1}{2(1 + \nu_{12})}$$

in this case, and we are left with only five independent components.

Finally, for a totally isotropic material,

$$E_1 = E_2 = E_3 = E$$

$$\nu_{13} = \nu_{23} = \nu_{12} = \nu$$

$$G_{13} = G_{23} = G_{12} = \frac{E}{2(1 + \nu)}$$

and the relationship between the elastic moduli and the constants E and ν is given by Equation 19.91.

22.1. Three-dimensional local problem

It was shown in Section 19 that in order to evaluate the local stress distributions as given by Equations 19.74, and to determine the effective stiffness moduli involved in the homogenized relations, Equations 19.88, the corresponding local problem must be solved for the functions $b_{ij}^{\lambda\mu}(\xi, z)$ and $b_{ij}^{*\lambda\mu}(\xi, z)$ assumed to be periodic in ξ_1 and ξ_2 with respective periods A_1 and A_2. There are two groups of elastic problems that arise from the preceding local problems and are set in the orthotropic region. One of these—to be referred to as the (b) group—is expressed by Equations 19.55, 19.57, 19.58, and 19.60, and the other—the (b^*) group—by Equations 19.56, 19.57, 19.59, and 19.61.

If the material is homogeneous, simplification of these group of problems will result.

(b)-Type Local Problems

In each of three $(b\lambda\mu)$ problems $(\lambda\mu = 11, 22, 12)$ we determine the functions $U_1^{\lambda\mu}(\xi, z)$, $U_2^{\lambda\mu}(\xi, z)$, and $U_3^{\lambda\mu}(\xi, z)$, which are periodic in ξ_1 and ξ_2 (with respective periods A_1 and A_2) and satisfy the system of equations

$$\frac{1}{h_\beta}\frac{\partial \tau_{i\beta}^{\lambda\mu}}{\partial \xi_\beta} + \frac{\partial \tau_{i3}^{\lambda\mu}}{\partial z} = 0 \qquad (22.6)$$

and

$$\tau_{11}^{\lambda\mu} = \frac{1}{h_1}c_{11}\frac{\partial U_1^{\lambda\mu}}{\partial \xi_1} + \frac{1}{h_2}c_{12}\frac{\partial U_2^{\lambda\mu}}{\partial \xi_2} + c_{13}\frac{\partial U_3^{\lambda\mu}}{\partial z}$$

$$\tau_{22}^{\lambda\mu} = \frac{1}{h_1}c_{21}\frac{\partial U_1^{\lambda\mu}}{\partial \xi_1} + \frac{1}{h_2}c_{22}\frac{\partial U_2^{\lambda\mu}}{\partial \xi_2} + c_{23}\frac{\partial U_3^{\lambda\mu}}{\partial z}$$

$$\tau_{33}^{\lambda\mu} = \frac{1}{h_1}c_{31}\frac{\partial U_1^{\lambda\mu}}{\partial \xi_1} + \frac{1}{h_2}c_{32}\frac{\partial U_2^{\lambda\mu}}{\partial \xi_2} + c_{33}\frac{\partial U_3^{\lambda\mu}}{\partial z}$$

$$\tau_{23}^{\lambda\mu} = c_{44}\left(\frac{1}{h_2}\frac{\partial U_3^{\lambda\mu}}{\partial \xi_2} + \frac{\partial U_2^{\lambda\mu}}{\partial z}\right)$$

$$\tau_{13}^{\lambda\mu} = c_{55}\left(\frac{1}{h_1}\frac{\partial U_3^{\lambda\mu}}{\partial \xi_1} + \frac{\partial U_1^{\lambda\mu}}{\partial z}\right)$$

$$\tau_{12}^{\lambda\mu} = c_{66}\left(\frac{1}{h_2}\frac{\partial U_1^{\lambda\mu}}{\partial \xi_2} + \frac{1}{h_1}\frac{\partial U_2^{\lambda\mu}}{\partial \xi_1}\right) \qquad (22.7)$$

together with the outer surface boundary conditions

$$t_1^{11} = -c_{11}\frac{n_1}{h_1} \qquad t_2^{11} = -c_{21}\frac{n_2}{h_2} \qquad t_3^{11} = c_{31}n_3$$

$$t_1^{22} = -c_{12}\frac{n_1}{h_1} \qquad t_2^{22} = -c_{22}\frac{n_2}{h_2} \qquad t_3^{22} = -c_{32}n_3 \qquad (22.8)$$

$$t_1^{12} = -c_{66}\frac{n_2}{h_2} \qquad t_2^{12} = c_{66}\frac{n_1}{h_1} \qquad t_3^{12} = 0$$

where

$$t_1^{\lambda\mu} = \tau_{11}^{\lambda\mu}\frac{n_1}{h_1} + \tau_{12}^{\lambda\mu}\frac{n_2}{h_2} + \tau_{13}^{\lambda\mu}n_3$$

$$t_2^{\lambda\mu} = \tau_{12}^{\lambda\mu}\frac{n_1}{h_1} + \tau_{22}^{\lambda\mu}\frac{n_2}{h_2} + \tau_{23}^{\lambda\mu}n_3 \qquad (22.9)$$

$$t_3^{\lambda\mu} = \tau_{13}^{\lambda\mu}\frac{n_1}{h_1} + \tau_{23}^{\lambda\mu}\frac{n_2}{h_2} + \tau_{33}^{\lambda\mu}n_3$$

where $n_1\,n_2, n_3$ are the components of the unit vector normal to the bounding surface of the region Ω.

The functions that are determined from the group (b) problems are

$$b_{11}^{11} = \tau_{11}^{11} + c_{11} \qquad b_{12}^{11} = \tau_{12}^{11} \qquad\qquad b_{22}^{11} = \tau_{22}^{11} + c_{21}$$

$$b_{11}^{22} = \tau_{11}^{22} + c_{21} \qquad b_{12}^{22} = \tau_{12}^{22} \qquad\qquad b_{22}^{22} = \tau_{22}^{22} + c_{22}$$

$$b_{11}^{12} = \tau_{11}^{12} \qquad\qquad b_{12}^{12} = \tau_{12}^{12} + c_{66} \qquad b_{22}^{12} = \tau_{22}^{12} \qquad\qquad (22.10)$$

$$b_{33}^{11} = \tau_3^{11} + c_{31} \qquad b_{33}^{22} = \tau_{33}^{22} + c_{32} \qquad b_{3\beta}^{\lambda\mu} = \tau_{\beta3}^{\lambda\mu}$$

(b*)-Type Local Problems

From the functions (b*11), (b*22), and (b*12), one determines the functions $V_1^{\lambda\mu}(\xi, z)$, $V_2^{\lambda\mu}(\xi, z)$, and $V_3^{\lambda\mu}(\xi, z)$, also periodic in ξ_1 and ξ_2 and satisfying the equations

$$\frac{1}{h_\beta}\frac{\partial \bar\tau_{i\beta}^{\lambda\mu}}{\partial \xi_\beta} + \frac{\partial \bar\tau_{i3}^{\lambda\mu}}{\partial z} = -c_{33\lambda\mu}\delta_{i3} \qquad (22.11)$$

and

$$\bar{\tau}_{11}^{\lambda\mu} = \frac{1}{h_1}c_{11}\frac{\partial V_1^{\lambda\mu}}{\partial \xi_1} + \frac{1}{h_2}c_{12}\frac{\partial V_2^{\lambda\mu}}{\partial \xi_2} + c_{13}\frac{\partial V_3^{\lambda\mu}}{\partial z}$$

$$\bar{\tau}_{22}^{\lambda\mu} = \frac{1}{h_1}c_{21}\frac{\partial V_1^{\lambda\mu}}{\partial \xi_1} + \frac{1}{h_2}c_{22}\frac{\partial V_2^{\lambda\mu}}{\partial \xi_2} + c_{23}\frac{\partial V_3^{\lambda\mu}}{\partial z}$$

$$\bar{\tau}_{33}^{\lambda\mu} = \frac{1}{h_1}c_{31}\frac{\partial V_1^{\lambda\mu}}{\partial \xi_1} + \frac{1}{h_2}c_{32}\frac{\partial V_2^{\lambda\mu}}{\partial \xi_2} + c_{33}\frac{\partial V_3^{\lambda\mu}}{\partial z}$$

$$\bar{\tau}_{23}^{\lambda\mu} = c_{44}\left(\frac{1}{h_2}\frac{\partial V_3^{\lambda\mu}}{\partial \xi_2} + \frac{\partial V_2^{\lambda\mu}}{\partial z} \right) \qquad (22.12)$$

$$\bar{\tau}_{13}^{\lambda\mu} = c_{55}\left(\frac{1}{h_1}\frac{\partial V_3^{\lambda\mu}}{\partial \xi_1} + \frac{\partial V_1^{\lambda\mu}}{\partial z} \right)$$

$$\bar{\tau}_{12}^{\lambda\mu} = c_{66}\left(\frac{1}{h_2}\frac{\partial V_1^{\lambda\mu}}{\partial \xi_2} + \frac{1}{h_1}\frac{\partial V_2^{\lambda\mu}}{\partial \xi_1} \right)$$

and the boundary conditions on the outer surfaces of Ω,

$$\bar{t}_1^{11} = -zc_{11}\frac{n_1}{h_1} \qquad \bar{t}_2^{11} = -zc_{21}\frac{n_2}{h_2} \qquad \bar{t}_3^{11} = -zc_{31}n_3$$

$$\bar{t}_1^{22} = -zc_{12}\frac{n_1}{h_1} \qquad \bar{t}_2^{22} = -zc_{22}\frac{n_2}{h_2} \qquad \bar{t}_3^{22} = -zc_{32}n_3 \qquad (22.13)$$

$$\bar{t}_1^{12} = -zc_{66}\frac{n_2}{h_2} \qquad \bar{t}_2^{12} = -zc_{66}\frac{n_1}{h_1} \qquad \bar{t}_1^{12} = 0$$

where

$$\bar{t}_1^{\lambda\mu} = \bar{\tau}_{11}^{\lambda\mu}\frac{n_1}{h_1} + \bar{\tau}_{12}^{\lambda\mu}\frac{n_2}{h_2} + \bar{\tau}_{13}^{\lambda\mu}n_3$$

$$\bar{t}_2^{\lambda\mu} = \bar{\tau}_{12}^{\lambda\mu}\frac{n_1}{h_1} + \bar{\tau}_{22}^{\lambda\mu}\frac{n_2}{h_2} + \bar{\tau}_{23}^{\lambda\mu}n_3 \qquad (22.14)$$

$$\bar{t}_3^{\lambda\mu} = \bar{\tau}_{13}^{\lambda\mu}\frac{n_1}{h_1} + \bar{\tau}_{23}^{\lambda\mu}\frac{n_2}{h_2} + \bar{\tau}_{33}^{\lambda\mu}n_3$$

Having solved the $(b*)$-type problems, we are in a position to obtain the functions

$$b_{11}^{*11} = \bar{\tau}_{11}^{11} + zc_{11} \qquad b_{12}^{*11} = \bar{\tau}_{12}^{11} \qquad b_{22}^{*11} = \bar{\tau}_{22}^{11} + zc_{21}$$

$$b_{11}^{*22} = \bar{\tau}_{11}^{22} + zc_{21} \qquad b_{12}^{*22} = \bar{\tau}_{12}^{22} \qquad b_{22}^{*22} = \bar{\tau}_{22}^{22} + zc_{22}$$

$$b_{11}^{*12} = \bar{\tau}_{11}^{12} \qquad b_{12}^{*12} = \bar{\tau}_{12}^{12} + zc_{66} \qquad b_{22}^{*12} = \bar{\tau}_{22}^{12} \qquad (22.15)$$

$$b_{33}^{*11} = \bar{\tau}_{33}^{11} + zc_{31} \qquad b_{33}^{*22} = \bar{\tau}_{33}^{22} + zc_{32} \qquad b_{3\beta}^{*\lambda\mu} = \bar{\tau}_{\beta 3}^{\lambda\mu}$$

22.2. Local problems in a two-dimensional formulation

Consider a shell strengthened by a set of ribs or cavities located along one of the coordinate axes, say α_1. The functions $U_i^{\lambda\mu}$ and $V_i^{\lambda\mu}$ will then depend on the variables ξ_2 and z, and the local problems of the group $(b\lambda\mu)$ $(\lambda\mu = 11, 22, 12)$ reduce to the determination of the functions $U_2^{\lambda\mu}(\xi_2, z)$, $U_3^{\lambda\mu}(\xi_2, z)$ $(\lambda\mu = 11, 22)$, and $U_1^{12}(\xi_2, z)$ from the system of equations

$$\frac{1}{h_2} \frac{\partial \tau_{i2}^{\lambda\mu}}{\partial \xi_2} + \frac{\partial \tau_{i3}^{\lambda\mu}}{\partial z} = 0 \qquad (i = 2, 3, \lambda\mu = 11, 22) \qquad (22.16)$$

$$\frac{1}{h_2} \frac{\partial \tau_{12}^{12}}{\partial \xi_2} + \frac{\partial \tau_{13}^{12}}{\partial z} = 0 \qquad (22.17)$$

$$\tau_{11}^{\lambda\mu} = \frac{1}{h_2} c_{12} \frac{\partial U_2^{\lambda\mu}}{\partial \xi_2} + c_{13} \frac{\partial U_3^{\lambda\mu}}{\partial z}$$

$$\tau_{22}^{\lambda\mu} = \frac{1}{h_2} c_{22} \frac{\partial U_2^{\lambda\mu}}{\partial \xi_2} + c_{23} \frac{\partial U_3^{\lambda\mu}}{\partial z}$$

$$\tau_{33}^{\lambda\mu} = \frac{1}{h_2} c_{32} \frac{\partial U_2^{\lambda\mu}}{\partial \xi_2} + c_{33} \frac{\partial U_3^{\lambda\mu}}{\partial z}$$

$$\tau_{23}^{\lambda\mu} = c_{44} \left(\frac{1}{h_2} \frac{\partial U_3^{\lambda\mu}}{\partial \xi_2} + \frac{\partial U_2^{\lambda\mu}}{\partial z} \right) \qquad (\lambda\mu = 11, 22)$$

$$\tau_{13}^{12} = c_{55} \frac{\partial U_1^{12}}{\partial z}$$

$$\tau_{12}^{12} = c_{66} \frac{1}{h_2} \frac{\partial U_1^{12}}{\partial \xi_2} \qquad (22.18)$$

subjected to the following conditions on the outer surface of Ω:

$$t_2^{11} = -c_{21}\frac{n_2}{h_2} \qquad t_3^{11} = -c_{31}n_3$$

$$t_2^{22} = -c_{22}\frac{n_2}{h_2} \qquad t_3^{22} = -c_{32}n_3 \qquad\qquad (22.19)$$

$$t_1^{12} = -c_{66}\frac{n_2}{h_2}$$

where $(\lambda\mu = 11, 22)$

$$t_2^{\lambda\mu} = \tau_{22}^{\lambda\mu}\frac{n_2}{h_2} + \tau_{23}^{\lambda\mu}n_3$$

$$t_3^{\lambda\mu} = \tau_{23}^{\lambda\mu}\frac{n_2}{h_2} + \tau_{33}^{\lambda\mu}n_3 \qquad\qquad (22.20)$$

$$t_1^{12} = \tau_{12}^{12}\frac{n_2}{h_2} + \tau_{13}^{12}n_3$$

The above problems yield

$$b_{11}^{11} = \tau_{11}^{11} + c_{11} \qquad\qquad b_{22}^{11} = \tau_{22}^{11} + c_{21}$$

$$b_{11}^{22} = \tau_{11}^{22} + c_{21} \qquad\qquad b_{22}^{22} = \tau_{22}^{22} + c_{22}$$

$$\qquad\qquad\qquad\qquad\qquad\qquad\qquad\qquad\qquad (22.21)$$

$$b_{12}^{12} = \tau_{12}^{12} + c_{66} \qquad\qquad b_{33}^{11} = \tau_{33}^{11} + c_{31}$$

$$b_{33}^{22} = \tau_{33}^{22} + c_{32} \qquad\qquad b_{3\beta}^{\lambda\mu} = \tau_{\beta3}^{\lambda\mu}$$

The important point to be made about the two-dimensional analysis is that each of the local problems divides itself into a plane strain problem, Equation 22.16, and the antiplane strain problem, Equation 22.17, either one being solved independently of the other. A similar situation exists with regard to the group (b^*) problems, whose two-dimensional form is

$$\frac{1}{h_2}\frac{\partial \bar{\tau}_{i2}^{\lambda\mu}}{\partial \xi_2} + \frac{\partial \bar{\tau}_{i3}^{\lambda\mu}}{\partial z} = -c_{33\lambda\mu}\delta_{i3} \qquad (i = 2, 3, \ \lambda\mu = 11, 22) \quad (22.22)$$

$$\frac{1}{h_2}\frac{\partial \bar{\tau}_{12}^{12}}{\partial \xi_2} + \frac{\partial \bar{\tau}_{13}^{12}}{\partial z} = 0 \qquad\qquad (22.23)$$

$$\bar{\tau}_{11}^{\lambda\mu} = \frac{1}{h_2}c_{12}\frac{\partial V_2^{\lambda\mu}}{\partial \xi_2} + c_{13}\frac{\partial V_3^{\lambda\mu}}{\partial z}$$

$$\bar{\tau}_{22}^{\lambda\mu} = \frac{1}{h_2}c_{22}\frac{\partial V_2^{\lambda\mu}}{\partial \xi_2} + c_{23}\frac{\partial V_3^{\lambda\mu}}{\partial z}$$

$$\bar{\tau}_{33}^{\lambda\mu} = \frac{1}{h_2}c_{32}\frac{\partial V_2^{\lambda\mu}}{\partial \xi_2} + c_{33}\frac{\partial V_3^{\lambda\mu}}{\partial z} \qquad (22.24)$$

$$\bar{\tau}_{23}^{\lambda\mu} = c_{44}\left(\frac{1}{h_2}\frac{\partial V_3^{\lambda\mu}}{\partial \xi_2} + \frac{\partial V_2^{\lambda\mu}}{\partial z}\right) \qquad (\lambda\mu = 11, 22)$$

$$\bar{\tau}_{13}^{12} = c_{55}\frac{\partial V_1^{12}}{\partial z} \qquad \bar{\tau}_{12}^{12} = c_{66}\frac{1}{h_2}\frac{\partial V_1^{12}}{\partial \xi_2}$$

To these we must adjoin the following boundary conditions

$$\bar{t}_2^{11} = -zc_{21}\frac{n_2}{h_2} \qquad \bar{t}_3^{11} = -zc_{31}n_3$$

$$\bar{t}_2^{22} = -zc_{22}\frac{n_2}{h_2} \qquad \bar{t}_3^{22} = -zc_{32}n_3 \qquad (22.25)$$

$$\bar{t}_1^{12} = -zc_{66}\frac{n_2}{h_2}$$

where

$$\bar{t}_2^{\lambda\mu} = \bar{\tau}_{22}^{\lambda\mu}\frac{n_2}{h_2} + \bar{\tau}_{23}^{\lambda\mu}n_3$$

$$\bar{t}_3^{\lambda\mu} = \bar{\tau}_{23}^{\lambda\mu}\frac{n_2}{h_2} + \bar{\tau}_{33}^{\lambda\mu}n_3 \qquad (\lambda\mu = 11, 22) \quad (22.26)$$

$$\bar{t}_1^{12} = \bar{\tau}_{12}^{12}\frac{n_2}{h_2} + \bar{\tau}_{13}^{12}n_3$$

Having found the solutions of the (b^*) group problems, we then calculate the functions

$$b_{11}^{*11} = \bar{\tau}_{11}^{11} + zc_{11} \qquad b_{22}^{*11} = \bar{\tau}_{22}^{11} + zc_{21}$$

$$b_{11}^{*22} = \bar{\tau}_{11}^{22} + zc_{21} \qquad b_{22}^{*22} = \bar{\tau}_{22}^{22} + zc_{22}$$

$$b_{12}^{*12} = \tau_{12}^{12} + zc_{66} \qquad b_{3}^{*11} = \bar{\tau}_{31}^{11} + zc_{3i} \qquad (22.27)$$

$$b_{33}^{*22} = \bar{\tau}_{33}^{22} + zc_{32} \qquad b_{3\beta}^{*\lambda\mu} = \bar{\tau}_{\beta 3}^{\lambda\mu}$$

$$b_{11}^{*12} = b_{22}^{*12} = b_{21}^{*11} = b_{12}^{*22} = 0$$

22.3. General features of the effective elastic moduli of ribbed shell and plate structures

Based on the analysis of the two-dimensional local problems given by Equations 22.16 through 22.20 and 22.22 through 22.26 in combination with Relations 22.21 and 22.27, some general predictions concerning the effective elastic properties of ribbed plates or shells can be made that are independent of the specific shape of the unit cell. In particular, it turns out that in the case of a homogeneous isotropic material we have

$$\langle b_{11}^{11} \rangle = v_{21}^2 \langle b_{22}^{22} \rangle + E_1 \langle 1 \rangle \qquad \langle b_{11}^{22} \rangle = \langle b_{22}^{11} \rangle = v_{21} \langle b_{22}^{22} \rangle$$

$$\langle b_{11}^{*11} \rangle = v_{21}^2 \langle b_{22}^{*22} \rangle + E_1 \langle z \rangle \qquad \langle b_{11}^{*22} \rangle = \langle b_{22}^{*11} \rangle = v_{21} \langle b_{22}^{*22} \rangle \quad (22.28)$$

$$\langle zb_{11}^{*11} \rangle = v_{21}^2 \langle zb_{22}^{*22} \rangle + E_1 \langle z^2 \rangle \quad \langle zb_{11}^{*22} \rangle = \langle zb_{22}^{*11} \rangle = v_{21} \langle zb_{22}^{*22} \rangle$$

The proof of these relations is as follows. Using expressions for $\tau_{22}^{\lambda\mu}$ and $t_{33}^{\lambda\mu}$ from Equations 22.18, it is easily verified that

$$\frac{1}{h_2} \frac{\partial U_2^{\lambda\mu}}{\partial \xi_2} = \frac{c_{33}\tau_{22}^{\lambda\mu} - c_{23}\tau_{33}^{\lambda\mu}}{c_{23}c_{33} - c_{23}^2}$$

$$\frac{\partial U_3^{\lambda\mu}}{\partial z} = \frac{c_{22}\tau_{33}^{\lambda\mu} - c_{23}\tau_{22}^{\lambda\mu}}{c_{22}c_{33} - c_{23}^2} \qquad (22.29)$$

which when substituted into Equations 22.18 for $\tau_{11}^{\lambda\mu}$ gives

$$\tau_{11}^{\lambda\mu} = v_{21}\tau_{22}^{\lambda\mu} + v_{31}\tau_{33}^{\lambda\mu} \qquad (22.30)$$

with the help of expressions for v_{22} and v_{31} in Equations 22.4).

From Equations 22.24, proceeding in a precisely similar manner, we find

$$\bar{\tau}_{11}^{\lambda\mu} = \nu_{21}\bar{\tau}_{22}^{\lambda\mu} + \nu_{31}\bar{\tau}_{33}^{\lambda\mu} \tag{22.31}$$

Now using Equations 22.21 and 22.30 along with Equations 22.4, we find that

$$b_{11}^{11} = c_{11} + \nu_{21}\tau_{22}^{11} + \nu_{31}\tau_{33}^{11}$$

$$= c_{11} - \nu_{21}c_{21} - \nu_{31}c_{31} + \nu_{21}b_{22}^{11} + \nu_{31}b_{33}^{11}$$

$$= E_1 + \nu_{21}b_{22}^{11} + \nu_{31}b_{33}^{11}$$

$$b_{11}^{22} = c_{12} + \nu_{21}\tau_{22}^{22} + \nu_{31}\tau_{33}^{22}$$

$$= c_{12} - \nu_{21}c_{22} - \nu_{31}c_{32} + \nu_{21}b_{22}^{22} + \nu_{31}b_{33}^{22}$$

$$= \nu_{21}b_{22}^{22} + \nu_{31}b_{33}^{22}$$

Using Equations 22.27 and 22.31 we find that, for the general case of a two-dimensional region Ω,

$$b_{11}^{11} = E_1 + \nu_{21}b_{22}^{11} + \nu_{31}b_{33}^{11} \tag{22.32}$$

$$b_{11}^{22} = \nu_{21}b_{22}^{22} + \nu_{31}b_{33}^{22} \tag{22.33}$$

$$b_{11}^{*11} = zE_1 + \nu_{21}b_{22}^{*11} + \nu_{31}b_{33}^{*11} \tag{22.34}$$

$$b_{11}^{*22} = \nu_{21}b_{22}^{*22} + \nu_{31}b_{33}^{*22} \tag{22.35}$$

Letting Equation 19.27 act on Equation 22.32 and using the general formula Equation 19.62 now gives, for a homogeneous material,

$$\langle b_{11}^{11} \rangle = \nu_{21}\langle b_{22}^{11} \rangle + E_1\langle 1 \rangle \qquad \langle b_{11}^{22} \rangle = \nu_{21}\langle b_{22}^{22} \rangle \tag{22.36}$$

which when combined with the general properties of the effective elastic moduli yields the first line in Equations 22.28.

Averaging Equation 22.33 after first multiplying by z, and using Equation 19.62 we find

$$\langle zb_{11}^{22} \rangle = \nu_{21}\langle zb_{22}^{22} \rangle$$

which, by Equations 19.76, may be written as

$$\langle b_{22}^{*11} \rangle = \nu_{21}\langle b_{22}^{*22} \rangle \tag{22.37}$$

Next average Equations 22.34 and 22.35 using Equations 19.76 again. This gives

$$\langle b_{11}^{*11}\rangle = \nu_{21}\langle b_{22}^{*11}\rangle + E_1\langle z\rangle \qquad \langle b_{11}^{*22}\rangle = \nu_{21}\langle b_{22}^{*22}\rangle \qquad (22.38)$$

and by combining Equations 22.37 and 22.38, the second line in Equations 22.28 is obtained.

The third and last line in Equations 22.28 follows on averaging Equations 22.24 and 22.35, after first multiplying them by z, and with the aid of Equations 18.62 and 19.76.

We note that Formulas 22.28 are perfectly general as far as two-dimensional problems are concerned and that they lead to considerable simplification in the homogenized-shell elastic relations (19.88).

22.4. Formulation of the local heat conduction and thermoelastic problems

The effective thermal conductivities occurring in the homogenized heat conduction relations (Equations 20.35) are found by solving the local heat conduction problems for the functions $l_{i\mu}(\xi, z)$ and $l_{i\mu}^*(\xi, z)$, which are periodic in ξ_1 and ξ_2 with respective periods A_1 and A_2. These local problems divide themselves into two groups of problems for the orthotropic region, (l) and (l^*), posed by Equations 20.24, 20.26, and 20.28 through 20.30 and Equations 20.24, 20.27 through 20.29, and 20.31, respectively.

The heat conductivity tensor of an orthotropic material is of the form

$$\lambda_{ij} = \begin{bmatrix} \lambda_2 & 0 & 0 \\ 0 & \lambda_2 & 0 \\ 0 & 0 & \lambda_3 \end{bmatrix} \qquad (22.39)$$

and from Equations 20.24 it therefore follows that

$$l_{11} = \lambda_1 = \frac{\lambda_1}{h_1}\frac{\partial W_1}{\partial \xi_1}, \qquad l_{21} = \frac{\lambda_2}{h_2}\frac{\partial W_1}{\partial \xi_2} \qquad l_{31} = \lambda_2\frac{\partial W_1}{\partial z}$$

$$l_{12} = \frac{\lambda_1}{h_1}\frac{\partial W_2}{\partial \xi_1} \qquad l_{22} = \lambda_2 + \frac{\lambda_2}{h_2}\frac{\partial W_2}{\partial \xi_2} \qquad l_{32} = \lambda_3\frac{\partial W_2}{\partial z} \qquad (22.40)$$

$$l_{11}^* = z\lambda_1 + \frac{\lambda_1}{h_1}\frac{\partial W_1^*}{\partial \xi_1} \qquad l_{21}^* = \frac{\lambda_2}{h_2}\frac{\partial W_1^*}{\partial \xi_2} \qquad l_{31}^* = \lambda_3\frac{\partial W_1^*}{\partial z}$$

$$l_{12}^* = \frac{\lambda_1}{h_1}\frac{\partial W_2^*}{\partial \xi_1} \qquad l_{22}^* = z\lambda_2 + \frac{\lambda_2}{h_2}\frac{\partial W_2^*}{\partial \xi_2} \qquad l_{32}^* = \lambda_3\frac{\partial W_2^*}{\partial z} \qquad (22.41)$$

Substituting Equations 22.40 into Equations 20.26 yields

$$\frac{\lambda_\beta}{h_\beta^2}\frac{\partial^2 W_1}{\partial \xi_\beta^2} + \lambda_3 \frac{\partial^2 W_1}{\partial z^2} = 0 \tag{22.42}$$

$$\frac{\lambda_\beta}{h_\beta^2}\frac{\partial^2 W_2}{\partial \xi_\beta^2} + \lambda_3 \frac{\partial^2 W_2}{\partial z^2} = 0 \tag{22.43}$$

and from Equations 20.29, 20.30, and 22.40, the conditions to be satisfied on the boundary surfaces are

$$\frac{\lambda_\beta}{h_\beta}\frac{\partial W_1}{\partial \xi_\beta}n_\beta + \lambda_3 \frac{\partial W_1}{\partial z}n_3 = -\frac{\lambda_1}{h_2}n_1 \tag{22.44}$$

$$\frac{\lambda_\beta}{h_\beta^2}\frac{\partial W_2}{\partial \xi_\beta}n_\beta + \lambda_3 \frac{\partial W_2}{\partial z}n_3 = -\frac{\lambda_2}{h_2}n_2 \tag{22.45}$$

The local problem (*l*) thus divides itself into two uncoupled problems: (*l1*) (given by Equations 22.42 and 22.44) and (*l2*) (given by Equations 22.43 and 22.45), the solutions of which determine the functions $W_1(\xi, z)$ and $W_2(\xi, z)$, which are periodic in ξ_1 and ξ_2 with respective periods A_1 and A_2.

Similarly, from Equations 22.41, 20.27, 20.29, and 20.31 we obtain the formulation of the problem (*l**) which also decouples into two independent problems: (*l*1*) for the function $W_1^*(\xi_1, z)$; (*l*2*) for $W_2(\xi, z)$. That is,

$$\frac{\lambda_\beta}{h_\beta^2}\frac{\partial^2 W_1^*}{\partial \xi_\beta^2} + \lambda_3 \frac{\partial^2 W_1^*}{\partial z^2} = 0$$

$$\frac{\lambda_\beta}{h_\beta^2}\frac{\partial W_1^*}{\partial \xi_\beta}n_\beta + \lambda_3 \frac{\partial W_1^*}{\partial z}n_3 = -z\frac{\lambda_1}{h_1}n_1 \tag{22.46}$$

$$\frac{\lambda_\beta}{h_\beta^2}\frac{\partial^2 W_2^*}{\partial \xi_\beta^2} + \lambda_3 \frac{\partial^2 W_2^*}{\partial z^2} = 0$$

$$\frac{\lambda_\beta}{h_\beta^2}\frac{\partial W_2^*}{\partial \xi_\beta}n_\beta + \lambda_3 \frac{\partial W_2^*}{\partial z}n_3 = -z\frac{\lambda_2}{h_2}n_2 \tag{22.47}$$

where $W_1^*(\xi, z)$ and $W_2^*(\xi, z)$ are also periodic in ξ_1 and ξ_2 with periods A_1 and A_2.

These local problem formulations should be complemented by Relations 20.28 in order to secure the uniqueness of their solutions.

The local thermoelastic problems given by Equations 21.9, 21.11, 21.15, and 21.16 give us the functions $s_{ij}(\xi, z)$ and $s_{ij}(\xi, z)$, from which the effective thermoelastic coefficients involved in the constitutive relation (Equation 21.33) are calculated. The local problems for these functions again form two groups of problems, (s) and (s^*), for the functions $S_m(\xi, z)$ and $S_m^*(\xi, z)$, respectively. These local problems may be represented in a form entirely analogous to the local elastic problems given by Equations 22.6 through 22.10 and 22.11 through 22.15, with appropriate modifications in Relations 19.41 and 19.53 for elastic problems and in Equations 21.9 for thermoelastic problems.

We remind the reader, with reference to Section 21, that the evaluation of the effective thermoelastic properties does not actually presuppose the solution of the preceding problems and can be carried out instead by Formulas 21.17, in which only the local elastic solutions are needed.

23. AN APPROXIMATE METHOD FOR DETERMINING THE EFFECTIVE PROPERTIES OF RIBBED AND WAFER SHELLS

23.1. Effective elastic moduli of ribbed and wafer shells of orthotropic material

The structure we are concerned with in this section is the eccentrically reinforced wafer type of shell (or plate), whose unit cell is shown schematically in Figure 23.1. Neglecting unavoidable interelemental fillets, as we do in the figure, it is seen that the unit cell consists of three mutually perpendicular plane elements, or plates.

An approximate solution of the local problems relevant to this kind of geometry may be found under the assumption that the thickness of each of the three elements is small as compared with its other two dimensions, that is,

$$t_1 \ll h_2 \qquad t_2 \ll h_1 \qquad H \sim h_1, h_2 \tag{23.1}$$

The local problems can then be approximately solved for each of the cell elements separately and, although the near-joint stress concentrations clearly remain unaccounted for in this approach, one feels confident in believing that the overall properties of the shell will be predicted accurately enough because the troublesome regions are highly localized at the joints and contribute little to integrals over the unit cell. The error

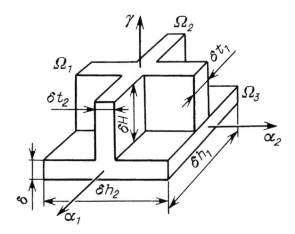

FIGURE 23.1. Unit cell for a wafer shell.

introduced in this way will be estimated in a more refined analysis (Section 25), with proper attention to all the interactions involved.

If we change to the coordinate system ξ_1, ξ_2, z, the unit cell of the problem will be shown as Figure 23.2, where

$$\delta_1 = \frac{t_1 A_1}{h_1} \qquad \delta_2 = \frac{t_2 A_2}{h_2} \tag{23.2}$$

and if Conditions 23.1 are fulfilled, the local problems of interest can be solved independently of the elements Ω_1, Ω_2, and Ω_3 shown in Figure 23.3. We begin our discussion with the problem $(b11)$ posed by Equations 22.6 through 22.9 and 19.57 with $\lambda\mu = 11$.

(a) In region Ω_3 defined by $|\xi_1| < A_1/2$, $|\xi_2| < A_2/2$, $|z| < 1/2$, the functions τ_{13}^{11}, τ_{23}^{11}, and τ_{33}^{11} are supposed to satisfy the following conditions on the surfaces $z = \pm(1/2)$:

Because of the periodicity in ξ_1 and ξ_2, this implies that

$$U_1^{11} = U_2^{11} = 0 \quad \text{and} \quad U_3^{11} = -\frac{C_{31}}{C_{33}} z \tag{23.3}$$

and hence

$$\tau_{11}^{11} = -\frac{C_{13}^2}{C_{33}} \qquad \tau_{22}^{11} = -\frac{C_{23} C_{13}}{C_{33}}$$

$$\tau_{33}^{11} = -C_{13} \qquad \tau_{23}^{11} = \tau_{13}^{11} = \tau_{12}^{11} = 0 \tag{23.4}$$

everywhere in Ω_3.

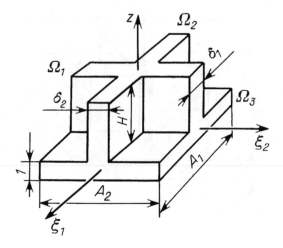

FIGURE 23.2. Unit cell of a wafer shell in the coordinate system ξ_1, ξ_2, z.

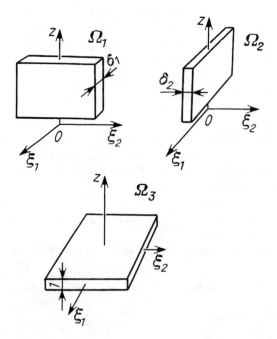

FIGURE 23.3. Individual elements of the unit cell of a wafer shell.

(b) In region Ω_1 defined by $|\xi_1| < \delta_1/2, |\xi_2| < A_2/2, 1/2 < z < 1/2 +$ H, the boundary conditions given by Equations 22.8 and 22.9 take the form

$$\tau_{11}^{11} = -c_{11} \qquad \tau_{12}^{11} = \tau_{13}^{11} = 0 \qquad (\xi_1 = \pm\delta_1/2)$$

$$\tau_{13}^{11} = \tau_{23}^{11} = 0 \quad \tau_{33}^{11} = -c_{13} \qquad (z = \tfrac{1}{2}, z = \tfrac{1}{2} + H)$$

giving

$$U_1^{11} = -h_1\xi_1 \quad \text{and} \quad U_2^{11} = U_3^{11} = 0 \tag{23.5}$$

in view of the periodicity in ξ_2, so that everywhere in Ω_1, we have

$$\tau_{11}^{11} = -c_{11} \qquad \tau_{22}^{11} = -c_{12} \qquad \tau_{33}^{11} - c_{13} \qquad \tau_{23}^{11} = \tau_{13}^{11} = \tau_{12}^{11} = 0 \tag{23.6}$$

(c) In region Ω_2 defined by $|\xi_1| < A_1/2,\ |\xi_2| < \delta_2/2,\ 1/2 < z <$ $1/2 + H$, the face conditions are

$$\tau_{12}^{11} = \tau_{23}^{11} = 0 \qquad \tau_{22}^{11} = -c_{12} \qquad (\xi_2 = \pm\delta_2/2),$$

$$\tau_{13}^{11} = \tau_{23}^{11} = 0 \qquad \tau_{33}^{11} = -c_{13} \quad (x = \tfrac{1}{2}, z = \tfrac{1}{2} + H)$$

and because of the periodicity in ξ_1

$$U_1^{11} = 0$$

$$U_2^{11} = h_2 = \frac{c_{13}c_{23} - c_{12}c_{33}}{c_{22}c_{33} - c_{23}^2}\xi_2$$

$$U_3^{11} = \frac{c_{12}c_{23} - c_{13}c_{22}}{c_{22}c_{33} - c_{23}^2}z \tag{23.7}$$

which means that

$$\tau_{11}^{11} = \frac{2c_{12}c_{13}c_{23} - c_{12}^2c_{33} - c_{13}^2c_{22}}{c_{22}c_{33} - c_{23}^2} \tag{23.8}$$

$$\tau_{22}^{11} = -c_{12} \qquad \tau_{33}^{11} = -c_{13} \quad \tau_{23}^{11} = \tau_{13}^{11} = \tau_{12}^{11} = 0$$

in region Ω_2.

Combining Solutions 23.4, 23.6, and 23.8, and using Equations 22.4 and 22.10, one finds

$$
b_{11}^{11} = \begin{cases} c_{11} - \dfrac{c_{13}^2}{c_{33}} = \dfrac{E_1}{1 - \nu_{12}\nu_{21}} & \text{in } \Omega_3 \\[3mm] 0 & \text{in } \Omega_1 \\[3mm] c_{11} + \dfrac{2c_{12}c_{13}c_{23} - c_{12}^2 c_{33} - c_{13}^2 c_{22}}{c_{22}c_{33} - c_{23}^2} = E_1 & \text{in } \Omega_2 \end{cases}
$$

$$\tag{23.9}$$

$$
b_{22}^{11} = \begin{cases} c_{12} - \dfrac{c_{23}c_{13}}{c_{33}} = \dfrac{\nu_{12}E_1}{1 - \nu_{12}\nu_{21}} = \dfrac{\nu_{21}E_2}{1 - \nu_{12}\nu_{21}} & \text{in } \Omega_3 \\[3mm] 0 & \text{in } \Omega_1, \Omega_2 \end{cases}
$$

$$
b_{12}^{11} = b_{33}^{11} = b_{32}^{11} = b_{31}^{11} = 0
$$

Solutions of the problems $(b22)$ and $(b12)$ are obtained in a precisely similar manner. We give the solution of the problem $(b22)$ first.

(a) In region Ω_3,

$$
U_1^{22} = U_2^{22} = 0 \qquad U_3^{22} = -\frac{c_{32}}{c_{33}}z
$$

$$
\tau_{11}^{22} = -\frac{c_{13}c_{23}}{c_{33}} \qquad \tau_{22}^{22} = -\frac{c_{23}}{c_{33}} \tag{23.10}
$$

$$
\tau_{33}^{22} = -c_{32} \qquad \tau_{23}^{22} = \tau_{13}^{22} = \tau_{12}^{22} = 0
$$

(b) In region Ω_1,

$$
U_1^{22} = h_1 \frac{c_{13}c_{23} - c_{12}c_{33}}{c_{11}c_{33} - c_{13}^2}\xi_1 \qquad U_2^{22} = 0
$$

$$
U_3^{22} = \frac{c_{13}c_{12} - c_{11}c_{23}}{c_{11}c_{33} - c_{13}^2}z
$$

$$\tag{23.11}$$

$$
\tau_{11}^{22} = -c_{12} \qquad \tau_{22}^{22} = \frac{2c_{12}c_{13}c_{23} - c_{12}^2 c_{33} - c_{11} - c_{23}^2}{c_{11}c_{33} - c_{13}^2}
$$

$$
\tau_{33}^{22} = -c_{23} \qquad \tau_{23}^{22} = \tau_{13}^{22} = \tau_{12}^{22} = 0
$$

(c) In region Ω_2,

$$U_1^{22} = U_3^{33} = 0 \qquad U_2^{22} = -h_2\xi_2$$

$$\tau_{11}^{22} = -c_{12} \qquad \tau_{22}^{22} = -c_{22} \qquad \tau_{33}^{22} = -c_{23} \qquad (23.12)$$

$$\tau_{23}^{22} = \tau_{13}^{22} = \tau_{12}^{22} = 0$$

Using Equations 23.10 through 23.12 together with Equations 22.3 and 22.10, we find that

$$b_{12}^{22} = \begin{cases} c_{12} - \dfrac{c_{13}c_{23}}{c_{33}} = \dfrac{\nu_{12}E_1}{1-\nu_{12}\nu_{21}} = \dfrac{\nu_{21}E_2}{1-\nu_{12}\nu_{21}} & \text{in } \Omega_3 \\[2mm] 0 & \text{in } \Omega_1, \Omega_2 \end{cases}$$

$$b_{22}^{22} = \begin{cases} c_{22} - \dfrac{c_{23}^2}{c_{33}} = \dfrac{E_2}{1-\nu_{12}\nu_{21}} & \text{in } \Omega_3 \\[2mm] c_{22} + \dfrac{2c_{12}c_{13}c_{23}-c_{12}^2c_{33}-c_{11}c_{23}^2}{c_{11}c_{33}-c_{13}^2} = E_2 & \text{in } \Omega_1 \\[2mm] 0 & \text{in } \Omega_2 \end{cases} \qquad (23.13)$$

$$b_{12}^{22} = b_{33}^{22} = b_{32}^{22} = b_{31}^{22} = 0$$

The solution of the problem (b12) is as follows:
(a) In region Ω_3,

$$U_1^{12} = U_2^{12} = U_3^{12} = 0$$

$$\tau_{11}^{12} = \tau_{22}^{12} = \tau_{33}^{12} = \tau_{23}^{12} = \tau_{13}^{12} = \tau_{12}^{12} = 0 \qquad (23.14)$$

(b) In region Ω_1,

$$U_1^{12} = U_3^{12} = 0 \qquad U_2^{12} = -h_1\xi_1$$

$$\tau_{12}^{12} = -c_{66} \qquad \tau_{11}^{12} = \tau_{22}^{12} = \tau_{33}^{12} = \tau_{23}^{12} = \tau_{13}^{12} = 0 \qquad (23.15)$$

(c) In region Ω_2,

$$U_1^{12} = -h_2\xi_2 \qquad U_2^{12} = U_3^{12} = 0$$

$$\tau_{12}^{12} = -c_{66} \qquad \tau_{11}^{12} = \tau_{22}^{12} = \tau_{33}^{12} = \tau_{23}^{12} = \tau_{13}^{12} = 0 \qquad (23.16)$$

Hence by Equations 22.4 and 22.10

$$b_{12}^{12} = \begin{cases} c_{66} = G_{12} & \text{in } \Omega_3 \\ 0 & \text{in } \Omega_1, \Omega_2 \end{cases} \tag{23.17}$$

$$b_{11}^{12} = b_{22}^{12} = b_{33}^{12} = b_{23}^{12} = b_{13}^{12} = 0$$

To proceed further it is necessary to construct approximate solutions for the local problems of the group ($b*$). Here again they are obtained by superposing the solutions for regions Ω_1, Ω_2, and Ω_3.

The approximate solution for the problem ($b*11$) is as follows:

(a) In region Ω_3,

$$V_1^{11} = V_2^{11} = 0 \qquad V_3^{11} = -\frac{c_{31}}{c_{33}}\frac{z^2}{2}$$

$$\bar\tau_{11}^{11} = -z\frac{c_{13}^2}{c_{33}} \qquad \bar\tau_{22}^{11} = -\frac{c_{23}c_{13}}{c_{33}}z \tag{23.18}$$

$$\bar\tau_{33}^{11} = -zc_{13} \qquad \bar\tau_{23}^{11} = \bar\tau_{13}^{11} = \bar\tau_{12}^{11} = 0$$

(b) In region Ω_1,

$$V_1^{11} = -h_1\xi_1 z \qquad V_2^{11} = 0 \qquad V_3^{11} = \frac{h_1^2}{2}\left(\xi_1^2 - \frac{\delta_1^2}{12}\right)$$

$$\bar\tau_{11}^{11} = -zc_{11} \qquad \bar\tau_{22}^{11} = -zc_{21} \tag{23.19}$$

$$\bar\tau_{33}^{11} = -zc_{31} \qquad \bar\tau_{23}^{11} = \bar\tau_{13}^{11} = \bar\tau_{12}^{11} = 0$$

(c) In region Ω_2,

$$V_1^{11} = 0 \qquad V_2^{11} = h_2\frac{c_{13}c_{23} - c_{12}c_{33}}{c_{22}c_{33} - c_{23}^2}z\xi_2$$

$$V_3^{11} = \frac{c_{12}c_{23} - c_{13}c_{22}}{c_{22}c_{33} - c_{23}^2}\frac{z^2}{2} - \frac{h_2^2}{2}\frac{c_{13}c_{23} - c_{12}c_{33}}{c_{22}c_{33} - c_{23}^2}\left(\xi_2^2 - \frac{\delta_2^2}{12}\right)$$

$$\bar\tau_{11}^{11} = \frac{2c_{12}c_{13}c_{23} - c_{12}^2c_{33} - c_{13}^2c_{22}}{c_{22}c_{33} - c_{23}^2}z \tag{23.20}$$

$$\bar\tau_{22}^{11} = -zc_{12} \qquad \bar\tau_{33}^{11} = -zc_{13} \qquad \bar\tau_{23}^{11} = \bar\tau_{13}^{11} = \bar\tau_{12}^{11} = 0$$

Using Equations 23.19 and 23.20 along with Equations 22.4 and 22.15 now gives expressions for the functions b_{ij}^{11}, which when compared with Equations 23.9 lead to the result

$$b_{ij}^{*11} = zb_{ij}^{11} \tag{23.21}$$

The solution of the problem $(b*22)$ is as follows:
(a) In region Ω_3,

$$V_1^{22} = V_2^{22} = 0 \qquad V_3^{22} = -\frac{c_{23}}{c_{33}} \frac{z^2}{2}$$

$$\bar{\tau}_{11}^{22} = -\frac{c_{13}c_{23}}{c_{33}} z \qquad \bar{\tau}_{22}^{22} = -\frac{c_{23}^2}{c_{33}} z \tag{23.22}$$

$$\bar{\tau}_{33}^{22} = -zc_{23} \qquad \bar{\tau}_{23}^{22} = \bar{\tau}_{13}^{22} = \bar{\tau}_{12}^{22} = 0$$

(b) In region Ω_1,

$$V_1^{22} = h_1 \frac{c_{13}c_{23} - c_{12}c_{33}}{c_{11}c_{33} - c_{13}^2} z\xi_1 \qquad V_2^{22} = 0$$

$$V_3^{22} = \frac{c_{13}c_{12} - c_{11}c_{23}}{c_{11}c_{33} - c_{13}^2} \frac{z^2}{2} - \frac{h_1^2}{2} \frac{c_{13}c_{23} - c_{12}c_{33}}{c_{11}c_{33} - c_{13}^2} \left(\xi_1^2 - \frac{\delta_1^2}{12} \right) \tag{23.23}$$

$$\bar{\tau}_{11}^{22} = zc_{12} \qquad \bar{\tau}_{22}^{22} = \frac{2c_{12}c_{13}c_{23} - c_{12}^2 c_{33} - c_{11}c_{23}^2}{c_{11}c_{33} - c_{13}^2} z$$

$$\bar{\tau}_{33}^{22} = -zc_{23} \qquad \bar{\tau}_{23}^{22} = \bar{\tau}_{13}^{22} = \bar{\tau}_{12}^{22} = 0$$

(c) In region Ω_2,

$$V_1^{22} = c \qquad V_2^{22} = -h_2\xi_2 z \qquad V_3^{22} = \frac{h_2^2}{2} \left(\xi_2^2 - \frac{\delta_2^2}{12} \right)$$

$$\bar{\tau}_{11}^{22} = -zc_{12} \qquad \bar{\tau}_{22}^{22} = -zc_{22} \tag{23.24}$$

$$\bar{\tau}_{22}^{22} = -zc_{23} \qquad \bar{\tau}_{23}^{22} = \bar{\tau}_{13}^{22} = \bar{\tau}_{12}^{22} = 0$$

From Equations 23.22 through 23.24, 22.4 and 22.15 one may now obtain expressions for the functions b_{ij}^{*22} and by using Equations 23.13 we find that

$$b_{ij}^{*22} = zb_{ij}^{22} \tag{23.25}$$

Turning now to the problem (b^*12) we must admit that its solution is not so elementary as we have seen to be the case with the problems (b^*11) and (b^*22).

(a) In region Ω_3 we have

$$V_1^{12} = V_2^{12} = V_3^{12} = 0$$

$$\overline{\tau}_{11}^{12} = \overline{\tau}_{22}^{12} = \overline{\tau}_{33}^{12} = \overline{\tau}_{23}^{12} = \overline{\tau}_{13}^{12} = \overline{\tau}_{12}^{12} = 0 \tag{23.26}$$

(b) In region Ω_1, taking into account the boundary conditions

$$\overline{\tau}_{11}^{12} = 0 \qquad \overline{\tau}_{12}^{12} = -zc_{66} \qquad \overline{\tau}_{13}^{12} = 0 \qquad (\xi_1 = \pm \delta_1/2)$$

$$\overline{\tau}_{13}^{12} = 0 \qquad \overline{\tau}_{23}^{12} = 0 \qquad \overline{\tau}_{33}^{12} = 0 \qquad (z = \tfrac{1}{2}, z = \tfrac{1}{2} + H)$$

and the periodicity in ξ_2, we may set

$$\overline{\tau}_{11}^{12} = \overline{\tau}_{33}^{12} = \overline{\tau}_{13}^{12} = 0 \quad \text{everywhere in } \Omega_1$$

But then $V_1^{12} = 0$, $V_2^{12} = V_2^{12}(\xi_1, z)$, $V_3^{12} = 0$, and the only nonzero functions are

$$\overline{\tau}_{12}^{12} = \frac{c_{66}}{h_1} \frac{\partial V_2^{12}(\xi_1, z)}{\partial \xi_1} \quad \text{and} \quad \overline{\tau}_{23}^{12} c_{44} \frac{\partial V_2^{12}}{\partial z} \tag{23.27}$$

which must be related by

$$\frac{1}{h_1} \frac{\partial \overline{\tau}_{12}^{12}}{\partial \xi_1} + \frac{\partial \overline{\tau}_{23}^{12}}{\partial z} = 0$$

in region $|\xi_1| < \delta_1/2, 1/2 < z < 1/2 + H$, and must satisfy the boundary conditions

$$\overline{\tau}_{12}^{12} = -zc_{66} \qquad (\xi_1 = \pm \delta_1/2)$$

$$\overline{\tau}_{23}^{12} = 0 \qquad (z = \tfrac{1}{2}, z = \tfrac{1}{2} + H)$$

A boundary value problem for $V_2^{12}(\xi_1, z)$, can now be obtained from Equations 23.27 and 22.4 as

$$\frac{G_{12}}{h_1^2}\frac{\partial^2 V_2^{12}}{\partial \xi_1^2} + G_{23}\frac{\partial^2 V_2^{12}}{\partial z^2} = 0$$

$$\frac{1}{h_1}\frac{\partial V_2^{12}}{\partial \xi_1} = -z \qquad (\xi_1 = \pm\delta_1/2) \qquad (23.28)$$

$$\frac{\partial V_2^{12}}{\partial z} = 0 \qquad \left(z = \frac{1}{2}, z = 1/2 + H\right)$$

and the solution satisfying Conditions 19.57 is found to be given by

$$V_2^{12} = -\frac{h_1}{2}(H+1)\xi_1$$

$$+ \sqrt{\frac{G_{12}}{G_{23}}\frac{2H^2}{\pi^3}}$$

$$\times \sum_{n=1}^{\infty} \frac{[1-(-1)^n]\sinh\left(\dfrac{\pi n h_1 \xi_1}{H}\sqrt{\dfrac{G_{23}}{G_{12}}}\right)}{n^3\cosh\left(\dfrac{\pi n h_1 \delta_1}{2H}\sqrt{\dfrac{G_{23}}{G_{12}}}\right)}\cos\frac{\pi n\left(z-\dfrac{1}{2}\right)}{H}$$

$$(23.29)$$

Finally, in region Ω_1 we have

$$\bar{\tau}_{11}^{12} = \bar{\tau}_{22}^{12} = \bar{\tau}_{33}^{12} = \bar{\tau}_{13}^{12} = 0$$

$$\bar{\tau}_{12}^{12} = -G_{12}\frac{H+1}{2} + G_{12}\frac{2H}{\pi^2}$$

$$\times \sum_{n=1}^{\infty} \frac{[1-(-1)^n]\cosh\left(\dfrac{\pi n h_1 \xi_1}{H}\sqrt{\dfrac{G_{23}}{G_{12}}}\right)}{n^2\cosh\left(\dfrac{\pi n h_2 \delta_1}{2H}\sqrt{\dfrac{G_{23}}{G_{12}}}\right)}\cos\frac{\pi n\left(z-\dfrac{1}{2}\right)}{H}$$

$$\bar{\tau}_{23}^{12} = -\sqrt{G_{23}G_{12}}\,\frac{2H}{\pi^2}\sum_{n=1}^{\infty}\frac{[1-(-1)^n]\sinh\left(\dfrac{\pi nh_1\xi_1}{H}\sqrt{\dfrac{G_{23}}{G_{12}}}\right)}{n^2\cosh\left(\dfrac{\pi nh_1\delta_1}{2H}\sqrt{\dfrac{G_{23}}{G_{12}}}\right)}$$

$$\times \sin\frac{\pi n\left(z-\dfrac{1}{2}\right)}{H} \tag{23.30}$$

(c) The solution for region Ω_2 is carried out in much the same way as for Ω_1. The result is

$$V_1^{12} = \frac{h_2}{2}(H+1)\xi_2 + \sqrt{\frac{G_{12}}{G_{13}}}\,\frac{2H^2}{\pi^3}$$

$$\times \sum_{n=1}^{\infty}\frac{[1-(-1)^n]\sinh\left(\dfrac{\pi nh_2\xi_2}{H}\sqrt{\dfrac{G_{13}}{G_{12}}}\right)}{n^3\cosh\left(\dfrac{\pi nh_2\delta_2}{2H}\sqrt{\dfrac{G_{13}}{G_{12}}}\right)}\cos\frac{\pi n\left(z-\dfrac{1}{2}\right)}{H}$$

$$V_2^{12} = 0 \qquad V_3^{12} = 0$$

$$\bar{\tau}_{11}^{12} = \bar{\tau}_{22}^{12} = \bar{\tau}_{33}^{12} = \bar{\tau}_{23}^{12} = 0$$

$$\bar{\tau}_{12}^{12} = -G_{12}\frac{H+1}{2} + G_{12}\frac{2H}{\pi^2}$$

$$\times \sum_{n=1}^{\infty}\frac{[1-(-1)^n]\cosh\left(\dfrac{\pi nh_2\xi_2}{H}\sqrt{\dfrac{G_{13}}{G_{12}}}\right)}{n^2\cosh\left(\dfrac{\pi nh_2\delta_2}{2H}\sqrt{\dfrac{G_{13}}{G_{12}}}\right)}\cos\frac{\pi n\left(z-\dfrac{1}{2}\right)}{H}$$

$$\tag{23.31}$$

$$\bar{\tau}_{13}^{12} = \sqrt{G_{13}G_{12}}\, \frac{2H}{\pi^2}$$

$$\times \sum_{n=1}^{\infty} \frac{[1 - (-1)^n]\sinh\left(\dfrac{\pi n h_2 \xi_2}{H}\sqrt{\dfrac{G_{13}}{G_{12}}}\right)}{n^2 \cosh\left(\dfrac{\pi n h_2 \delta_2}{2H}\sqrt{\dfrac{G_{13}}{G_{12}}}\right)} \sin\frac{\pi n\left(z - \dfrac{1}{2}\right)}{H}$$

From Equations 22.15, 23.26, 23.30, and 23.31 we find

$$b_{11}^{*12} = b_{22}^{*12} = b_{33}^{*12} = 0$$

$$
b_{12}^{*12} = \begin{cases}
zG_{12} & \text{in } \Omega_3 \\[4pt]
G_{12}\left(z - \dfrac{H+1}{2}\right) + G_{12}\dfrac{2H}{\pi^2} \\[6pt]
\quad \times \displaystyle\sum_{n=1}^{\infty} \frac{[1 - (-1)^n]}{n^2} \frac{\cosh\left(\dfrac{\pi n h_1 \xi_1}{H}\sqrt{\dfrac{G_{23}}{G_{12}}}\right)}{\cosh\left(\dfrac{\pi n h_1 \delta_1}{2H}\right)} \\[10pt]
\quad \times \cos\dfrac{\pi n\left(z - \dfrac{1}{2}\right)}{H} & \text{in } \Omega_1 \\[10pt]
G_{12}\left(z - \dfrac{H+1}{2}\right) + G_{12}\dfrac{2H}{\pi^2} \\[6pt]
\quad \times \displaystyle\sum_{n=1}^{\infty} \frac{[1 - (-1)^n]}{n^2} \frac{\cosh\left(\dfrac{\pi n h_2 \xi_2}{H}\sqrt{\dfrac{G_{13}}{G_{12}}}\right)}{\cosh\left(\dfrac{\pi n h_2 \delta_2}{2H}\sqrt{\dfrac{G_{13}}{G_{12}}}\right)} \\[10pt]
\quad \times \cos\dfrac{\pi n\left(z - \dfrac{1}{2}\right)}{H} & \text{in } \Omega_2
\end{cases}
$$

$$
b_{13}^{*12} = \begin{cases} \bar{\tau}_{13}^{12} & \text{in } \Omega_2 \\ 0 & \text{in } \Omega_1, \Omega_3 \end{cases}
\qquad
b_{23}^{*12} = \begin{cases} \bar{\tau}_{23}^{12} & \text{in } \Omega_1 \\ 0 & \text{in } \Omega_2, \Omega_3 \end{cases}
\qquad (23.32)
$$

Note that since $\langle \tau_{13}^{12} \rangle \Omega_2 = \langle \tau_{23}^{12} \rangle \Omega_1 = 0$ in the obvious notation, it follows that here we have not violated the truth of Relation 19.62, earlier proved in a general fashion.

Relations 23.9, 23.13, 23.17, 23.21, 23.25, and 23.32 enable one to obtain explicit expressions for the entire set of the effective elastic moduli of the wafer-type shell with the unit cell shown in Figure 23.1.

Noting that averaging in the sense of Equation 19.27 is actually integration over the coordinates y_1, y_2, and z, the following auxiliary formulas can be proved (c.f. Figure 23.1):

$$\langle 1 \rangle_{\Omega_1} = t_1 H/h_1 = F_1^{(b)}$$

$$\langle z \rangle_{\Omega_1} = (H^2 + H)t_1/(2h_1) = S_1^{(b)}$$

$$\langle z^2 \rangle_{\Omega_1} = t_1(4H^3 + 6H^2 + 3H)/(12h_1) = J_1^{(b)} \qquad (1 \leftrightarrow 2)$$

$$(23.33)$$

$$\langle 1 \rangle_{\Omega_3} = 1 \qquad \langle z \rangle_{\Omega_3} = 0 \qquad \langle z^2 \rangle_{\Omega_3} = \tfrac{1}{12}$$

where $F_1^{(b)}$ and $F_2^{(b)}$ are the cross-sectional areas, $S_1^{(b)}$ and $S_2^{(b)}$ are the static moments, and $J_1^{(b)}$ and $J_2^{(b)}$ are the inertia moments of the cross sections of the reinforcing elements Ω_1 and Ω_2 relative to the middle surface of the shell, are calculated in the coordinate system y_1, y_2, z.

If each of the elements Ω_k ($k = 1, 2, 3$) is made of an orthotropic material with the engineering properties $E_1^{(k)}$, $E_2^{(k)}$, $\nu_{21}^{(k)}$, $G_{12}^{(k)}$, $G_{13}^{(k)}$, and $G_{23}^{(k)}$, the nonzero effective moduli of the wafer-type shell of Figure 23.1 are given by the expressions

$$\langle b_{11}^{11} \rangle = \frac{E_1^{(3)}}{1 - \nu_{12}^{(3)} \nu_{21}^{(3)}} + E_1^{(2)} F_2^{(b)}$$

$$\langle b_{22}^{22} \rangle = \frac{E_2^{(3)}}{1 - \nu_{12}^{(3)} \nu_{21}^{(3)}} + E_2^{(1)} F_1^{(b)}$$

$$\langle b_{22}^{11} \rangle = \langle b_{11}^{22} \rangle = \frac{\nu_{12}^{(3)} E_1^{(3)}}{1 - \nu_{12}^{(3)} \nu_{21}^{(3)}} \qquad \langle b_{12}^{12} \rangle = G_{12}^{(3)}$$

$$\langle zb_{11}^{11} \rangle = \langle b_{11}^{*11} \rangle = E_1^{(2)} S_2^{(b)} \qquad \langle zb_{22}^{22} \rangle = \langle b_{22}^{*22} \rangle = E_2^{(1)} S_1^{(b)}$$

$$(23.34)$$

$$\langle zb_{11}^{*11} \rangle = \frac{E_1^{(3)}}{12(1 - \nu_{12}^{(3)} \nu_{21}^{(3)})} + E_1^{(2)} J_2^{(b)}$$

$$\langle zb_{22}^{*22} \rangle = \frac{E_2^{(3)}}{12(1 - v_{12}^{(3)}v_{21}^{(3)})} + E_2^{(1)}J_1^{(b)}$$

$$\langle zb_{22}^{*11} \rangle = \langle zb_{11}^{*22} \rangle = \frac{v_{12}^{(3)}E_1^{(3)}}{12(1 - v_{12}^{(3)}v_{21}^{(3)})}$$

$$\langle zb_{12}^{*12} \rangle = \frac{G_{12}^{(3)}}{12} + \frac{G_{12}^{(1)}}{12}\left(\frac{H^3 t_1}{h_1} - K_1 \right) + \frac{G_{12}^{(2)}}{12}\left(\frac{H^3 t_2}{h_2} - K_2 \right)$$

where K_1 and K_2 are defined by the equations

$$K_1 = \frac{96H^4}{\pi^5 A_1 h_1} \sqrt{\frac{G_{12}^{(1)}}{G_{23}^{(1)}}} \sum_{n=1}^{\infty} \frac{[1 - (-1)^n]}{n^5} \tanh\left(\frac{\pi n A_1 t_1}{2H} \sqrt{\frac{G_{23}^{(1)}}{G_{12}^{(1)}}} \right)$$

$$\tag{23.35}$$

$$K_2 = \frac{96H^4}{\pi^5 A_2 h_2} \sqrt{\frac{G_{12}^{(2)}}{G_{13}^{(2)}}} \sum_{n=1}^{\infty} \frac{[1 - (-1)^n]}{n^5} \tanh\left(\frac{\pi n A_2 t_2}{2H} \sqrt{\frac{G_{13}^{(2)}}{G_{12}^{(2)}}} \right)$$

Equations 23.34 may be employed for calculating the effective stiffness moduli of ribbed shells. In particular, if the strengthening ribs are directed along the coordinate axis α_1, the strengthening element Ω_1 will be absent in Figure 23.1, and in Equations 23.33 we set $t_1 = 0$, thereby also turning to zero the quantities of $F_1^{(b)}$, $S_1^{(b)}$, $J_i^{(b)}$, and K_1.

If the material is isotropic, Equations 23.34 take a simpler form and agree closely with results known from the structurally anisotropic theory of strengthened plates and shells (see, e.g., Korolev 1971, Elpatyevskii and Gavva 1983). A notable exception is the torsional stiffness $\langle zb_{12}^{*12} \rangle$, which in terms of the structural theory is given by

$$\langle zb_{12}^{*12} \rangle = \frac{E}{24(1 + v)}\left(1 + H\frac{t_1^3}{h_1} + H\frac{t_2^3}{h_2} \right) \tag{23.36}$$

The difference between Equations 23.34 and 23.35 on the one hand and Equation 23.36 on the other turns out to be considerable for high ribs. If, for example, we take $A_1 = A_2 = 1$, $v = 0.3$, $H = 20$, $h_1 = h_2 = 60$, and $t_1 = t_2 = 2$, then from Equations 23.34 and 23.35 we obtain $\langle zb_{12}^{*12} \rangle/E = 0.192$ whereas Equation 23.36 gives a result of 0.203, that is, 5.4% more. If $H = 10$, $h_1 = h_2 = 10$, and $t_1 = t_2 = 1$, then Formulas 23.34 and 23.35 yield $\langle zb_{12}^{*12} \rangle/E = 0.0921$, to be compared with 0.0962 from Equation 23.36. The percentage change is in this case 4.3%.

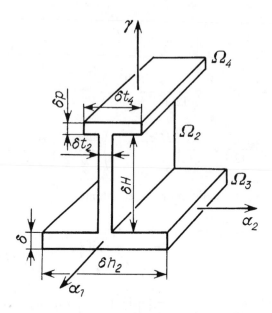

FIGURE 23.4.   Unit cell of a ribbed shell for the case of T-shaped ribs.

We shall now apply the above approximating scheme to the case of T-shaped ribs shown in Figure 23.4. Because the solutions for regions Ω_1 and Ω_2 are already known, clearly only the local problems associated with the element Ω_4 remain to be considered. By the same procedure used in the Ω_2 and Ω_3 problems, the nonzero moduli are found to be, for an isotropic material,

$$\langle b_{11}^{11} \rangle = \frac{E}{1-\nu^2} + E(F_2^{(b)} + F_4) \qquad \langle b_{22}^{22} \rangle = \frac{E}{1-\nu^2}$$

$$\langle b_{22}^{11} \rangle = \langle b_{11}^{22} \rangle = \frac{\nu E}{1-\nu^2} \qquad\qquad \langle b_{12}^{12} \rangle = G \qquad\qquad (23.37)$$

$$\langle zb_{11}^{11} \rangle = \langle b_{11}^{*11} \rangle = E(S_2^{(b)} + S_4) \qquad \langle zb_{11}^{*11} \rangle = \frac{E}{12(1-\nu^2)}$$

$$+ E(J_2^{(b)} + J_4)$$

$$\langle zb_{22}^{*22} \rangle = \frac{E}{12(1-\nu^2)} \qquad\qquad \langle zb_{22}^{*11} \rangle = \langle zb_{11}^{*22} \rangle = \frac{\nu E}{12(1-\nu^2)}$$

$$\langle zb_{12}^{*12} \rangle = \frac{G}{12} \left\{ 1 + \left(\frac{H^3 t_2}{h_2} - K_2 \right) \right.$$

$$\left. + \left[\frac{p^3 t_4}{h_2} + \frac{96 p^4}{\pi^5 A_2 h_2} \sum_{n=1}^{\infty} \frac{[1 - (-1)^n]}{n^5} \tanh \left(\frac{\pi n A_2 t_4}{2p} \right) \right] \right\}$$

where the quantities F_4, S_4, and J_4 are given by formulas analogous to Equations 23.33. For complete details of the derivation, see Parton et al. (1989b).

Changing to the coordinate system y_1, y_2, z (cf. Figure 23.4), we find

$$F_4 = \frac{p t_4}{h_2}$$

$$S_4 = \frac{p t_4 (1 + 2H + p)}{2 h_2}$$

$$J_4 = \frac{p t_4 [3(1 + 2H)(1 + 2H + 2p) + 4p^3]}{12 h_2}$$

Again with the exception of the torsional stiffness expression, Equations 23.37 are identical to the corresponding results of structurally anisotropic theory (Elpatyevskii and Gavva 1983) mentioned previously. The difference in the $\langle zb_{12}^{*12} \rangle$ expressions is associated with the square bracket term, which describes the contribution from the "shelf" (element Ω_4) and in which the second term (the sum) is absent in the structurally anisotropic theory. Calculations show, however, that this term changes the result substantially at large values of p and t_4. If, for example, we take $H = 20$, $h_2 = 60$, $t_2 = 2$, $t_4 = 15$, $p = 1.5$, and $A_2 = 1$, then for the value of $\langle zb_{12}^{*12} (G/12)^{-1}$ we have 4.510 from the structurally anisotropic theory and 4.289 (i.e., 4.9% less) from Equations 23.37. For $H = 30$, $h_2 = 80$, $t_2 = 3$, $t_4 = 20$, and $p = 2$, a 5.8% reduction is obtained, from 13.125 to 12.362.

23.2. Effective thermoelastic and heat conduction properties of wafer-type and ribbed shells of orthotropic materials

In this section we apply Equations 21.17 and the approximate solutions of the cell local problems to the calculation of the effective thermoelastic properties of the wafer shell of Figure 23.1 (Parton et al. 1986b, Kalamkarov et al. 1987d, Parton and Kalamkarov 1988a).

For an orthotropic material we have

$$
\alpha_{ij}^T = \begin{pmatrix} \alpha_1^T & 0 & 0 \\ 0 & \alpha_2^T & 0 \\ 0 & 0 & \alpha_3^T \end{pmatrix}
\tag{23.38}
$$

With this, using Equations 21.17, 23.9, 23.13, 23.17, 23.21, 23.25, and 23.32, the nonzero properties of the structure are found to be

$$
\delta\langle s_{11}\rangle = \frac{\alpha_1^{T(3)}E_1^{(3)} + \alpha_2^{T(3)}\nu_{12}^{(3)}E_1^{(3)}}{1 - \nu_{12}^{(3)}\nu_{21}^{(3)}} + \alpha_1^{T(2)}E_1^{(2)}F_2^{(b)}
$$

$$
\delta\langle s_{22}\rangle = \frac{\alpha_2^{T(3)}E_2^{(3)} + \alpha_1^{T(3)}\nu_{21}^{(3)}E_2^{(3)}}{1 - \nu_{12}^{(3)}\nu_{21}^{(3)}} + \alpha_2^{T(1)}E_2^{(1)}F_1^{(b)}
$$

$$
\delta\langle zs_{11}\rangle = \delta\langle s_{11}^*\rangle = \alpha_1^{T(2)}E_1^{(2)}S_2^{(b)}
$$

$$
\delta\langle zs_{22}\rangle = \delta\langle s_{22}^*\rangle = \alpha_2^{T(1)}E_2^{(1)}S_1^{(b)}
\tag{23.39}
$$

$$
\delta\langle zs_{11}^*\rangle = \frac{\alpha_1^{T(3)}E_1^{(3)} + \alpha_2^{T(3)}\nu_{12}^{(3)}E_1^{(3)}}{12(1 - \nu_{12}^{(3)}\nu_{21}^{(3)})} + \alpha_1^{T(2)}E_1^{(2)}J_2^{(b)}
$$

$$
\delta\langle zs_{22}^*\rangle = \frac{\alpha_2^{T(3)}E_2^{(3)} + \alpha_1^{T(3)}\nu_{21}^{(3)}E_2^{(3)}}{12(1 - \nu_{12}^{(3)}\nu_{21}^{(3)})} + \alpha_2^{T(1)}E_2^{(1)}J_1^{(b)}
$$

Turning now to heat conduction properties of the same shell, the local problems to be considered are

(l1). Equations 22.42 and 22.44
(l2). Equations 22.43 and 22.45
(l*1). Equations 22.46
(l*2). Equations 22.47,

all in combination with Equations 20.28.

Approximate solutions for these problems may be found for each of the elements Ω_1, Ω_2, and Ω_3 independently (see Figure 23.3) by proceeding in much the same way as for the local problems of elasticity. We omit the details of the solution and present here only the final expressions for the functions $l_{i\mu}$, and $l_{i\mu}^*$ as obtained from Equations 22.40 and 22.41. We have

$$l_{11} = \begin{cases} \lambda_1 & \text{in } \Omega_3, \Omega_2 \\ 0 & \text{in } \Omega_1 \end{cases} \qquad l_{22} = \begin{cases} \lambda_2 & \text{in } \Omega_3, \Omega_1 \\ 0 & \text{in } \Omega_2 \end{cases}$$

$$l_{12} = l_{21} = l_{31} = l_{32} = 0$$

$$l_{11}^* = \begin{cases} \lambda_1 z \\ \quad \text{in } \Omega_3, \Omega_2 \\ \lambda_1 z - \lambda_1 \dfrac{H+1}{2} + \lambda_1 \dfrac{2H}{\pi^2} \\ \quad \times \sum_{n=1}^{\infty} \dfrac{[1 - (-1)^n]\cosh\left(\dfrac{\pi n h_1 \xi_1}{H}\sqrt{\dfrac{\lambda_3}{\lambda_1}}\right)}{n^2 \cosh\left(\dfrac{\pi n h_1 \delta_1}{2H}\sqrt{\dfrac{\lambda_3}{\lambda_1}}\right)} \cos\dfrac{\pi n\left(z - \dfrac{1}{2}\right)}{H} \\ \quad \text{in } \Omega_1 \end{cases}$$

(23.40)

$$l_{22}^* = \begin{cases} \lambda_2 z \\ \quad \text{in } \Omega_3, \Omega_1 \\ \lambda_2 z - \lambda_2 \dfrac{H+1}{2} + \lambda_2 \dfrac{2H}{\pi^2} \\ \quad \times \sum_{n=1}^{\infty} \dfrac{[1 - (-1)^n]\cosh\left(\dfrac{\pi n h_2 \xi_2}{H}\sqrt{\dfrac{\lambda_3}{\lambda_2}}\right)}{n^2 \cosh\left(\dfrac{\pi n h_2 \delta_2}{2H}\sqrt{\dfrac{\lambda_3}{\lambda_2}}\right)} \cos\dfrac{\pi n\left(z - \dfrac{1}{2}\right)}{H} \\ \quad \text{in } \Omega_2 \end{cases}$$

$$l_{12}^* = l_{21}^* = 0$$

Note that although the functions l_{31}^* and l_{32}^* are also different from zero, they do turn into zero after the averaging operation, in full accord with Equations 20.36 (which we proved earlier for the general case).

From Equations 23.40, the nonzero effective heat conduction properties of the wafer shell of Figure 23.1 are found to be given by the equations

$$\langle l_{11} \rangle = \lambda_1^{(3)} + \alpha_1^{(2)} F_2^{(b)} \qquad \langle l_{22} \rangle = \lambda_2^{(3)} + \lambda_2^{(1)} F_1^{(b)}$$

$$\langle zl_{11} \rangle = \langle l_{11}^* \rangle = \lambda_1^{(2)} S_2^{(b)} \qquad \langle zl_{22} \rangle = \langle l_{22}^* \rangle = \lambda_2^{(1)} S_1^{(b)}$$

$$\langle zl_{11}^* \rangle = \frac{\lambda_1^{(3)}}{12} + \lambda_1^{(2)} J_2^{(b)} + \frac{\lambda_1^{(1)}}{12} \left(\frac{H^3 t_1}{h_1} - L_1 \right) \qquad (23.41)$$

$$\langle zl_{22}^* \rangle = \frac{\lambda_2^{(3)}}{12} + \lambda_2^{(1)} J_1^{(b)} + \frac{\lambda_2^{(2)}}{12} \left(\frac{H^3 t_2}{h_2} - L_2 \right)$$

which together with the definitions

$$L_1 = \sqrt{\frac{\lambda_1}{\lambda_2}} \frac{96 H^4}{\pi^5 A_1 h_1} \sum_{n=1}^{\infty} \frac{[1 - (-1)^n]}{n^5} \tanh\left(\frac{\pi n A_1 t_1}{2H} \sqrt{\frac{\lambda_3}{\lambda_1}} \right)$$

$$L_2 = \sqrt{\frac{\lambda_2}{\lambda_3}} \frac{96 H^4}{\pi^5 A_2 h_2} \sum_{n=1}^{\infty} \frac{[1 - (-1)^n]}{n^5} \tanh\left(\frac{\pi n A_2 t_2}{2H} \sqrt{\frac{\lambda_3}{\lambda_2}} \right)$$

(23.42)

determine all the coefficients occurring in the heat conduction constitutive relations of the homogenized shell, Equations 20.35. The system of the heat conduction governing equations of the homogenized shell, Equations 20.32 and 20.34, also involves the coefficients of $\mathscr{I}_m$ ($m = 0, 1, 2$), R_m, and Z_m ($m = 0, 1$), whose values are given by Equations 20.33 and 19.21. For the unit cell of Figure 23.1, we find

$$\mathscr{I}_0 = \alpha_S^+ \left[1 + \frac{2H(h_1 - t_1)}{h_1 h_2 A_2} + \frac{2H(h_2 - t_2)}{h_1 h_2 A_1} \right] + \alpha_S^-$$

$$\mathscr{I}_1 = \alpha_S^+ \left[\frac{1}{2} + \frac{Ht_2}{h_2} + \frac{H^2 + H}{h_2 A_2} \right] - \frac{\alpha_S^-}{2}$$

(23.43)

$$\mathscr{I}_2 = \alpha_S^+ \left[\frac{1}{4} + (H^2 + H)\left(\frac{t_2}{h_2} + \frac{t_1}{h_1} - \frac{t_1 t_2}{h_1 h_2} \right) \right.$$

$$\left. + \frac{(4H^3 + 6H^2 + 3H)}{6 h_1 h_2} \left(\frac{h_1 - t_1}{A_2} + \frac{h_2 - t_2}{A_1} \right) \right] + \frac{\alpha_S^-}{4}$$

Expressions for R_0 and R_1 are obtained from $\mathscr{I}_0$ and $\mathscr{I}_1$ by replacing $\alpha_S^\pm$ by $g_S^\pm$. Finally, $Z_0 = Z_1 = 0$ because in the case we consider the quantity $\lambda_{3i} N_i^\pm = \lambda_3 N_3^\pm$ differs from zero only on the surfaces parallel to the middle surface of the shell, but there we have $z^{\pm} = $ constant, and hence $\zeta^{\pm} = 0$.

Equations 23.41 may be useful in determining the effective heat conduction properties of ribbed shells. If the ribs are directed along the coordinate line α_1, for example, we set $t_1 = 0$ in E 23.41 and, accordingly, $F_1 = S_1 = J_1 = L_1 = 0$. The parameters $\mathscr{I}_m$ of the ribbed shell are given by

$$\mathscr{I}_0 = \alpha_S^+ \left(1 + \frac{2H}{h_2 A_2} \right) + \alpha_S^-$$

$$\mathscr{I}_1 = \alpha_S^+ \left(\frac{1}{2} + \frac{Ht_2}{h_2} + \frac{H^2 + H}{h_2 A_2} \right) - \frac{\alpha_S^-}{2} \qquad (23.44)$$

$$\mathscr{I}_2 = \alpha_S^+ \left[\frac{1}{4} + \frac{(H^2 + H)t_2}{h_2} + \frac{4H^3 + 6H^2 + 3H}{6h_2 A_2} \right] + \frac{\alpha_S^-}{4}$$

To determine R_0 and R_1, we again replace $\alpha_S^{\pm}$ by $g_S^{\pm}$ in $\mathscr{I}_0$ and $\mathscr{I}_1$, respectively. Note that Equations 23.44 cannot be deduced from Equations 23.43 by merely setting $t_1 = 0$.

24. EFFECTIVE PROPERTIES OF A THREE-LAYERED SHELL WITH A HONEYCOMB FILLER

24.1. Effective properties of a honeycomb shell of tetrahedral structure

The problem we consider here is that of a three-layered shell (or plate) composed of a honeycomb filler of tetrahedral structure sandwiched between two carrying layers as shown in Figure 24.1.

The effective properties of this system may be evaluated by employing the approximate method based on the independent solution of the local problems for each of the elements of the unit cell of the system $(\Omega_1, \Omega_2, \Omega_3, \Omega_4$; see Figure 24.2). One finds that the solutions of the local problems (b), (b^*11), (b^*22), and (l), obtained previously for the elements Ω_1, Ω_2, and Ω_3 of the wafer-type shell of Figure 23.3, may be utilized without any modifications in the analysis of the honeycomb structure, whereas solutions for Ω_4 simply coincide with those of the corresponding problems for Ω_3. The same is true for the local problems (b^*12), (l^*1), and (l^*2) for the elements Ω_3 and Ω_4. As regards the elements Ω_1 and Ω_2, however, some modification is needed, associated with the change in the orientation of the elements with respect to the z axis (see Figures 23.3 and 24.2).

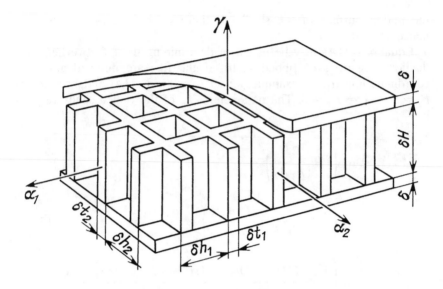

FIGURE 24.1.　Three-layered shell with a tetrahedral honeycomb filler.

Omitting all the intermediate steps of the solution, the final results are

$$
b_{12}^{*12} = \begin{cases}
zG_{12} \\
\quad \text{in } \Omega_3, \Omega_4 \\
zG_{12} + G_{12}\dfrac{2H}{\pi^2} \\
\quad \times \displaystyle\sum_{n=1}^{\infty} \dfrac{[1-(-1)^n]}{n^2} \dfrac{\cosh\left(\dfrac{\pi n h_1 \xi_1}{H}\sqrt{\dfrac{G_{23}}{G_{12}}}\right)}{\cosh\left(\dfrac{\pi n h_1 \delta_1}{2H}\sqrt{\dfrac{G_{23}}{G_{12}}}\right)} \cos\dfrac{\pi n\left(z+\dfrac{H}{2}\right)}{H} \\
\quad \text{in } \Omega_1 \\
zG_{12} + G_{12}\dfrac{2H}{\pi^2} \\
\quad \times \displaystyle\sum_{n=1}^{\infty} \dfrac{[1-(-1)^n]}{n^2} \dfrac{\cosh\left(\dfrac{\pi n h_2 \xi_2}{H}\sqrt{\dfrac{G_{13}}{G_{12}}}\right)}{\cosh\left(\dfrac{\pi n h_2 \delta_2}{2H}\sqrt{\dfrac{G_{13}}{G_{12}}}\right)} \cos\dfrac{\pi n\left(z+\dfrac{H}{2}\right)}{H} \\
\quad \text{in } \Omega_2
\end{cases}
$$

$$(24.1)$$

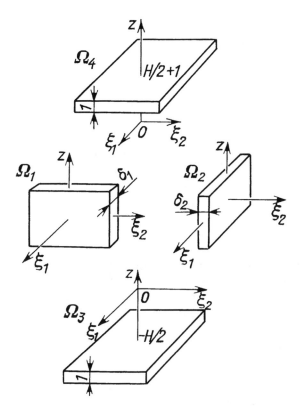

FIGURE 24.2. Unit cell elements of a three-layered shell with a tetrahedral honeycomb filler.

$$b_{11}^{*12} = b_{22}^{*12} = b_{33}^{*12} = 0$$

$$l_{11}^* = \begin{cases} \lambda_1 z \\ \qquad \text{in } \Omega_2, \Omega_3, \Omega_4 \\ \lambda_1 z + \lambda_1 \dfrac{2H}{\pi^2} \\ \qquad \times \sum_{n=1}^{\infty} \dfrac{[1 - (-1)^n]\cosh\left(\dfrac{\pi n h_1 \xi_1}{H}\sqrt{\dfrac{\lambda_3}{\lambda_1}}\right)}{n^2 \cosh\left(\dfrac{\pi n h_1 \delta_1}{2H}\sqrt{\dfrac{\lambda_3}{\lambda_1}}\right)} \cos\dfrac{\pi n\left(z + \dfrac{H}{2}\right)}{H} \\ \qquad \text{in } \Omega_1 \end{cases}$$

$$(24.1)$$

$$
l_{22}^* = \begin{cases} \lambda_2 z \\[4pt] \quad \text{in } \Omega_1, \Omega_3, \Omega_4 \\[6pt] \lambda_2 z + \lambda_2 \dfrac{2H}{\pi^2} \\[6pt] \quad \times \displaystyle\sum_{n=1}^{\infty} \dfrac{[1-(-1)^n]\cosh\!\left(\dfrac{\pi n h_2 \xi_2}{H}\sqrt{\dfrac{\lambda_3}{\lambda_2}}\right)}{n^2 \cosh\!\left(\dfrac{\pi n h_2 \delta_2}{2H}\sqrt{\dfrac{\lambda_3}{\lambda_2}}\right)} \cos \dfrac{\pi n\!\left(z+\dfrac{H}{2}\right)}{H} \\[14pt] \quad \text{in } \Omega_2 \end{cases}
$$

$$
l_{12}^* = l_{21}^* = 0 \tag{24.1}
$$

With these solutions, we are in a position to derive the entire set of the nonzero effective properties of the three-layered shell (plate) with a honeycomb filling element of tetrahedral structure (Figure 24.1). The following results are reproduced from a paper by Parton and Kalamkarov (1989).

Effective moduli:

$$
\langle b_{11}^{11} \rangle = \frac{E_1^{(3)}}{1 - \nu_{12}^{(3)}\nu_{21}^{(3)}} + \frac{E_1^{(4)}}{1 - \nu_{12}^{(4)}\nu_{21}^{(4)}} + E_1^{(2)}F_2^{(c)} \qquad (1 \leftrightarrow 2)
$$

$$
\langle b_{11}^{22} \rangle = \langle b_{22}^{11} \rangle = \frac{\nu_{12}^{(3)}E_1^{(3)}}{1 - \nu_{12}^{(3)}\nu_{21}^{(3)}} + \frac{\nu_{12}^{(4)}E_1^{(4)}}{1 - \nu_{12}^{(4)}\nu_{21}^{(4)}}
$$

$$
\langle b_{12}^{12} \rangle = G_{12}^{(3)} + G_{12}^{(4)} \qquad \langle b_{12}^{*12} \rangle = (G_{12}^{(3)} - G_{12}^{(4)})S_3^{(c)}
$$

$$
\langle zb_{11}^{11} \rangle = \langle b_{11}^{*11} \rangle = \left(\frac{E_1^{(3)}}{1 - \nu_{12}^{(3)}\nu_{21}^{(3)}} - \frac{E_1^{(4)}}{1 - \nu_{12}^{(4)}\nu_{21}^{(4)}} \right) S_3^{(c)} \qquad (1 \leftrightarrow 2)
$$

$$
\langle zb_{11}^{22} \rangle = \langle zb_{22}^{11} \rangle = \langle b_{11}^{*22} \rangle = \langle b_{22}^{*11} \rangle
$$

$$
= \left(\frac{\nu_{12}^{(3)}E_1^{(3)}}{1 - \nu_{12}^{(3)}\nu_{21}^{(3)}} - \frac{\nu_{12}^{(4)}E_1^{(4)}}{1 - \nu_{12}^{(4)}\nu_{21}^{(4)}} \right) S_3^{(c)} \tag{24.2}
$$

$$
\langle zb_{11}^{*11} \rangle = \left(\frac{E_1^{(3)}}{1 - \nu_{12}^{(3)}\nu_{21}^{(3)}} + \frac{E_1^{(4)}}{1 - \nu_{12}^{(4)}\nu_{21}^{(4)}} \right) J_3^{(c)} + E_1^{(2)}J_2^{(c)} \qquad (1 \leftrightarrow 2)
$$

$$\langle zb_{22}^{*11} \rangle = \langle zb_{11}^{*22} \rangle = \left(\frac{\nu_{12}^{(3)} E_1^{(3)}}{1 - \nu_{12}^{(3)} \nu_{21}^{(3)}} + \frac{\nu_{12}^{(4)} E_1^{(4)}}{1 - \nu_{12}^{(4)} \nu_{21}^{(4)}} \right) J_3^{(c)}$$

$$\langle zb_{12}^{*12} \rangle = (G_{12}^{(3)} + G_{12}^{(4)}) J_3^{(c)} + \frac{G_{12}^{(1)}}{12} \left(\frac{H^3 t_1}{h_1} - K_1 \right)$$

$$+ \frac{G_{12}^{(2)}}{12} \left(\frac{H^3 t_2}{h_2} - K_2 \right)$$

Effective thermoelastic properties:

$$\langle s_{11} \rangle = \frac{\alpha_1^{T(3)} E_1^{(3)} + \alpha_2^{T(3)} \nu_{12}^{(3)} E_1^{(3)}}{1 - \nu_{12}^{(3)} \nu_{21}^{(3)}}$$

$$+ \frac{\alpha_1^{T(4)} E_1^{(4)} \alpha_2^{T(4)} \nu_{12}^{(4)} E_1^{(4)}}{1 - \nu_{12}^{(4)} \nu_{21}^{(4)}} + \alpha_1^{T(2)} E_1^{(2)} F_2^{(c)} \quad (1 \leftrightarrow 2)$$

$$\langle zs_{12} \rangle = \langle s_{11}^* \rangle = \left(\frac{\alpha_1^{T(3)} E_1^{(3)} + \alpha_2^{T(3)} \nu_{12}^{(3)} E_1^{(3)}}{1 - \nu_{12}^{(3)} \nu_{21}^{(3)}} \right. \tag{24.3}$$

$$\left. - \frac{\alpha_1^{T(4)} E_1^{(4)} + \alpha_2^{T(4)} \nu_{12}^{(4)} E_1^{(4)}}{1 - \nu_{12}^{(4)} \nu_{21}^{(4)}} \right) S_3^{(c)} \quad (1 \leftrightarrow 2)$$

$$\langle zs_{11}^* \rangle = \frac{\alpha_1^{T(3)} E_1^{(3)} + \alpha_2^{T(3)} \nu_{12}^{(3)} E_1^{(3)}}{1 - \nu_{12}^{(3)} \nu_{21}^{(3)}} J_3^{(c)} + \frac{\alpha_1^{T(4)} E_1^{(4)} + \alpha_2^{T(4)} \nu_{12}^{(4)} E_1^{(4)}}{1 - \nu_{12}^{(4)} \nu_{21}^{(4)}} J_4^{(c)}$$

$$+ \alpha_1^{T(2)} E_1^{(2)} J_2^{(c)} \quad (1 \leftrightarrow 2)$$

Effective heat conduction coefficients:

$$\langle l_{11} \rangle = \lambda_1^{(3)} + \lambda_1^{(4)} + \lambda_1^{(2)} F_2^{(c)} \qquad \langle l_{22} \rangle = \lambda_2^{(3)} + \lambda_2^{(4)} + \lambda_2^{(1)} F_2^{(c)}$$

$$\langle l_{11}^* \rangle = (\lambda_1^{(3)} - \lambda_1^{(4)}) S_3^{(c)} \qquad \langle l_{22}^* \rangle = (\lambda_2^{(3)} - \lambda_2^{(4)}) S_3^{(c)}$$

$$\langle zl_{11}^* \rangle = \lambda_1^{(2)} J_2^{(c)} + \lambda_1^{(3)} J_3^{(c)} + \lambda_1^{(4)} J_4^{(c)} + \frac{\lambda_1^{(1)}}{12} \left(\frac{H^3 t_1}{h_1} - L_1 \right)$$

$$\langle zl_{22}^* \rangle = \lambda_2^{(1)} J_1^{(c)} + \lambda_2^{(3)} J_3^{(c)} + \lambda_2^{(4)} J_4^{(c)} + \frac{\lambda_2^{(2)}}{12} \left(\frac{H^3 t_2}{h_2} - L_2 \right) \tag{24.4}$$

In Equations 24.4, the elements of the unit cell are referenced by the appropriate superscript on material properties, and the quantities K_1, K_2, L_1, and L_2 are defined in Equations 23.35 and 23.42. We have also introduced the notation

$$\langle 1 \rangle_{\Omega_1} = \frac{t_1 H}{h_1} = F_1^{(c)}$$

$$\langle z^2 \rangle_{\Omega_1} = \frac{t_1 H^3}{12 h_1} = J_1^{(c)} \qquad (1 \leftrightarrow 2)$$

$$\langle z \rangle_{\Omega_3} = \frac{-(H+1)}{2} = S_3^{(c)} \tag{24.5}$$

$$\langle z \rangle_{\Omega_4} = \frac{H+1}{2} = -S_3^{(c)}$$

$$\langle z^2 \rangle_{\Omega_3} = \langle z^2 \rangle_{\Omega_4} = \frac{3H^2 + 6H + 4}{12} = J_3^{(c)} = J_4^{(c)}$$

and recalled that, for the honeycomb shell of Fig. 24.1,

$$\langle 1 \rangle_{\Omega_3} = \langle 1 \rangle_{\Omega_4} = 1 \quad \text{and} \quad \langle z \rangle_{\Omega_1} = \langle z \rangle_{\Omega_2} = 0$$

It should be noted that because the honeycomb shell we are considering is geometrically symmetric with respect to its middle surface, the presence of nonzero skew-symmetric properties in Equations 24.2 through 24.4 (that is, $\langle b_{11}^{*11} \rangle$, $\langle b_{22}^{*22} \rangle$, $\langle s_{11}^* \rangle$, $\langle s_{22}^* \rangle$, $\langle l_{11}^* \rangle$, and $\langle l_{22}^* \rangle$) is due entirely to the difference in the material properties of the upper and lower carrying layers. Clearly, all these coefficients turn to zero if the corresponding properties are the same in the elements Ω_3 and Ω_4.

Note also that if $t_1 = 0$ (and hence $F_1^{(c)} = S_1^{(c)} = J_1^{(c)} = K_1 = L_1 = 0$), what Equations 24.2 through 24.4 yield is the effective properties of a shell containing within itself a system of grooves of square cross section directed along the coordinate α_1 (see Figure 24.1 for $t_1 = 0$).

Finally, the coefficients in the heat conduction equations, Equations 20.32 and 20.34, should be written down for the case of a honeycomb shell:

$$\mathscr{I}_0 = \alpha_S^+ + \alpha_S^- \qquad \mathscr{I}_1 = \left(\frac{H}{2} + 1\right)(\alpha_S^+ - \alpha_S^-)$$

$$\mathscr{I}_2 = \left(\frac{H}{2} + 1\right)^2 (\alpha_S^+ + \alpha_S^-)$$

$$R_0 = g_S^+ + g_S^- \qquad R_1 = \left(\frac{H}{2} + 1\right)(g_S^+ - g_S^-)$$

$$z_0 = z_1 = 0$$

(24.6)

Expressions 24.2 for the effective elastic moduli of a honeycomb shell agree quite well with the corresponding results of Aleksandrov (1965) obtained for an isotropic material by applying the energy balance approach. Note, however, that the expression for the torsional stiffness $\langle z b_{12}^{*12} \rangle$ as given by Equations 24.2 is more accurate than its Aleksandrov counterpart derived neglecting the contribution from the honeycomb filler. This neglect assumption is justifiable in many practical situations when the elastic properties of the carrying layers and the filler are of the same order of magnitude; the contribution from the filler, as estimated by Equations 24.4, is then about only one-tenth of a percent of the contributions from the carrying layers. The error may be significant, however, if the filler material is much stiffer than that of the layers.

24.2. Effective elastic moduli of a three-layered shell with a honeycomb filler of hexagonal structure

Referring to Figure 24.3, which shows the geometry of the problem, we assume that

$$t_0 \ll H \quad \text{and} \quad t \ll H \tag{24.7}$$

that is, the vertical dimension of the honeycomb filler is much larger than the thickness of the carrying layers and of the honeycomb cells themselves. These conditions, as well as those in Conditions 23.1, are typically fulfilled in practical honeycomb plate and shell structures (Panin 1982, Endogur et al. 1986) and provide sufficient justification for the approximating scheme of the previous sections, in which the pertinent local problems are solved independently for each of the unit cell elements. We confine our attention to the case in which $A_1 = A_2 = 0$.

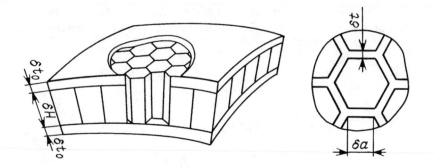

FIGURE 24.3. Three-layered shell with a hexagonal honeycomb filler.

If we change to the coordinates ξ_1, ξ_2, z in terms of which the local problems are formulated, the unit cell of our structure deforms into the shape shown in Figure 24.4, and, by the rules of transformation of coordinates,

$$h_1 = 3a \qquad h_2 = \sqrt{3}\,a \qquad t_1 = \frac{t}{a\sqrt{3}} \qquad t_2 = \frac{2t}{a\sqrt{30}} \qquad \sin \alpha = \frac{3}{\sqrt{10}}$$

The additional difficulty that arises in this particular example is that we have to perform a rotation of coordinate axes when considering the local problems for those of the cell elements that make and angle α with the ξ_1 axis (see Figure 24.4).

The calculation of the nonvanishing effective elastic moduli of the shell of Figure 24.3 thus includes the solution of the local problems on each of the cell elements and is somewhat lengthy to be reproduced here, so we only quote the final results of the calculation. For an isotropic material both in the carrying layers (E_0, ν_0) and in the filler foil (E, ν), we have (Parton et al. 1989b)

$$\langle b_{11}^{11} \rangle = \langle b_{22}^{22} \rangle = \frac{2E_0 t_0}{1 + \nu_0^2} + \frac{\sqrt{3}}{4}\frac{EHt}{a}$$

$$\langle b_{11}^{22} \rangle = \langle b_{22}^{11} \rangle = \frac{2\nu_0 E_0 t_0}{1 - \nu_0^2} + \frac{\sqrt{3}}{12}\frac{EHt}{a}$$

$$\langle b_{12}^{12} \rangle = \frac{E_0 t_0}{1 + \nu_0} + \frac{\sqrt{3}}{12}\frac{EHt}{a}$$

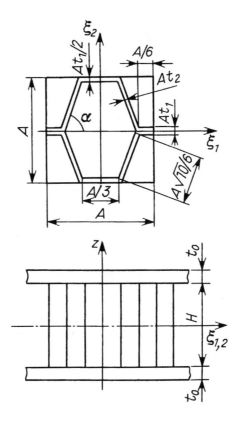

FIGURE 24.4. Unit cell of a three-layered shell with a hexagonal honeycomb filler, in the coordinate system ξ_1, ξ_2, z.

$$\langle zb_{22}^{*11} \rangle = \langle zb_{22}^{*22} \rangle = \frac{E_0}{1 - v_0^2} \left(\frac{H^2 t_0}{2} + H t_0^2 + \frac{2}{3} t_0^3 \right) + \frac{\sqrt{3}}{48} \frac{EH^3 t}{a}$$

$$\langle zb_{11}^{*22} \rangle = \langle zb_{22}^{*11} \rangle = \frac{v_0 E_0}{1 - v_0^2} \left(\frac{H^2 t_0}{2} + H t_0^2 + \frac{2}{3} t_0^3 \right) + \frac{\sqrt{3}}{144} \frac{EH^3 t}{a}$$

$$\langle zb_{12}^{*12} \rangle = \frac{E_0}{2(1 + v_0)} \left(\frac{H^2 t_0}{2} + H t_0^2 + \frac{2t_0^3}{3} \right)$$

$$+ \frac{EH^3 t}{12(1 + v)a} \left\{ \frac{3 + v}{4\sqrt{3}} - \frac{128 H}{\sqrt{3} \, \pi^5 At} \right.$$

$$\left. \times \sum_{n=1}^{\infty} \frac{\tanh[\pi(2n - 1)At/(2H)]}{(2n - 1)^5} \right\} \tag{24.8}$$

TABLE 24.1
Effective Elastic Moduli of Honeycomb Shells
of Hexagonal Structure (Figure 25.3)
$$H = 25, \; a = 2.5, \; t_0 = 1, \; v_0 = v = 0.3, \; A = 1$$

Effective elastic moduli	Total contribution from carrying layers	Contribution from honeycomb filler	
		$t = 0.1$	$t = 0.5$
$\langle b_{11}^{11} \rangle$	$2.1978 \times E_0$	$0.4330 \times E$	$2.1651 \times E$
$\langle b_{22}^{11} \rangle$	$0.6593 \times E_0$	$0.1443 \times E$	$0.7217 \times E$
$\langle b_{12}^{12} \rangle$	$0.7692 \times E_0$	$0.1443 \times E$	$0.7217 \times E$
$\langle zb_{11}^{*11} \rangle$	$371.6117 \times E_0$	$22.5527 \times E$	$112.7637 \times E$
$\langle zb_{22}^{*11} \rangle$	$111.4835 \times E_0$	$7.5176 \times E$	$37.5897 \times E$
$\langle zb_{12}^{*12} \rangle$	$130.0641 \times E_0$	$3.6643 \times E$	$18.3540 \times E$

where the first terms everywhere describe the contribution from the carrying layers, and the second terms describe that from the filler. It is seen that the latter contribution may be made comparable to, or even greater than, the former by appropriately varying the parameters E, H, t, and a. This is clearly borne out by Table 24.1 where the results based on Equations 24.8 are listed.

It is of interest to compare Equations 24.8 with the elastic moduli results obtained by different methods in the earlier work on the subject. Comparison with the work of Aleksandrov (1965) (see Table 24.2) shows the greatest corrections occur in the elastic moduli $\langle b_{11}^{22} \rangle$ and $\langle zb_{11}^{*22} \rangle$. Although the formulas of Aleksandrov et al. (1960) are more accurate in comparison with those of Aleksandrov (1965) (very much at the expense of simplicity), calculations show that even in this case the corrections given by Equations 24.8 are quite appreciable, as illustrated by Table 24.3 for $\langle b_{11}^{22} \rangle$.

25. ELASTIC MODULI AND LOCAL STRESSES IN WAFER-TYPE PLATES AND SHELLS, INCLUDING THE INTERACTION BETWEEN CELL ELEMENTS

We remind the reader that the elastic analysis of wafer-type plate and shell structures (Figure 23.1)—and hence the derivation of the elastic properties of such structures (Equations 23.34)—were performed in Section 23 under the assumption of there being no interaction between the individual elements of the unit cell of the problem, Ω_1, Ω_2, and Ω_3 (see Figures 23.1 and 23.3). The removal of this assumption in this section will enable us to obtain more accurate solutions as well as to estimate the

TABLE 24.2
Comparison of Computational Formulas for the Elastic Properties
of a Hexagonal Honeycomb Filler

	Formulas from Aleksandrov (1965)	Formulas 24.8
$\langle b_{11}^{11}\rangle$	$\dfrac{2.5}{3\sqrt{3}}\dfrac{EtH}{a}\approx 0.481$	$\dfrac{EtH}{a}-\dfrac{\sqrt{3}}{4}\dfrac{EthH}{a}\approx 0.433\dfrac{EtH}{a}$
$\langle b_{22}^{22}\rangle$	$0.5\dfrac{EtH}{a}$	$\dfrac{\sqrt{3}}{4}\dfrac{EtH}{a}\approx 0.433\dfrac{EtH}{a}$
$\langle b_{11}^{22}\rangle=\langle b_{22}^{11}\rangle$	$\dfrac{2.5}{\sqrt{15\sqrt{3}}}\sqrt{\nu_0}\dfrac{EtH}{a}\approx 0.49\sqrt{\nu_0}\dfrac{EtHE}{a}$ for $\nu_0=0.3$: $0.268\dfrac{EtH}{a}$ for $\nu_0=0.25$: $0.245\dfrac{Eth}{a}$	$\dfrac{\sqrt{3}}{12}\dfrac{EtH}{a}\approx 0.144\dfrac{EtH}{a}$
$\langle b_{12}^{12}\rangle$	$\dfrac{\sqrt{3}}{12}\dfrac{EtH}{a}$	$\dfrac{\sqrt{3}}{12}\dfrac{EtH}{a}$
$\langle zb_{11}^{*11}\rangle$	$\dfrac{2.5}{36\sqrt{3}}\dfrac{EtH^3}{a}\approx 0.04\dfrac{EtH^3}{a}$	$\dfrac{\sqrt{3}}{48}\dfrac{EtH^3}{a}\approx 0.036\dfrac{EtH^3}{a}$
$\langle zb_{22}^{*22}\rangle$	$\dfrac{1}{24}\dfrac{EtH^3}{a}\approx 0.042\dfrac{EtH^3}{a}$	$\dfrac{\sqrt{3}}{48}\dfrac{EtH^3}{a}\approx 0.036\dfrac{EtH^3}{a}$
$\langle zb_{22}^{*11}\rangle$ $=\langle zb_{11}^{*22}\rangle$	$\dfrac{2.5\sqrt{\nu_0}}{12\sqrt{15\sqrt{3}}}\dfrac{EtH^3}{a}$ $\approx\sqrt{\nu_0}\,0.041\dfrac{EtH^3}{a}$ for $\nu_0=0.3$: $0.022\ EtH^3/a$ for $\nu_0=0.25$: $0.021\ EtH^3/a$	$\dfrac{\sqrt{3}}{144}\dfrac{EtH^3}{a}\approx 0.012\dfrac{EtH^3}{a}$
$\langle zb_{12}^{*12}\rangle$	$\dfrac{\sqrt{3}}{288}\dfrac{EtH^3}{a}\approx 0.006\dfrac{EtH^3}{a}$	for $H=25$, $\nu=0.3$, $A=1$ if $t=0.1$: $0.00586\dfrac{EtH^3}{a}$ if $t=0.5$: $0.00587\dfrac{EtH^3}{a}$

accuracy of the solutions given by Equations 23.34. We again assume that Conditions 23.1 are satisfied and we limit ourselves to shells made of isotropic materials.

Based on Conditions 23.1, the pertinent local problems will be solved for each cell element separately. The functions sought will be averaged over the thickness of an element, and the interaction between the ele-

TABLE 24.3

$$\langle b_{11}^{22} \rangle (EtH / a)^{-1}$$

$H = 25,\ a = 2.5,\ A = 1$	Accurate formulas due to Aleksandrove et al. (1960)	Formula 24.8
$t = 0.1$:		
$\nu = 0.3$	0.179	0.144
$\nu = 0.25$	0.713	0.144
$t = 0.5$:		
$\nu = 0.3$	0.225	0.144
$\nu = 0.25$	0.218	0.144

ments will be modeled by introducing certain stress singularities, which have to be found in the course of solution. In particular, referring to Figure 25.1 and in accordance with the boundary local problems posed by Equations 22.8 and 22.9 and by Equations 22.13 and 22.14, the boundary conditions to be satisfied along Ω_1–Ω_2 junctions are

$$
\begin{cases}
\dfrac{1}{h_1} \tau_{12}^{\lambda\mu} = \pm \dfrac{1}{2}\delta(\xi_2)\tau_3^{\lambda\mu}(z) \\[2mm]
\dfrac{1}{h_2} \bar\tau_{13}^{\lambda\mu} = \pm \dfrac{1}{2}(\xi_2)\bar\tau_3^{\lambda\mu}(z)
\end{cases}
\qquad
\left(\xi_1 = \pm \dfrac{\delta_1}{2} \right)
$$

$$(25.1)$$

$$
\begin{cases}
\dfrac{1}{h_2} \tau_{23}^{\lambda\mu} = \mp \dfrac{1}{2}\delta(\xi_1)\tau_3^{\lambda\mu}(z) \\[2mm]
\dfrac{1}{h_2} \bar\tau_{23}^{\lambda\mu} = \mp \dfrac{1}{2}\delta(\xi_1)\bar\tau_3^{\lambda\mu}(z)
\end{cases}
\qquad
\left(\xi_2 = \pm \dfrac{\delta_2}{2} \right)
$$

Along junctions between the strengthening elements Ω_1 and Ω_2 and the carrying element Ω_3, that is, at $z = \frac{1}{2}$, we have the conditions (cf. Figure 25.1)

$$\tau_{23}^{\lambda\mu} = \delta(\xi_1)\tau_2^{\lambda\mu}(\xi_2) \qquad \bar\tau_{23}^{\lambda\mu} = \delta(\xi_1)\bar\tau_2^{\lambda\mu}(\xi_2)$$

$$\tau_{13}^{\lambda\mu} = \delta(\xi_2)\tau_1^{\lambda\mu}(\xi_1) \qquad \bar\tau_{13}^{\lambda\mu} = \delta(\xi_2)\bar\tau_1^{\lambda\mu}(\xi_1)$$

$$(25.2)$$

The stresses (Equations 25.1 and 25.2) occurring at the junctions of the elements are found by requiring that material displacements remain con-

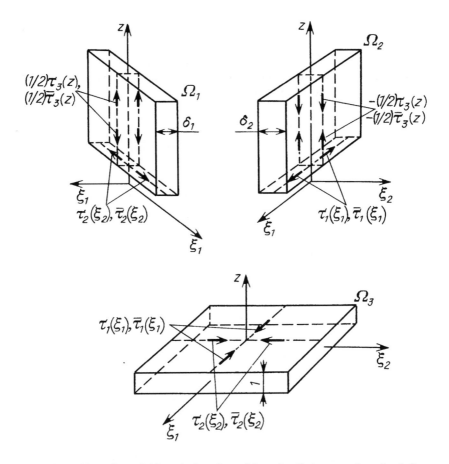

FIGURE 25.1. Stress fields at the junctions of the unit cell elements of a wafer shell.

tinuous as we pass from one element to another, that is,

$$\tilde{U}_1^{\lambda\mu}\Big|_{z=1/2}^{\Omega_2} = \tilde{U}_1^{\lambda\mu}\Big|_{\xi_2=0}^{\Omega_3} \qquad \tilde{U}_2^{\lambda\mu}\Big|_{z=1/2}^{\Omega_1} = \tilde{U}_2^{\lambda\mu}\Big|_{\xi_1=0}^{\Omega_3}$$

$$\tilde{U}_3^{\lambda\mu}\Big|_{\xi_2=0}^{\Omega_1} = \tilde{U}_3^{\lambda\mu}\Big|_{\xi_1=0}^{\Omega_2} \qquad \left(\tilde{U}_i^{\lambda\mu} \leftrightarrow \tilde{V}_i^{\lambda\mu}\right)$$

(25.3)

where a tilde denotes thickness average over the corresponding element.

As shown by Parton et al. (Parton et al. 1989a, Parton et al. 1989b) all six types of local problems $(b\lambda\mu)$ and $(b^*\lambda\mu)$ may be solved analytically by applying the preceding approximating scheme and expanding the solutions in trigonometric series in the coordinates ξ_1 and ξ_2 (which secures in a very natural fashion the periodicity conditions in these coordinates; as regards the coordinate z, which is "nonperiodic," the local problems associated with the elements Ω_1 and Ω_2 were solved by expanding in terms of the variable $z' = z - \frac{1}{2}$ on the segment $[O, H]$.

For the stresses along the junctions of the elements, taking into account the symmetry of each particular problem for different cell elements, the necessary trigonometric expansions may be written as

$$\tau_1^{\lambda\mu}(\xi_1) = \sum_{m=1}^{\infty} \tau_{1m}^{\lambda\mu} \sin\left(\frac{2\pi m}{A_1}\xi_1\right)$$

$$\tau_2^{\lambda\mu}(\xi_2) = \sum_{m=1}^{\infty} \tau_{2m}^{\lambda\mu} \sin\left(\frac{2\pi m}{A_2}\xi_2\right)$$

$$\bar{\tau}_1^{\lambda\mu}(\xi_1) = \sum_{m=1}^{\infty} \bar{\tau}_{1m}^{\lambda\mu} \sin\left(\frac{2\pi m}{A_1}\xi_1\right)$$

(25.4)

$$\bar{\tau}_2^{\lambda\mu}(\xi_2) = \sum_{m=1}^{\infty} \bar{\tau}_{2m}^{\lambda\mu} \sin\left(\frac{2\pi m}{A_2}\xi_2\right)$$

$$\tau_3^{\lambda\mu}(z) = \sum_{k=1}^{\infty} \tau_{3k}^{\lambda\mu} \cos\left(\frac{\pi k z'}{H}\right)$$

$$\bar{\tau}_3^{\lambda\mu}(z) = \sum_{k=1}^{\infty} \bar{\tau}_{3k}^{\lambda\mu} \cos\left(\frac{\pi k z'}{H}\right)$$

The solution of the entire set of the local problems is tedious, and we shall only quote the results here. The nonzero effective elastic moduli are as follows (Parton et al. 1989a, Parton et al. 1989b):

$$\langle b_{11}^{11} \rangle = \frac{E}{1 - \nu^2} + E(F_2^{(b)} + K\Sigma_1) \qquad \langle b_{22}^{22} \rangle = \frac{E}{1 - \nu^2}$$

$$+ E(F_1^{(b)} + K\Sigma_1)$$

$$\langle b_{11}^{22} \rangle = \langle b_{22}^{11} \rangle = \frac{\nu E}{1 - \nu^2} - EK\Sigma_1 \qquad \langle b_{12}^{12} \rangle = G$$

$$\langle zb_{11}^{11} \rangle = \langle b_{11}^{*11} \rangle = E(S_2^{(b)} + K\Sigma_2) \qquad \langle zb_{22}^{22} \rangle = \langle b_{22}^{*22} \rangle = E(S_1^{(b)} + K\Sigma_2)$$

$$\langle zb_{11}^{22} \rangle = \langle zb_{22}^{11} \rangle = \langle b_{22}^{*22} \rangle = \langle b_{22}^{*11} \rangle = -EK\Sigma_2$$

$$\langle zb_{11}^{*11} \rangle = \frac{E}{12(1 - \nu^2)} + E(J_2^{(b)} + K\Sigma_3)$$

$$\langle zb_{22}^{*22} \rangle = \frac{E}{12(1 - \nu^2)} + E(J_1^{(b)} + K\Sigma_3)$$

$$\langle zb_{22}^{*11} \rangle = \langle zb_{11}^{*22} \rangle = \frac{\nu E}{12(1 - \nu^2)} - EK\Sigma_3 \qquad (25.5)$$

Formula 23.34 for the torsional stiffness $\langle zb_{12}^{*12} \rangle$ remains unchanged and is not repeated here.

In Equations 25.5 we have defined, for $A_1 = A_2 = A$, $h_1 = h_2 = h$,

$$K = \frac{8\nu^2 H^2 t_1 t_2}{\pi^3(1 + \nu)Ah^2(t_1 + t_2)} \qquad \Sigma_1 = \sum_{k=1}^{\infty} \frac{1 - (-1)^k}{k^2} x_k$$

$$(25.6)$$

$$\Sigma_2 = \sum_{k=1}^{\infty} \frac{1 - (-1)^k}{k^2} \bar{x}_k \qquad \Sigma_3 = \sum_{k=1}^{\infty} \frac{[1 - (-1)^k]/2 - H(-1)^k}{k^2} \bar{x}_k$$

Here the coefficients x_k are determined from the following infinite system of simultaneous algebraic equations (Parton et al. 1989b):

$$+ \frac{16At_1 t_2 n}{\pi(1 + \nu)h^2(t_1 + t_2)} \sum_{m=1}^{\infty} r_{nm} y_m = \frac{1 - (-1)^n}{n}$$

$$(25.7)$$

$$p_{0n} y_n + \sum_{m=1}^{\infty} p_{mn} y_m + \frac{8H^2(t_1 + t_2)n}{\pi(1 + \nu)^2 A^3 h t_1 t_2} \sum_{k=1}^{\infty} r_{kn} x_k = 0$$

(for $n = 1, 2, 3, \ldots$), where

$$q_{0n} = (3 - \nu)\coth\frac{\pi n}{a} - \frac{(1 + \nu)\pi n}{a \sinh^2(\pi n/a)} \qquad a = \frac{2H}{Ah}$$

$$q_{kn} = \frac{16a^2}{\pi^2(1 + \nu)}$$

$$\times \sum_{m=1}^{\infty} \frac{mn(\nu k^2 - a^2 m^2)(\nu n^2 - a^2 m^2)[\cosh(\pi am) - (-1)^n]}{(k^2 + a^2 m^2)^2(n^2 + a^2 m^2)^2[\sinh(\pi am) - (-1)^n \pi am]}$$

$$r_{nm} = \frac{(\nu n^2 - a^2 m^2)[\cosh(\pi am) - (-1)^n]}{(n^2 + a^2 m^2)^2[\sinh(\pi am) - (-1)^n \pi am]}$$

$$p_{0n} = \frac{3 - \nu}{4(1 + \nu)} \coth(\pi n) - \frac{\pi n}{4 \sinh^2(\pi n)}$$

$$+ \frac{1[\sinh(\pi na)\cosh(\pi na) - \pi na]}{(1 + \nu)^2 h[\sinh^2(\pi na) - (\pi na)^2]}$$

$$p_{mn} = \frac{mn^2}{\pi(n^2 + m^2)^2}$$

The infinite system of equations for the coefficients x_k in Equations 25.6 differs from Equations 25.7 only in having

$$\left[(1 - (-1)^n)/2 - H(-1)^n\right] n^{-1}$$

instead of $(1 - (-1)^n)/n$ in the right-hand side of the first equation and, clearly, we must replace x_k by $\bar{x}_k$ and y_k by $\bar{y}_k$ everywhere in Equations 25.7.

The solutions of the preceding systems of equations and the coefficients in the expansions 25.4 for the stresses along the junctions (see Figures 25.1 and 25.2) are related by

$$\tau_{3k}^{11} = EK'x_k \qquad \tau_{1m}^{11} - \tau_{2m}^{11} = EK'y_m$$

$$\bar{\tau}_{3k}^{11} = EK'\bar{x}_k \qquad \bar{\tau}_{1m}^{11} - \bar{\tau}_{2m}^{11} = EK'\bar{y}_m$$

$$\tau_{3k}^{22} = -\tau_{3k}^{11} \qquad \tau_{3k}^{12} = 0 \qquad \tau_{1m}^{22} = -\tau_{1m}^{11} \qquad (\tau \leftrightarrow \bar{\tau})$$

$$\tau_{2m}^{22} = -\tau_{2m}^{11} \qquad \tau_{1m}^{12} = \tau_{2m}^{12} = 0 \qquad (\tau \leftrightarrow \bar{\tau})$$ (25.8)

$$K' = \frac{8\nu A t_1 t_2}{\pi(1 + \nu)h^2(t_1 + t_2)}$$

It should also be noted that

$$\tau_{2m}^{\lambda\mu} = -\tau_{1m}^{\lambda\mu} \qquad \text{and} \qquad \bar{\tau}_{2m}^{\lambda\mu} = -\bar{\tau}_{1m}^{\lambda\mu}$$

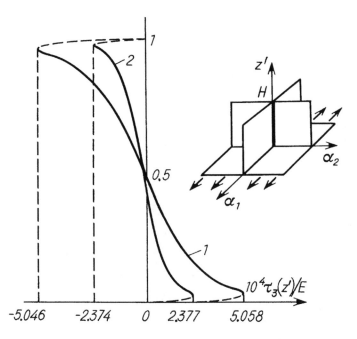

FIGURE 25.2. Stress fields at the junctions of the strengthening elements of a wafer plate or shell under tension.

For the more general case in which $A_1 = A_2$ and $h_1 = h_2$, the solution follows exactly the same pattern except for the fact that the system of Equations 25.7 will be augmented by one more group of equations and the corresponding coefficients will be more complex in form.

Expressions 25.5 for the effective elastic moduli of the wafer shell are a refined version of Equations 23.34 and can be reduced back to these latter by dropping the terms of the type $K\Sigma_i$ ($i = 1, 2, 3$). The magnitude of the corrections was estimated by a series of calculations for different values of the parameters determining the dimensions of the unit cell of Figure 23.1. The authors showed numerically that, unless an accuracy of better than 10^{-4} is required, there is no need to consider more than 40 equations in the system of Equations 25.7.

In Table 25.1 we present both the numerical results obtained from Equations 25.5 (rows labeled I) and from Equations 23.34 (rows labeled II) for $A_1 = A_2 = 1$, $\nu = 0.3$. The magnitudes of the percentage changes are shown in the rows labeled III. Based on the analysis of the table, the following conclusions are made:

1. As found from Formulas 23.34, the elastic moduli $\langle b_{11}^{11} \rangle$, $\langle b_{22}^{22} \rangle$, $\langle b_{11}^{*11} \rangle$, $\langle b_{22}^{*22} \rangle$, $\langle zb_{11}^{*11} \rangle$, and $\langle zb_{22}^{*22} \rangle$ show a percentage change of no more than 1% from the values given by Equations 25.5.

TABLE 25.1

Effective Elastic Moduli of Wafer Plates and Shells; $A_1 = A_2 = 1$, $\nu = 0.3$

Rows labeled I, Formulas 25.5; rows labeled II, Formulas 23.34; rows labeled III, percentage changes

Wafer shell	$\dfrac{\langle b_{11}^{11}\rangle}{E}$	$\dfrac{\langle b_{22}^{22}\rangle}{E}$	$\dfrac{\langle b_{22}^{11}\rangle}{E}$	$\dfrac{\langle b_{11}^{*11}\rangle}{E}$	$\dfrac{\langle b_{22}^{*22}\rangle}{E}$	$\dfrac{\langle b_{22}^{*11}\rangle}{E}$	$\dfrac{\langle zb_{11}^{*11}\rangle}{E}$	$\dfrac{\langle zb_{22}^{*22}\rangle}{E}$	$\dfrac{\langle zb_{22}^{*11}\rangle}{E}$	$\dfrac{\langle zb_{12}^{*12}\rangle}{E}$
(1) $H = 10$, $h = 10$, $t_1 = 1$, $t_2 = 1$										
I	2.1118	2.1118	0.3168	5.5648	5.5648	−0.0648	39.0728	39.0728	−0.3705	0.0921
II	2.0989	2.0989	0.3297	5.5	5.5	0	38.6749	38.6749	0.0275	0.0921
III	0.6%	0.6%	−3.9%	1.2%	1.2%	—	1.03%	1.03%	—	—
(2) $H = 10$, $h = 10$, $t_1 = 1$, $t_2 = 0.5$										
I	1.6075	2.1075	0.3211	2.7932	5.5432	−0.0432	19.6485	38.9402	−0.2378	0.0600
II	1.5989	2.0989	0.3297	2.75	5.5	0	19.3833	38.6749	0.0275	0.0660
III	0.54%	0.41%	−2.62%	1.57%	0.78%	—	1.37%	0.69%	—	—
(3) $H = 10$, $h = 20$, $t_1 = 0.5$, $t_2 = 0.5$										
I	1.3505	1.3505	0.3281	1.3828	1.3828	−0.0078	9.7856	9.7856	−0.0207	0.0359
II	1.3489	1.3489	0.3297	1.375	1.375	0	9.7374	9.7374	0.0275	0.0359
III	0.12%	0.12%	−0.48%	0.57%	0.57%	—	0.49%	0.49%	—	—
(4) $H = 20$, $h = 20$, $t_1 = 0.5$, $t_2 = 0.5$										
I	1.6053	1.0653	0.3232	5.3148	5.3148	−0.0648	72.6514	72.6514	−0.7406	0.0400
II	1.5989	1.5989	0.3297	5.25	5.25	0	71.8832	71.8832	0.0275	0.0400
III	0.4%	0.4%	−1.96%	1.23%	1.23%	—	1.07%	—	—	—

(5) $H = 2, h = 25, t_1 = 0.7, t_2 = 0.7$

I	1.1550	1.1550	0.3296	0.0841	0.0841	-0.0001	0.2363	0.2363	0.0274	0.0334
II	1.1549	1.1549	0.3297	0.084	0.084	0	0.2362	0.2362	0.0275	0.0334
III	0.005%	0.005%	-0.03%	0.07%	0.07%	—	0.055%	0.055%	-0.33%	—

(6) $H = 5, h = 20, t_1 = 0.8, t_2 = 0.8$

I	1.300	1.300	0.329	0.6016	0.6016	-0.1561	2.3134	2.3134	0.0223	0.0394
II	1.299	1.299	0.3297	0.6	0.6	0	2.3082	2.3082	0.0275	0.0394
III	0.05%	0.05%	-0.2%	0.26%	0.26%	—	0.22%	0.22%	-18.75%	—

(7) $H = 8, h = 30, t_1 = 0.8, t_2 = 0.8$

I	1.3129	1.3129	0.3290	0.9628	0.9628	-0.0028	5.5636	5.5636	0.0132	0.0403
II	1.3122	1.3122	0.3297	0.96	0.96	0	5.5494	5.5494	0.0275	0.0403
III	0.06%	0.06%	-0.22%	0.3%	0.3%	—	0.26%	0.26%	-51.96%	—

(8) $H = 8, h = 30, t_1 = 0.8, t_2 = 0.4$

I	1.2060	1.3127	0.3292	0.4819	0.9619	-0.0019	2.8300	5.5589	0.0180	0.0367
II	1.2056	1.3122	0.3297	0.48	0.96	0	2.8205	5.5494	0.0275	0.0367
III	0.04%	0.04%	-0.15%	0.39%	0.2%	—	0.34%	0.17%	-34.62%	—

(9) $H = 20, h = 60, t_1 = 2, t_2 = 2$

I	1.7683	1.7683	0.3269	7.0278	7.0278	-0.0277	96.1443	96.1442	-0.3030	0.1922
II	1.7656	1.7656	0.3297	7.0	7.0	0	95.8138	95.8138	0.0275	0.1922
III	0.15%	0.15%	-0.85%	0.4%	0.4%	—	0.34%	0.34%	—	—

2. For the modulus $\langle b_{22}^{11} \rangle$, somewhat greater, but also reasonable percentage changes are obtained (for example, 3.9% in the first example in Table 25.1).

3. For the moduli $\langle b_{22}^{*11} \rangle$ and $\langle zb_{22}^{*11} \rangle$, more significant percentage changes occur. Note that although the change for $\langle zb_{22}^{*11} \rangle$ is only of a quantitative nature, that for $\langle b_{22}^{*11} \rangle$ is also interesting at a qualitative level because this modulus vanishes, in view of Equations 24.34, and is assumed to be zero in the framework of the structurally anisotropic theory of strengthened plates and shells (Korolev 1971, Elpatyevskii and Gavva 1983).

One more point to be made is that the percentage changes invariably increase with the height of the ribs (parameter H) and decrease with the distance between the ribs (parameter h).

From Equations 19.74, 22.10, 22.15, 25.1, 25.2, 25.4, and 25.8, using the solutions of the system of Equations 25.7, we are in a position to obtain the local stress distributions along the junctions of the cell elements (Figures 25.2 and 25.3).

In particular, if we take $h_1 = h_2 = h$, then, in the case of a simple tension ($\varepsilon_1 \neq 0$), the junctions of reinforcing elements Ω_1 and Ω_2 will be subjected to the stresses

$$\left.\frac{\sigma_{13}}{\varepsilon_1}\right|_{\substack{\alpha_1 = \pm \delta t/2 \\ \alpha_2 = 0}} = \pm \frac{h}{2} \tau_3(z')$$

and in the case of simple bending ($\kappa_1 \neq 0$)

$$\left.\frac{\sigma_{13}}{\delta\kappa_1}\right|_{\substack{\alpha_1 = \pm \delta t/2 \\ \alpha_2 = 0}} = \pm \frac{h}{2} \bar{\tau}_2(z')$$

The functions $\tau_3(z')$ and $\bar{\tau}_2(z')$ are shown graphically in Figure 25.2 (for tension) and Figure 25.3 (for bending). The curves marked 1 and 2 in these figures correspond, respectively, to the cases 4 and 9 in Table 25.1. Note that the preceding stresses turn to zero at $z' = H$ and $z' = 0$, but the Fourier cosine representation that we assumed for $\tau_2(z')$ and $\bar{\tau}_3(z')$ in Equations 25.4 makes it difficult to approximate with reasonable accuracy the stress field in the immediate vicinity of these points.

26. TENSION OF PLATES OR SHELLS REINFORCED BY A REGULAR SYSTEM OF THIN SURFACE STRIPS

In the previous sections we have obtained approximate solutions of the cell local problems for the case(s) when the thickness of the reinforcing element is small as compared with its height. In this section the opposite

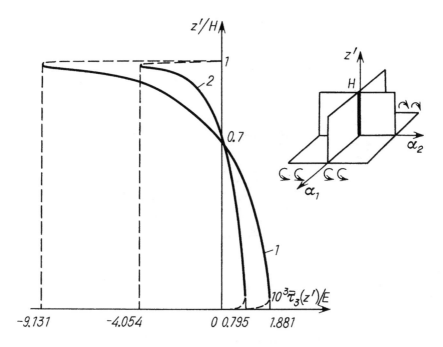

FIGURE 25.3. Stress fields at the junctions of the strengthening elements of a wafer plate or shell under bending.

extreme will be considered. We will be concerned, namely, with shells (or plates) strengthened with a regular system of thin ribs arranged symmetrically with respect to the middle surface of the shell (see Figure 26.1), and we will assume that the height of a rib is much smaller than its length. It will be seen that with thisassumption the local problem admits of being treated analytically even in the case of a nonvanishing interaction between the ribs and the carrying surfaces (Parton et al. 1988b).

We assume that the ribs and the carrying surfaces of the composite are made of isotropic materials (which may or may not be dissimilar); also that the strengthening elements within the unit cell of the structure are thin plates having no flexural stiffness and working in tension–compression only.

26.1. The solution of the local problems (b11) and (b22)

The shape of the unit cell of our problem in the coordinate system ξ_2, z is shown in Figure 26.2, in which $2\alpha = lA_2/h_2$ by the rules of coordinate transformation.

We begin with a discussion of the local problem (b11), posed by Equations 22.16, and 22.18 through 22.20 for $\lambda\mu = 11$; first we consider the region $\Omega_2 = \{|z| \leq \frac{1}{2}, |\xi_2| \leq A_2/2\}$ on the carrying surface. In view of

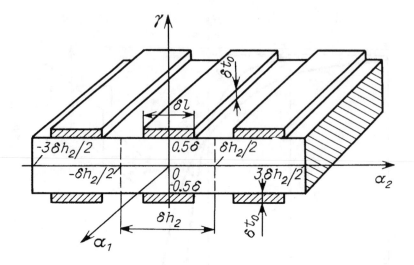

FIGURE 26.1. Plate or shell strengthened by a regular system of thin surface strips.

the symmetry with respect to z and taking into account the periodicity of the functions U_2^{11} and U_3^{11} in ξ_2 with period A_2, the solution of the plane elastic problem that arises may be represented in the form of the series expansions

$$U_2^{11} = 2 \sum_{n=1}^{\infty} [B_n \cosh(\lambda_n z) + C_n \sinh(\lambda_n z)]\sin(\lambda_n h_2 \xi_2)$$

$$U_3^{11} = \Phi_0 z + 2 \sum_{n=1}^{\infty} [B_n \sinh(\lambda_n z) \tag{26.1}$$

$$+ z C_n \cosh(\lambda_n z)]\cos(\lambda_n h_2 \xi_2)$$

where $\lambda_n = 2\pi n/(h_2 A_2)$.

The symmetry with respect to z now permits us to consider the half-plane $z > 0$, in which we mentally remove the ribs occupying the region $\Omega_1 = \{\frac{1}{2} \leq z \leq \frac{1}{2} + t_0, |\xi_2| < \alpha\}$ and replace their effect by a certain unknown displacement field. The assumptions we have introduced about the surface strips then make it possible to write the boundary conditions in the form

$$\tau_{33}^{11} = -C_{12} \qquad (z = \tfrac{1}{2})$$

$$\tau_{23}^{11} = 0 \qquad (z = \tfrac{1}{2}, -A_2/2 \leq \xi_2 < -\alpha, \ \alpha < \xi_2 \leq A_2/2) \tag{26.2}$$

$$U_2^{11} = \Psi(\xi_2) \qquad (z = \tfrac{1}{2}, |\xi_2| < \alpha)$$

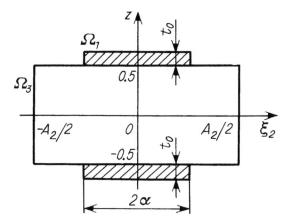

FIGURE 26.2. Unit cell for a plate or shell strengthened by a regular system of thin surface strips.

where τ_{33}^{11} and τ_{23}^{11} are given in 22.18 in terms of the displacements U_2^{11} and U_2^{11}, and the function $\Psi(\xi_2)$ is as yet unknown and must be found in the course of solution.

From the first of Conditions 26.2, we find that

$$\Phi_0 = \frac{c_{12}}{c_{11}} \quad \text{and} \quad B_n = -C_n\lambda_n^{-1}\left[\frac{1-\nu}{1-2\nu} + \frac{\lambda_n}{2}\tanh\left(\frac{\lambda_n}{\lambda}\right)\right]$$

whereas the remaining conditions lead to dual series relations that reduce after some calculation to

$$\sum_{n=1}^{\infty} F_n \sin(\lambda_n h_2 \xi_2) = \frac{1-2\nu}{1-\nu}\Psi(\xi_2) + \sum_{n=1}^{\infty} F_n R_n \sin(\lambda_n h_2 \xi_2)$$

$$(|\xi_2| < \alpha) \qquad\qquad (26.3)$$

$$\sum_{n=1}^{\infty} \lambda_n F_n \sin(\lambda_n h_2 \xi_2) = 0 \quad \left(-\frac{A_2}{2} \le \xi_2 < \alpha, \ \alpha < \xi_2 \le \frac{A}{2}\right)$$

provided

$$F_n = C_n \frac{2(1-2\nu)\lambda_n - \sinh(\lambda_n)}{\lambda_n \cosh(\lambda_n/2)}$$

$$R_n = \frac{2\cosh^2(\lambda_n/2) - \sinh(\lambda_n) + 2(1-2\nu)\lambda_n}{2(1-2\nu)\lambda_n - \sinh(\lambda_n)} \qquad \lim_{n\to\infty} R_n = 0$$

Now if we introduce an auxiliary function $p(\xi_2)$ for which

$$\sum_{n=1}^{\infty} \beta_n F_n \sin(\beta_n \xi_2) \begin{cases} p(\xi_2) & (|\xi_2| < \alpha) \\ 0 & \left(-\dfrac{A_2}{2} \le \xi_2 < -\alpha, \; \alpha < \xi_2 \le \dfrac{A_2}{2} \right) \end{cases}$$

(26.4)

where $\beta_n = \lambda_n h_2$, then

$$F_n = \frac{1}{\beta_n} \int_{-\alpha}^{\alpha} p(t)\sin(\beta_n t)\, dt \tag{26.5}$$

and we note that the function $p(\xi_2)$ and the contact stress $\tau_{23}^{11}|_{z=1/2}$ are related by

$$\tau_{23}^{11}|_{z=1/2} = p^*(\xi_2) = \frac{c_{44}}{(1 - 2\nu)h_2} p(\xi_2) \tag{26.6}$$

Now if we substitute Equation 26.5 into the first of Equations 26.3, we obtain the integral equation

$$\sum_{n=1}^{\infty} \frac{1}{\beta_n} \left(\int_{-\alpha}^{\alpha} p(t)\sin(\beta_n t)\, dt \right) \sin(\beta_n \xi_2)$$

$$= \frac{1 - 2\nu}{1 - \nu} \Psi(\xi_2) + \sum_{n=1}^{\infty} \frac{R_n}{\beta_n} \left(\int_{-\alpha}^{\alpha} p(t)\sin(\beta_n t)\, dt \right) \sin(\beta_n \xi_2)$$

$$(|\xi_2| < \alpha) \tag{26.7}$$

by means of which to calculate the function $p(t)$. The solution, in terms of the Chebyshev polynomial of the first kind, $T_{2m+1}(t)$ may be represented in the form

$$p(t) = \frac{1}{\sqrt{\alpha^2 - t^2}} \sum_{m=0}^{\infty} A_{2m+1} T_{2m+1}\left(\frac{t}{d}\right) \tag{26.8}$$

and using the fact that

$$\int_{-\alpha}^{\alpha} \frac{\sin(\beta_n t) T_{2m+1}(t/d)\, dt}{\sqrt{\alpha^2 - t^2}} = (-1)^m \pi J_{2m+1}(\beta_n \alpha)$$

Equation 26.7 can be put into the form

$$
\pi \sum_{n=1}^{\infty} \frac{1 - R_n}{\beta_n} \left[\sum_{m=0}^{\infty} (-1)^m A_{2m+1} J_{2m+1}(\beta_n \alpha) \right] \sin(\beta_n \xi_2)
$$

$$
= \frac{1 - 2\nu}{1 - \nu} \Psi(\xi_2) \qquad (|\xi_2| < \alpha) \tag{26.9}
$$

A system of equations for determining the coefficients A_{2m+1} occurring in Equation 26.8 can now be obtained by expressing $\Psi(\xi_2)$ in the form of an infinite series,

$$
\Phi(\xi_2) = \sum_{k=0}^{\infty} d_{2k+1} T_{2k+1}\left(\frac{\xi_2}{\alpha} \right) \tag{26.10}
$$

and recalling that

$$
\sin(\beta_n \xi_2) = 2 \sum_{k=0}^{\infty} (-1)^k J_{2k+1}(\beta_n \alpha) T_{2k+1}\left(\frac{\xi_2}{\alpha} \right)
$$

Substituting into Equation 26.9 then results in

$$
\sum_{m=0}^{\infty} (-1)^{m+k} a_{2m+1} \alpha_{2m+1, 2k+1} = d_{2k+1} \tag{26.11}
$$

where

$$
\alpha_{2m+1, 2k+1} = \sum_{n=1}^{\infty} \frac{1 - R_n}{n} J_{2m+1}(\beta_n \alpha) J_{2k+1}(\beta_n \alpha)
$$

$$
a_{2m+1} = \frac{A_1(1 - \nu)}{1 - 2\nu} A_{2m+1}
$$

To solve Equation 26.11 we assume

$$
a_{2m+1} = \sum_{k=0}^{\infty} a_{2k+1}^{(k)} d_{2k+1} \tag{26.12}
$$

where $a_{2m+1}^{(k)}$ is the solution of a system whose left-hand side is the same as in Equations 26.1 and whose right-hand side has 1 at the kth position and zeros at the others.

Substituting into Equation 26.8 now yields

$$p(t) = \frac{1 - 2\nu}{A_2(1 - \nu)} \frac{1}{\sqrt{\alpha^2 - t^2}} \sum_{k=0}^{\infty} d_{2k+1} \left[\sum_{m=0}^{\infty} a_{2m+1}^{(k)} T_{2m+1}\left(\frac{t}{\alpha}\right) \right] \quad (26.13)$$

To proceed with the analysis, the coefficients d_{2k+1} must be determined, for which purpose we turn to the local problem ($b11$), in region Ω_1 (see Figure 26.2) and make use of the strip–surface interface condition. From the boundary conditions given by Equations 22.19 and 22.20 for Ω_1 we have

$$\tau_{22}^{11} = -c_{12}^S \qquad \tau_{23}^{11} = 0 \qquad (\xi_2 = \pm\alpha)$$

$$\tau_{33}^{11} = -c_{12}^S \qquad \tau_{23}^{11} = 0 \qquad (z = \tfrac{1}{2} + t_0) \tag{26.14}$$

where (and further on) a superscript S is used to refer to the strip material.

Along the line of contact between Ω_1 and Ω_3, in accordance with Equations 19.60, 22.21, 26.2, and 26.6, we obtain

$$\tau_{33}^{11} = -c_{12}^S \qquad \tau_{23}^{11} = p^*(\xi_2) \qquad (z = \tfrac{1}{2}, |\xi_2| < \alpha) \tag{26.15}$$

Further, from Equations 26.14, using the assumed small height-to-length ratio $t_0/2\alpha$, one gets

$$\tau_{33}^{11} = -c_{12}^S \quad \text{in } \Omega_1 \tag{26.16}$$

which when substituted into Equations 22.18 for τ_{33}^{11} gives

$$\frac{\partial U_3^{11}}{\partial z} = -\frac{c_{12}^S}{c_{11}^S} - \frac{1}{h_2} \frac{c_{12}^S}{c_{11}^S} \frac{\partial U_2^{11}}{\partial \xi_2}$$

and using this in turn in the expression for τ_{22}^{11} in Equations 22.18, we have

$$\tau_{22}^{11} = \frac{(c_{12}^S)^2}{c_{11}^S} + \frac{1}{h_2}\left(c_{11}^S - \frac{(c_{12}^S)^2}{c_{11}^S}\right)\frac{\partial U_2^{11}}{\partial \xi_2} \tag{26.17}$$

If one averages Equation 22.16 with $i = 2$ over the thickness of the strip, one finds

$$\frac{1}{h_2} \frac{d\bar{\tau}_{22}^{11}}{\partial \xi_2} + \tau_{23}^{11}|_{z=1/2+t_0} - \tau_{23}^{12}|_{z=1/2} = 0 \tag{26.18}$$

and hence, by Equations 26.14 and 26.15,

$$\frac{d\tilde{\tau}_{22}^{11}}{d\xi_2} = h_2 p^*(\xi_2) \tag{26.19}$$

which when integrated yields

$$\tilde{\tau}_{22}^{11} = -c_{12}^S t_0 + h_2 \int_{-\alpha}^{\xi_2} p^*(u)\, du \tag{26.20}$$

after using Equations 26.14.

On the other hand, averaging Equation 26.17 over the thickness of the strip we have

$$\tilde{\tau}_{22}^{11} = -\frac{(c_{12}^S)^2 t_0}{c_{11}^S} + \frac{1}{h_2}\left(c_{11}^S - \frac{(c_{12}^S)^2}{c_{11}^S}\right) t_0 \frac{dU_2^{11}}{d\xi_2} \tag{26.21}$$

From Equations 26.20 and 26.21, with Equations 26.6 and 26.13 and the standard integral

$$\int_{-\alpha}^{\xi_2} \frac{T_{2m+1}(u/\alpha)\, du}{\sqrt{\alpha^2 - u^2}} = -\frac{\sqrt{1 - (\xi_2/\alpha)^2}}{2_{m+1}} U_{2m}\left(\frac{\xi_2}{\alpha}\right)$$

where U_{2m} is the Chebyshev polynomial of the second kind, we obtain

$$\frac{dU_2^{11}}{d\xi_2} = -\frac{\left(c_{12}^S - \dfrac{(c_{12}^S)^2}{c_{11}^S}\right) h_2}{c_{11}^S - \dfrac{(c_{12}^S)^2}{c_{11}^S}}$$

$$-\frac{c_{44} h_2}{(1-\nu)\left[c_{11}^S - \dfrac{(c_{12}^S)^2}{c_{11}^S}\right] t_0 A_2} \sum_{k=0}^{\infty} d_{2k+1} \sum_{m=0}^{\infty} a_{2m+1}^{(k)}$$

$$\times \frac{U_{2m}\left(\dfrac{\xi_2}{\alpha}\right)}{2_{m+2}} \sqrt{1 - \left(\frac{\xi_2}{\alpha}\right)^2} \tag{26.22}$$

This is now integrated using $U_2^{11}(0) = 0$ (which clearly follows from the symmetry of the problem), expressions for the elastic coefficients of an isotropic material, and the integral

$$\int_0^{\xi_2} \sqrt{1 - (\xi_2/\alpha)^2}\, U_{2m}\left(\frac{\xi_2}{2}\right) d\xi_2 = \frac{a}{2m+1} T_{2m+1}\left(\frac{\xi_2}{\alpha}\right)$$

The result is

$$U_2^{11} = -\nu_S h_2 \xi_2 - \frac{E(1 - \nu_S^2)h_2 \alpha}{2E_H(1 - \nu^2)t_0 A_2} \sum_{k=0}^{\infty} d_{2k+1}$$

$$\times \sum_{m=0}^{\infty} \frac{a_{2m+1}^{(k)}}{(2m+1)^2} T_{2m+1}\left(\frac{\xi_2}{\alpha}\right) \tag{26.23}$$

so that using Equations 26.2 and noting that U_2^{11} is continuous across the strip–surface interface, in view of Equations 19.60, we arrive at the following system of equations for the coefficients d_{2k+1}:

$$d_{2m+1} + \frac{E(1 - \nu_H^2)h_2 \alpha}{2E_H(1 - \nu^2)t_0 A_2} \sum_{k=0}^{\infty} a_{2m+1}^{(k)} \frac{d_{2k+1}}{(2m+1)^2} = -\nu_H h_2 \alpha \delta_{2m+1,1}$$

$$\tag{26.24}$$

where $\delta_{2m+1,1}$ is unity for $m = 0$ and zero for $m \geq 1$.

The local problem $(b22)$ can be treated in a similar fashion, and to obtain its solution, the double index {11} must be replaced by {22} everywhere in the preceding formulas. In Conditions 26.14 we write

$$\tau_{22}^{11} - c_{11}^S \quad (\xi_2 = \pm\alpha)$$

In this case and, accordingly, c_{11}^S replaces c_{12}^S in Equations 26.20 and 26.21.

One further change is that in the first term in Equation 26.23 the factor ν_S will disappear and, accordingly, it must also be dropped in the right-hand side of the system of Equations 26.24 (which is solved for d_{2k+1} this time rather than for $d_{2k+1} = d_{2k+1}^{(1)}$ as in the $(b11)$ case). We have therefore

$$d_{2k+1}^{(1)} = \nu_S d_{2k+1}^{(2)}$$

Having solved the local problems ($b11$) and ($b22$) we are now ready to calculate the effective elastic moduli of the shell (plate) reinforced by a regular system of thin surface strips (see Figure 26.1). Using Equations 22.21 we find that

$$\langle b_{22}^{11} \rangle = \langle b_{11}^{22} \rangle = \frac{\nu E}{1 - \nu^2}\left(1 - \frac{1}{\nu}\frac{l}{h_2}\Sigma\right)$$

(26.25)

$$\langle b_{22}^{22} \rangle = \frac{E}{1 - \nu^2}\left(1 - \frac{1}{\nu_S}\frac{l}{h_2}\Sigma\right)$$

where we have written as a shorthand

$$\Sigma = \sum_{k=0}^{\infty} d_{2k+1} \sum_{m=0}^{\infty} \frac{a_{2m+1}^{(k)}}{(2m + 1)^2}$$

If $E_s = E$ and $\nu_S = \nu$, calculations show that the elastic moduli given by Equations 26.25 satisfy Relations 22.28 which we proved in the general case for the effective moduli of homogeneous ribbed shells.

26.2. The effective elastic moduli and the local stresses

A direct comparison between Equations 26.25 and the corresponding results of the structurally anisotropic model (Birger 1961) shows that the latter differ from the former by the lack of terms containing Σ. The role of this correction is illustrated in Table 26.1, which shows the effective elastic moduli computed for homogeneous plates and shells for $h_2 = 10$, $\nu = 0.3$, $A_2 = 1$, and various values of α and t_0 from Equations 26.25 (rows labeled I) and by the method of Birger (rows labeled II); the magnitude of the percentage changes is given in the rows labeled III. It is seen that although the percentage changes for the modulus $\langle b_{11}^{11} \rangle$ are small, those for $\langle b_{22}^{11} \rangle$ are quite considerable.

The method of determining the distribution of local stress will be illustrated by considering the tension of a homogeneous strip reinforced by a set of surface strips when one end of the strip is clamped and the other subjected to an external force of magnitude Q normal to surface strips. The homogenized problem, that is, that of the tension of a strip with the effective moduli given by Equations 26.25, is solved in an elementary fashion to give

$$N_{22} = Q \quad \text{and} \quad \varepsilon_{22} = Q/(\delta\langle b_{22}^{22} \rangle)$$

TABLE 26.1
Effective Elastic Moduli of Plates and Shells
Strengthened by Thin Surface Strips; $v = 0.3$, $h_2 = 10$, $A_2 = 1$

	$\alpha = 0.4$				$t_0 = 0.05$			
	$t_0 = 0.005$	$t_0 = 0.01$	$t_0 = 0.05$	$t_0 = 0.1$	$\alpha = 0.475$	$\alpha = 0.45$	$\alpha = 0.35$	$\alpha = 0.3$
$\dfrac{\langle b_{11}^{11}\rangle}{E/(1-v^2)}$								
1	1.0082	1.0163	1.0811	1.1615	1.0954	1.0907	1.0708	1.0605
2	1.0073	1.0146	1.0728	1.1456	1.0865	1.0819	1.0637	1.0546
3	0.009%	0.17%	0.77%	1.39%	0.82%	0.81%	0.67%	0.56%
$\dfrac{\langle b_{22}^{11}\rangle}{vE/(1-v^2)}$								
1	1.0097	1.0194	1.0926	1.1770	1.0998	1.0974	1.0786	1.0654
2	1.0000	1.0000	1.0000	1.0000	1.0000	1.0000	1.0000	1.0000
3	0.97%	1.94%	9.26%	17.7%	9.98%	9.74%	7.86%	6.54%

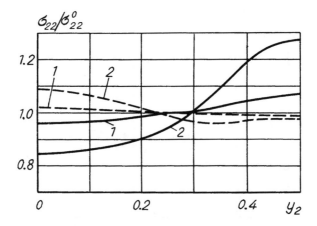

FIGURE 26.3. Stress distribution $\sigma_{22}/\sigma_{22}^0$ along the axis $z = 0$ for $\alpha = 0.4$, $t_0 = 0.1$, and $t_0 = 0.5$ (solid curves 1 and 2) and along the line $z = 0.5$ (dashed curves).

For the stress quantity σ_{22}, using Equation 19.74 and the solution of the problem $(b22)$, we have

$$\sigma_{22}/\sigma_{22}^0 = b_{22}^{22}/\langle b_{22}^{22} \rangle$$

$$= -\frac{\pi}{h_2} \sum_{n=1}^{\infty} G_n \sum_{k=0}^{\infty} d_{2k+1}^{(2)} \sum_{m=0}^{\infty} (-1)^m a_{2m+1}^{(k)} J_{2m+1}$$

$$\times (2\pi n\alpha)\cos(2\pi ny_2) \qquad (26.26)$$

where $\sigma_{22}^{(0)} = Q/\delta$, and G_n are unknown functions.

The results obtained from Equation 26.26 for $z = 0$ (the axis of the strip) and $\alpha = 0.4$ (respectively $\alpha = 0.45$) are shown in Figure 26.3 (respectively in Figure 26.4) for $t_0 = 0.01$ (curves 1) and $t_0 = 0.05$ (curves 2). The dashed curves in both figures refer to the case $z = 0.5$. All the calculations were performed for $\nu = 0.3$ and $h_2 = 10$. Note that the line $\sigma_{22}/\sigma_{22}^0$ corresponds to the stress distribution in the absence of surface strips.

27. PLATES AND SHELLS WITH CORRUGATED SURFACES OF REGULAR STRUCTURE

We will be concerned with different profiles of corrugated surfaces in this section, such as trapezoidal, sinusoidal, and so on (see Figures 27.2, 27.7, 27.11, and 27.12), and it is clear from the outset that because these profiles have all their dimensions of the same order of magnitude, the

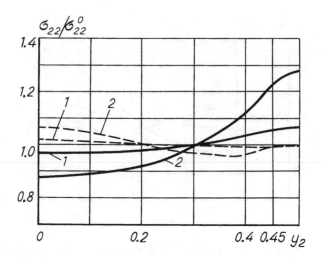

FIGURE 26.4. Stress distribution $\sigma_{22}/\sigma_{22}^0$ along the axis $z = 0$ for $\alpha = 0.45$, $t_0 = 0.01$, and $t_0 = 0.05$ (solid curves 1 and 2) and along the line $z = 0.5$ (dashed curves).

approximate methods we have thus far developed will be of no use for the local problems that arise. The trigonometric series expansions that we employ as an alternative approach will enable us to calculate the effective moduli of the structures under study and to evaluate the stress fields in them under different types of macrodeformations. It is assumed throughout that the bounding contour of the corrugated surface is nowhere perpendicular to the middle surface of the shell.

27.1. The solution in terms of trigonometric expansions

We begin by considering the local problem $(b11)$ (Equations 22.16 and 22.18 through 22.20 with $\lambda\mu = 11$) for the unit cell shown in Figure 27.1. We assume that the profile of the rib is symmetric with respect to the axis $0z$ so that $z^+(\xi_2)$ is an even function of ξ_2. It is also assumed that the shell material is isotropic and its properties are E and ν.

Considered together with Conditions 19.57, the local problem $(b11)$ belongs to the class of the plane problems of the theory of elasticity, and its solution can, in this case, be represented in the form

$$U_2^{11} = \sum_{n=1}^{\infty} \{A_n^{11} \sinh(\lambda_n z) + B_n^{12} \cosh(\lambda_n z) + C_n^{11}$$

$$\times [\cosh(\lambda_n z) + z\lambda_n \sinh(\lambda_n z)]$$

$$\times + D_n^{11} [\sinh(\lambda_n z) + z\lambda_n \cosh(\lambda_n z)]\} \sin(\lambda_n h_2 \xi_2)$$

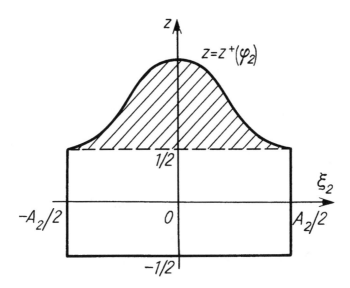

FIGURE 27.1. Unit cell for a shell with a corrugated surface in the coordinate system ξ_2, z.

$$U_3^{11} = A_0^{11}z + \sum_{n=1}^{\infty} \{-A_n^{11}\cosh(\lambda_n z) - B_n^{11}\sinh(\lambda_n z)$$

$$+ C_n^{12}[2(1 - 2\nu)\sinh(\lambda_n z) - z\lambda_n \cosh(\lambda_n z)]$$

$$\times + D_n^{12}[2(1 - 2\nu)\cosh(\lambda_n z) - z\lambda_n \sinh(\lambda_n z)]\}\cos(\lambda_n h_2 \xi_2)$$

$$\lambda_n = \frac{2\pi n}{h_2 A_2} \tag{27.1}$$

Using Equations 27.1 in Equations 22.18 yields the corresponding expansions for the functions τ_{22}^{11}, τ_{23}^{11}, and τ_{33}^{11} involved in the boundary conditions given by Equations 22.19 and 22.20 on the contour $z = z^+(\xi_2)$ and at $z = -\frac{1}{2}$ (see Figure 27.1). From the boundary conditions at $z = -\frac{1}{2}$, one can show that

$$A_0^{11} = \frac{\nu}{1 - \nu} \qquad A_n^{11} = C_n^{11}\left(\mathrm{sc} + \frac{\lambda_n}{2}\right) - D_n^{11}(2\nu + \mathrm{s}^2)$$

$$\tag{27.2}$$

$$B_n^{11} = C_n^{11}(1 - 2\nu + \mathrm{s}^2) + D_n^{11}\left(-\mathrm{sc} + \frac{\lambda_n}{2}\right)$$

where

$$s = \sinh\left(\frac{\lambda_n}{2}\right) \quad \text{and} \quad c = \cosh\left(\frac{\lambda_n}{2}\right)$$

From the boundary conditions on the contour $z = z^+(\xi_2)$, expanding once more in terms of trigonometric functions, and after some calculation, we obtain the following system of algebraic equations for determining the coefficients C_n^{11} and D_n^{11}:

$$\sum_{n=1}^{\infty} (\alpha_{nm} X_n + \beta_{nm} Y_n) = \alpha_m \qquad (m = 1, 2, \dots)$$

(27.3)

$$\sum_{n=1}^{\infty} (\gamma_{nm} X_n + \kappa_{nm} Y_n) = 0 \qquad (m = 0, 1, 2, \dots)$$

and

$$\lambda_n C_n^{11} = \frac{\nu}{1 - \nu} X_n \qquad \lambda_n D_n^{11} = \frac{\nu}{1 - \nu} Y_n \qquad (27.4)$$

where

$$\alpha_{nm} = \frac{4}{A_2} \int_0^{A_2/2} \left\{ -\left[\left(sc + \frac{\lambda_n}{2} \right) \sinh(\lambda_n z^+(\xi_2)) \right. \right.$$

$$\left. + z^+(\xi_2) \lambda_n \sinh(\lambda_n z^+(\xi_2)) + (2 + s^2)\cosh(\lambda_n z^+(\xi_2)) \right]$$

$$\times (z^+(\xi_2))' \cos(\lambda_n h_2 \xi_2)$$

$$+ \left[c^2 \sinh(\lambda_n z^+(\xi_2)) + \left(sc + \frac{\lambda_n}{2} \right) \cosh(\lambda_n z^+(\xi_2)) \right.$$

$$\left. + z^+(\xi_2) \lambda_n \cosh(\lambda_n z^+(\xi_2)) \right]$$

(27.5)

$$\left. \times h_2 \sin(\lambda_n h_2 \xi_2) \right\} \sin(\lambda_m h_2 \xi_2) \, d\xi_2$$

$$
\beta_{nm} = \frac{4}{A_2} \int_0^{A_2/2} \left\{ -\left[(1 - s^2)\sinh(\lambda_n z^+) + \left(\frac{\lambda_n}{2} - sc \right)\cosh(\lambda_n z^+) \right.\right.
$$

$$
+ z^+ \lambda_n \cosh(\lambda_n z^+) \Big] (z^+(\xi_2))' \sin(\lambda_n h_2 \xi_2)
$$

$$
- \left[\left(\frac{\lambda_n}{2} - sc \right)\sinh(\lambda_n z^+) + z^+ \lambda_n \sinh(\lambda_n z^+) + s^2 \cosh(\lambda_n z^+) \right] h_2
$$

$$
\left. \times \cos(\lambda_n h_2 \xi_2) \right\} \cos(\lambda_m h_2 \xi_2) \, d\xi_2
$$

$$
\gamma_{nm} = \frac{4}{A_2} \int_0^{A_2/2} \left\{ -\left[c^2 \sinh(\lambda_n z^+) + \left(sc + \frac{\lambda_n}{2} \right)\cosh(\lambda_n z^+) \right.\right.
$$

$$
+ z^+ \lambda_n \cosh(\lambda_n z^+) \Big]
$$

$$
\times (z^+(\xi_2))' \sin(\lambda_n h_2 \xi_2) - \left[\left(\frac{\lambda_n}{2} + sc \right)\sinh(\lambda_n z^+) \right.
$$

$$
+ z^+ \lambda_n \sinh(\lambda_n z^+) + s^2 \cosh(\lambda_n z^+) \Big]
$$

$$
\left. \times h_2 \cos(\lambda_n h_2 \xi_2) \right\} \cos(\lambda_m h_2 \xi_2) \, d\xi_2
$$

$$
\kappa_{nm} = \frac{4}{A_2} \int_0^{A_2/2} \left\{ -\left[\left(\frac{\lambda_n}{2} - sc \right)\sinh(\lambda_n z^+) + z^+ \lambda_n \sinh(\lambda_n z^+) \right.\right.
$$

$$
- s^2 \cosh(\lambda_n z^+) \Big]
$$

$$
\times (z^+(\xi_2))' \sin(\lambda_n h_2 \xi_2) + \left[c^2 \sinh(\lambda_n z^+) + \left(sc - \frac{\lambda_n}{2} \right) \right.
$$

$$
\times \cosh(\lambda_n z^+) - z^+ \lambda_n \cosh(\lambda_n z^+) \Big]
$$

$$
\left. \times h_2 \cos(\lambda_n h_2 \xi_2) \right\} \cos(\lambda_m h_2 \xi_2) \, d\xi_2
$$

$$\alpha_m = \frac{4}{A_2} \int_0^{A_2/2} (z^+(\xi_2))' \sin(\lambda_m h_2 \xi_2) \, d\xi_2 \tag{27.6}$$

with a prime indicating an ordinary derivative with respect to ξ_2.

Using Equations 22.18 and 22.21 along with Equations 27.1 and 27.2, the functions we are interested in are found to be related by

$$b_{11}^{11} = \nu(\tau_{22}^{11} + \tau_{33}^{11}) + c_{11} \qquad b_{22}^{11} = \tau_{22}^{11} + c_{12}$$

$$b_{33}^{11} = \tau_{33}^{11} + c_{12} \qquad b_{23}^{11} = \tau_{23}^{11}$$

$$\tau_{22}^{11} = -\frac{c_{12}^2}{c_{11}} + (c_{11} - c_{12}) \sum_{n=1}^{\infty} \left\{ C_n^{11} \left[\left(sc + \frac{\lambda_n}{2} \right) \sinh(\lambda_n z) \right. \right.$$

$$\left. + z\lambda_n \sinh(\lambda_n z) + (2 + s^2)\cosh(\lambda_n z) \right]$$

$$+ D_n^{11} \left[(1 - s^2)\sinh(\lambda_n z) + \left(\frac{\lambda_n}{2} - sc \right) \right.$$

$$\left. \times \cosh(\lambda_n z) + z\lambda_n \cosh(\lambda_n z) \right] \right\}$$

$$\times \lambda_n \cos(\lambda_n H_2 \xi_2)$$

$$\tau_{33}^{11} = -c_{12} + (c_{11} - c_{12}) \sum_{n=1}^{\infty} \left\{ -C_n^{11} \left[\left(\frac{\lambda_n}{2} + sc \right) \sinh(\lambda_n z) \right. \right.$$

$$\left. + z\lambda_n \sinh(\lambda_n z) + s^2 \cosh(\lambda_n z) \right]$$

$$+ D_n^{11} \left[c^2 \sinh(\lambda_n z) + \left(sc - \frac{\lambda_n}{2} \right) \cosh(\lambda_n z) \right.$$

$$\left. - z\lambda_n \cosh(\lambda_n z) \right] \right\} \lambda_n \cos(\lambda_n h_2 \xi_2)$$

$$\tau_{23}^{11} = (c_{11} - c_{12}) \sum_{n=1}^{\infty} \left\{ C_n^{11} \left[c^2 \sinh(\lambda_n z) + \left(sc + \frac{\lambda_n}{2} \right) \cosh(\lambda_n z) \right. \right.$$

$$\left. + z\lambda_n \cosh(\lambda_n z) \right] + D_n^{11} \left[\left(\frac{\lambda_n}{2} - sc \right) \sinh(\lambda_n z) + z\lambda_n \sinh(\lambda_n z) \right.$$

$$\left. \left. - s^2 \cosh(\lambda_n z) \right] \right\} \lambda_n \sin(\lambda_n h_2 \sigma_2) \tag{27.7}$$

The solution of the local problem ($b22$) (Equations 22.16 and 22.18 through 22.20 with $\lambda\mu = 22$) may be obtained along lines similar to those for ($b11$), with the superscript pair $\{11\}$ replaced by $\{22\}$ in Equations 27.1 and 27.2. It can be shown that the coefficients c_n^{22} and D_n^{22} are found from Equation 27.3 because

$$\lambda_n C_n^{22} = \frac{1}{1-\nu} X_n \quad \text{and} \quad \lambda_n D_n^{22} = \frac{1}{1-\nu} Y_n \tag{27.8}$$

Now if the expansions employed for the solution of the ($b11$) problem are compared with those for the ($b22$) problem, then, with the aid of Equations 27.3 and 27.8, the relations

$$b_{11}^{11} = \nu b_{11}^{22} + E \qquad b_{22}^{11} = \nu b_{22}^{22}$$

$$\tag{27.9}$$

$$b_{33}^{11} = \nu b_{33}^{22} \qquad b_{23}^{11} = \nu b_{23}^{22}$$

are obtained, which supplement the general conditions given by Equations 22.32 and 22.33 and are valid for any homogeneous ribbed shell and plates made of an orthotropic material. Because of the relationships established in Equations 27.9 between the functions b_{ij}^{22} and b_{ij}^{11}, it suffices to solve only the local problem ($b11$) for the determination of these functions.

We consider next the local problems (b^*11) and (b^*22) set by Equations 22.22 and 22.34 through 22.26 with, respectively, $\lambda\mu = 11$ and 22 on the unit cell shown in Figure 27.1. These problems also belong to the class of the plane elastic problems and are solved in the same manner as the problems ($b11$) and ($b22$).

The exact analytical solutions of the problem $(b*11)$ is of the form

$$V_2^{11} = \sum_{n=1}^{\infty} \left\{ \overline{C}_n^{11} \left[\left(\frac{\lambda_n}{2} + sc \right) \sinh(\lambda_n z) + \lambda_n z \sinh(\lambda_n z) \right.\right.$$

$$+ (s^2 + 2 - 2\nu)\cosh(\lambda_n z) \Big]$$

$$+ D_n^{11} \left[(1 - 2\nu - s^2)\sinh(\lambda_n z) + \left(\frac{\lambda_n}{2} - sc \right)\cosh(\lambda_n z) \right.$$

$$\left.\left. + z\lambda_n \cosh(\lambda_n z) \right] \right\} \sin(\lambda_n h_2 \xi_2)$$

$$V_3^{11} = -\frac{\nu}{1-\nu} \frac{z^2}{2} + \sum_{n=1}^{\infty} \left\{ \overline{C}_n^{11} \left[(1 - 2\nu - s^2)\sinh(\lambda_n z) \right.\right.$$

$$- \left(\frac{\lambda_n}{2} + sc \right)\cosh(\lambda_n z) - z\lambda_n \cosh(\lambda_n z) \Big]$$

$$+ \overline{D}_n^{11} \left[\left(sc - \frac{\lambda_n}{2} \right)\sinh(\lambda_n z) - z\lambda_n \sinh(\lambda_n z) \right.$$

$$\left.\left. + (2 - 2\nu + s^2)\cosh(\lambda_n z) \right] \right\} \cos(\lambda_n h_2 \xi_2) \qquad (27.10)$$

and for the functions b_{ij}^{11} we have

$$b_{11}^{*11} = \nu\left(\overline{\tau}_{22}^{11} + \overline{\tau}_{33}^{11} \right) + zc_{11} \qquad b_{22}^{*11} = \overline{\tau}_{22}^{11} + zc_{12}$$

$$b_{33}^{*11} = \overline{\tau}_{33}^{11} + zc_{12} \qquad\qquad\qquad b_{23}^{*11} = \overline{\tau}_{23}^{11}$$

$$\overline{\tau}_{22}^{11} = -\frac{c_{12}^2}{c_{11}}z + (c_{11} - c_{12}) \sum_{n=1}^{\infty} \left\{ \overline{C}_n^{11} \left[\left(sc + \frac{\lambda_n}{2} \right)\sinh(\lambda_n z) \right.\right.$$

$$\left. + z\lambda_n \sinh(\lambda_n z) + (2 + s^2)\cosh(\lambda_n z) \right]$$

$$+ \overline{D}_n^{11} \left[(1 - s^2)\sinh(\lambda_n z) + \left(\frac{\lambda_n}{2} - sc \right)\cosh(\lambda_n z) \right.$$

$$\left.\left. + z\lambda_n \cosh(\lambda_n z) \right] \right\} \lambda_n \cos(\lambda_n h_2 \xi_2)$$

$$\overline{\tau}_{33}^{11} = -c_{12}z + (c_{11} - c_{12}) \sum_{n=1}^{\infty} \left\{ -\overline{C}_n^{12} \left[\left(\frac{\lambda_n}{2} + sc \right) \sinh(\lambda_n z) \right.\right.$$

$$\left. + z\lambda_n \sinh(\lambda_n z) + s^2 \cosh(\lambda_n z) \right]$$

$$+ \overline{D}_n^{11} \left[c^2 \sinh(\lambda_n z) + \left(sc - \frac{\lambda_n}{2} \right) \cosh(\lambda_n z) \right.$$

$$\left.\left. - z\lambda_n \cosh(\lambda_n z) \right] \right\} \lambda_n \cos(\lambda_n h_2 \xi_2)$$

$$\overline{\tau}_{23}^{11} = (c_{11} - c_{12}) \sum_{n=1}^{\infty} \left\{ \overline{C}_n^{11} \left[c^2 \sinh(\lambda_n z) + \left(sc + \frac{\lambda_n}{2} \right) \cosh(\lambda_n z) \right.\right.$$

$$\left. + z\lambda_n \cosh(\lambda_n z) \right] + \overline{D}_n^{11} \left[\left(\frac{\lambda_n}{2} - sc \right) \sinh(\lambda_n z) \right.$$

$$\left.\left. + z\lambda_n \sinh(\lambda_n z) - s^2 \cosh(\lambda_n Z) \right] \right\} \lambda_n \sin(\lambda_n h_2 \xi_2) \qquad (27.11)$$

The coefficients C_n^{11} and D_n^{11} in Expansions 27.10 and 27.11 are determined from the equations

$$\lambda_n \overline{C}_n^{11} = \frac{\nu}{1 - \nu} \overline{X}_n \quad \text{and} \quad \lambda_n \overline{D}_n^{11} = \frac{\nu}{1 - \nu} \overline{Y}_n \qquad (27.12)$$

where X_n and Y_n are the solutions of the algebraic system

$$\sum_{n=1}^{\infty} \left(\alpha_{nm} \overline{X}_n + \beta_{nm} \overline{Y}_m \right) = \beta_m \qquad (m = 1, 2, \dots)$$

$$\qquad (27.13)$$

$$\sum_{n=1}^{\infty} \left(\gamma_{nm} \overline{X}_n + \kappa_{nm} \overline{Y}_n \right) = 0 \qquad (m = 0, 1, 2, \dots)$$

in which the coefficients are defined by Equations 27.5 and the right-hand side of the first equation is

$$\beta_m = \frac{4}{A_2} \int_0^{A_2/2} z^+(\xi_2)(z^+(\xi_2))' \sin(\lambda_m h_2 \xi_2) \, d\xi_2 \qquad (27.14)$$

The solutions of the local problem (b^*22) are obtained from Equations 27.10 by merely replacing the superscript pair $\{11\}$ by $\{22\}$; similarly for the group (b) local problems, it is readily shown that

$$\overline{C}_n^{11} = \nu \overline{C}_n^{22} \quad \text{and} \quad \overline{D}_n^{11} = \nu \overline{D}_n^{22} \tag{27.15}$$

Using this and comparing the series representations for the solutions of the (b^*11) and (b^*22) problems, we arrive at the following relations analogous to (27.9) and complementary to (22.34) and (22.35) (which express, we recall, the general properties of homogeneous ribbed and shells):

$$b_{11}^{*11} = \nu b_{11}^{*22} + zE \qquad b_{22}^{*11} = \nu b_{22}^{*22}$$

$$\tag{27.16}$$

$$b_{33}^{*11} = \nu b_{33}^{*22} \qquad b_{23}^{*11} = \nu b_{23}^{*22}$$

By means of Equations 27.15 and 27.16, we can express the solutions of the problem (b^*22) through the corresponding solutions of the problem (b^*11).

The local problem $(b12)$ that we consider next is set on the unit cell of Figure 27.1 by Equations 22.17 through 22.20 with $\lambda \mu = 12$. The problem belongs to the class of antiplane elastic problems and its solution may be taken in the form

$$U_1^{12} = \sum_{n=1}^{\infty} B_n^{12}[t \sinh(\lambda_n z) = \cosh(\lambda_n z)]\sin(\lambda_n h_2 \xi_2)$$

$$\tag{27.17}$$

$$t = \tanh\left(\frac{\lambda_n}{2}\right)$$

where the coefficients B_n^{12} are found from the system of algebraic equations

$$\sum_{n=1}^{\infty} (\lambda_n B_n^{12})\theta_{nm} = \alpha_m \qquad (m = 1, 2, \dots) \tag{27.18}$$

$$\theta_{nm} = \frac{4}{A_2} \int_0^{A_2/2} \{-[\cosh(\lambda_n z^+) + t \sinh(\lambda_n z^+)](z^+(\xi_2))'\cos(\lambda_n h_2 \xi_2)$$

$$+ [\sinh(\lambda_n z^+) + t \cosh(\lambda_n z^+)]h_2 \sin(\lambda_n h_2 \xi_2)\}\sin(\lambda_n h_2 \xi_2)\,d\xi_2$$

$$\tag{27.19}$$

with α_m defined in Equations 27.6.

Having solved the problem (b12), the functions

$$b_{12}^{12} = G + G \sum_{n=1}^{\infty} \lambda_n B_n^{12}[t \sinh(\lambda_n z) + \cosh(\lambda_n z)]\cos(\lambda_n h_2 \xi_2)$$

(27.20)

$$b_{13}^{12} = G \sum_{n=1}^{\infty} \lambda_n B_n^{12}[t \cosh(\lambda_n z) + \sinh(\lambda_n z)]\sin(\lambda_n h_2 \xi_2)$$

are determined.

The solution of the remaining local problem (b^*12), posed by Equations 22.23 through 22.26 with $\lambda\mu = 12$, is obtained in the same way as for the problem (b12) and may be represented in the form

$$V_1^{12} = \sum_{n=1}^{\infty} \bar{B}_n^{12}[t \sinh(\lambda_n z) + \cosh(\lambda_n z)]\sin(\lambda_n h_2 \xi_2)$$

$$b_{12}^{*12} = zG + G \sum_{n=1}^{\infty} \lambda_n \bar{B}_n^{12}[t \sinh(\lambda_n z) + \cosh(\lambda_n z)]\cos(\lambda_n h_2 \xi_2) \quad (27.21)$$

$$b_{13}^{*12} = G \sum_{n=1}^{\infty} \lambda_n \bar{B}_n^{12}[t \cosh(\lambda_n z) + \sinh(\lambda_n z)]\sin(\lambda_n h_2 \xi_2)$$

where the coefficients B_n^{12} are determined from the system of algebraic equations,

$$\sum_{n=1}^{\infty} (\lambda_n \bar{B}_n^{11}) \theta_{nm} = \beta_m \qquad (m = 1, 2 \ldots) \qquad (27.22)$$

in which θ_{nm} are defined by Equation 27.19 and β_m by Equation 27.14.

27.2. The effective elastic moduli and the local stresses

The relevant local problems having been solved, we can now proceed to the numerical analysis of the effective moduli and the local stress distributions in plate and shell structures with corrugated surfaces of various types. The question that immediately presents itself is, given the desired accuracy of calculation, where the (theoretically infinite) system of Equations 27.3, 27.13, 27.18, and 27.22 should be truncated? In order to get a feeling for the role of the size of the system, results obtained from 5 × 5 and 8 × 8 systems were compared. It was found that although the effective moduli are virtually insensible to the order of the system, the changes in local stress distribution are quite substantial, especially in the vicinity of

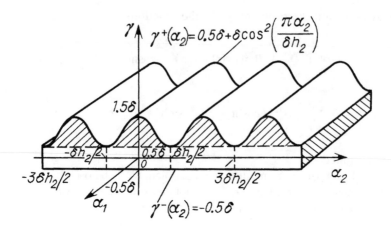

FIGURE 27.2.   A shell with a corrugated surface of cosinusoidal profile.

the edge of rib (up to 20%) for b_{22}^{22}), although they become progressively less away from the edge and for no stress component exceed 1 or 2% at the center of the composite shell. It was decided therefore that a 20 × 20 system should provide a sufficient computational accuracy. In the following, the results obtained for a few specific profiles of the reinforcing element are discussed.

For the case of a cosinusoidal profile (see Figure 27.2), the unit cell referred to the coordinate system ξ_2, z is shown in Figure 27.3 and the effective moduli calculated by the preceding scheme for $\nu = 0.3$ and as $A_2 = 1$ are summarized in Table 27.1. The table also shows the corresponding results of the structurally anisotropic model (Korolev 1971), both in absolute terms and in terms of percentage changes. These latter are seen to be quite considerable for some effective moduli.

The local distributions of the functions $b_{\alpha\beta}^{\lambda\mu}$ and $b_{\alpha\beta}^{*\lambda\mu}$ were calculated for a number of important cross sections of the unit cell of Figure 27.3 with $h_2 = 3$, $\nu = 0.3$, and $A_2 = 1$ and are shown in Figures 27.4 and 27.5 (for $\xi_2 = 0$) and in Figure 27.6 (for $\xi_2 = 0.5$). By means of Equation 19.74, these functions determine the local distribution of stress for various macrodeformations of the shell.

Figure 27.7 shows a shell-like structure reinforced by ribs of trapezoidal profile; Figure 27.8 shows its unit cell referred to the coordinates ξ_2, z. The shape of a rib is determined by the parameters H, a, and b ($0 \leq a \leq b \leq 0.5$); the dimensions of the rib as referred to the coordinate system α_1, γ, are δH, $\delta a h_2$, and $\delta b h_1$, the quantity δh_2 representing the distance between two neighboring ribs.

In the rows labeled I in Table 27.2 we show the effective stiffness properties of such structures obtained for $\nu = 0.3$, $A_2 = 1$, and for various

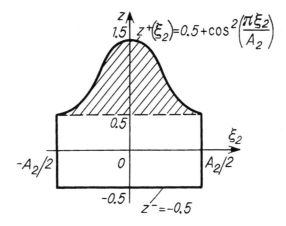

FIGURE 27.3. Unit cell for a shell with a corrugated surface of cosinusoidal profile, in coordinate system ξ_2, z.

TABLE 27.1
Effective Elastic Moduli of Plates and Shells
with a Corrugated Surface of a Cosinusoidal Profile; $\nu = 0.3$, $A_2 = 1$

Effective elastic moduli	Cases considered			Structurally anisotropic theory	Percentage changes (%)
	(1) $h_2 = 5$	(2) $h_2 = 2\pi$	(3) $h_2 = 3\pi$		
$\langle b_{11}^{11} \rangle / E$	1.618	1.619	1.616	1.599	1.1–1.3%
$\langle b_{22}^{22} \rangle / E$	1.311	1.322	1.289	1.099	17–20%
$\langle b_{22}^{11} \rangle / E$	0.393	0.397	0.387	0.330	17–20%
$\langle b_{12}^{12} \rangle / G$	1.345	1.368	1.391	1.000	35–39%
$\langle b_{11}^{*11} \rangle / E$	0.448	0.451	0.449	0.438	2–3%
$\langle b_{22}^{*22} \rangle / E$	0.117	0.150	0.128	0	—
$\langle b_{22}^{*11} \rangle / E$	0.035	0.045	0.038	0	—
$\langle b_{12}^{*12} \rangle / G$	0.283	0.306	0.330	0.219	29–51%
$\langle z b_{11}^{*11} \rangle / E$	0.515	0.518	0.517	0.508	1–2%
$\langle z b_{22}^{*22} \rangle / E$	0.167	0.200	0.189	0.092	82–117%
$\langle z b_{22}^{*11} \rangle / E$	0.050	0.060	0.057	0.028	82–117%
$\langle z b_{12}^{*12} \rangle / G$	0.344	0.366	0.391	—	—

values of the parameters h, a, b, and h_2. The row labeled II and III present, respectively, the values computed from the structurally anisotropic theory (Korolev 1971) and the differences between the results given in the first two rows. Again, the percentage changes prove to be quite large for many stiffness moduli.

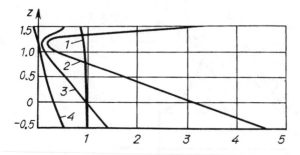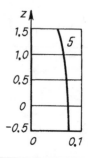

FIGURE 27.4. Local stress distributions over the cross section $\xi_2 = 0$ of the unit cell of Figure 27.3, for $h_2 = 3\pi$, $\nu = 0.3$, and $A_2 = 1$: curve 1, $b_{11}^{11}/[E/(1 - \nu^2)]$; curve 2, $b_{22}^{22}/[E/(1 - \nu^2)]$; curve 3, $b_{11}^{11}/[\nu E/(1 - \nu^2)]$; curve 4, $b_{22}^{11}/[\nu E/(1 - \nu^2)]$; curve 5, b_{12}^{12}/G.

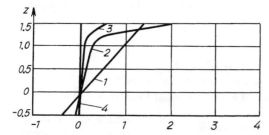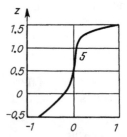

FIGURE 27.5. Local stress distributions over the cross section $\xi_2 = 0$ of the unit cell of Figure 27.3, for $h_2 = 3\pi$, $\nu = 0.3$, and $A_2 = 1$: curve 1, $b_{11}^{*11}/[E/(1 - \nu^2)]$; curve 2, $b_{22}^{*22}/[E/(1 - \nu^2)]$; curve 3, $b_{11}^{*22}/[\nu E/(1 - \nu^2)]$; curve 4, $b_{22}^{*11}/[\nu E/(1 - \nu^2)]$; curve 5, b_{12}^{*12}/G.

For case II of Table 27.2, the local distributions of the functions $b_{\alpha\beta}^{\lambda\mu}$ and $b_{\alpha\beta}^{*\lambda\mu}$ over some of the important cross sections of the unit cell in Figure 27.8 are shown in Figures 27.9 and 27.10 for $\xi_2 = 0$ and for $\xi_2 = 0.5$, respectively.

It should be noted that at small a/b ratios (high trapezoids) or in the limiting case when $a = 0$ (triangle), the violent oscillations and singularities that appear in solutions in the vicinity of rib edges complicate the numerical work considerably in terms of both increased grid density and, as a consequence, increased computer time.

It is also of interest to estimate the effect that fillets between the ribs and the carrying surfaces have on the effective properties of, and local stress fields in, the composite. We consider as an example the trapezoidal profile of Figures 27.7 and 27.8, choosing case III of Table 27.2 as a reference case and introducing a rounded-off fillet as shown in Figure

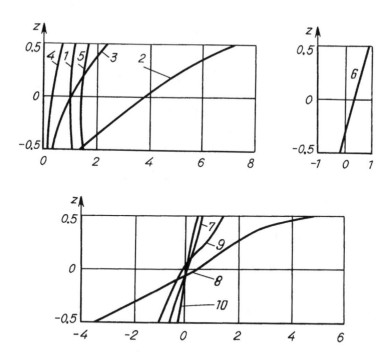

FIGURE 27.6. Local stress distributions over the cross section $\xi_2 = 0.5$ of the unit cell of Figure 27.3 for $h_2 = 3\pi$, $\nu = 0.3$, and $A_2 = 1$: curve 1, $b_{11}^{11}/[E/(1 - \nu^2)]$; curve 2, $b_{22}^{22}/[E/(1 - \nu^2)]$; curve 3, $b_{11}^{22}/[\nu E/(1 - \nu^2)]$; curve 4, $b_{12}^{11}/[\nu E/(1 - \nu^2)]$; curve 5, b_{12}^{12}/G; curve 6, b_{12}^{*12}/G; curve 7, $b_{11}^{*11}/[E/(1 - \nu^2)]$; curve 8, $b_{22}^{*22}/[E/(1 - \nu^2)]$; curve 9, $b_{11}^{*22}/[\nu E/(1 - \nu^2)]$; curve 10, $b_{22}^{*11}/[\nu E/(1 - \nu^2)]$.

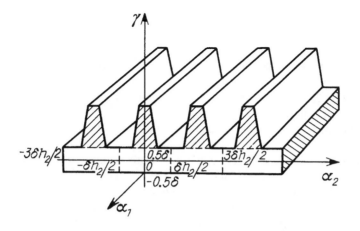

FIGURE 27.7. Shell with strengthening ribs of trapezoidal profile.

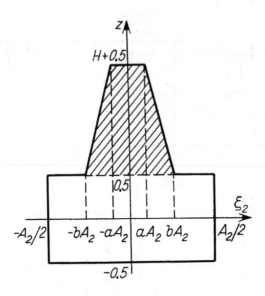

FIGURE 27.8. Unit cell for a shell with strengthening ribs of trapezoidal profile in the coordinate system ξ_2, z.

27.11. The values of the effective moduli calculated for $\nu = 0.3$ and $A_2 = 1$ are found to be

$$\frac{\langle b_{11}^{11} \rangle}{E} = 1.651 \qquad \frac{\langle b_{11}^{*11} \rangle}{E} = 0.723 \qquad \frac{\langle zb_{11}^{*11} \rangle}{E} = 1.220$$

$$\frac{\langle b_{12}^{12} \rangle}{G} = 1.874 \qquad \frac{\langle b_{12}^{*12} \rangle}{G} = 1.054 \qquad \frac{\langle zb_{12}^{*12} \rangle}{G} = 0.800$$

showing a 5–16% decrease in $\langle b_{11}^{11} \rangle$, $\langle b_{11}^{*11} \rangle$, $\langle zb_{11}^{*11} \rangle$, and a 22–40% increase in $\langle b_{12}^{12} \rangle$, $\langle b_{12}^{*12} \rangle$, and $\langle zb_{12}^{*12} \rangle$. The local distributions of the functions $b_{\alpha\beta}^{\lambda\mu}$ and $b_{\alpha\beta}^{*\lambda\mu}$ over the contour of the unit cell of Figure 27.11 are shown in Table 27.3.

As the final example we consider a shell (or plate) with a system of parallel grooves of circular cross section (see Figure 27.12). The values of the effective moduli of such structures were computed for the case in which $\nu = 0.3$ and $A_2 = 1$ and are listed in Table 27.4 (column I); the corresponding results obtained from the structurally anisotropic model of Korolev (1971) are also shown (in column II). The values of $b_{\alpha\beta}^{\lambda\mu}$ and $b_{\alpha\beta}^{*\lambda\mu}$ at some points on the external contour of the unit cell of Figure 27.12 are listed in Table 27.5.

TABLE 27.2
Effective Elastic Moduli of Plates and Shells
with Strengthening Ribs of Trapezoidal Profile; $v = 0.3$, $A_2 = 1$

Effective Elastic moduli		Cases considered		
		(1) $H = 0.5$, $h_2 = 1$, $a = 0.25$, $b = 0.45$	(2) $H = 1.5$, $h_2 = 1$, $a = 0.1$, $b = 0.2$	(3) $H = 2$, $h_2 = 3$, $a = 0.1$, $b = 1/6$
$\langle b_{11}^{11} \rangle / E$	I	1.4835	1.5923	1.7440
	II	1.4489	1.5489	1.6319
	III	2.4%	2.8%	6.9%
$\langle b_{22}^{22} \rangle / E$	I	1.4833	1.5811	2.3444
	II	1.0989	1.0989	1.0989
	III	35%	43.9%	113.3%
$\langle b_{22}^{11} \rangle / E$	I	0.4450	0.4743	0.7033
	II	0.3297	0.3297	0.3297
$\langle b_{12}^{12} \rangle / G$	I	1.344	1.190	1.539
	II	1.000	1.000	1.000
	III	34.4%	19%	53.9%
$\langle b_{11}^{*11} \rangle / E$	I	0.2802	0.5758	0.8626
	II	0.254	0.525	0.755
	III	10.3%	9.7%	14.3%
$\langle b_{22}^{*22} \rangle / E$	I	0.2911	0.5644	1.1956
	II	0	0	0
$\langle b_{22}^{*11} \rangle / E$	I	0.0873	0.1693	0.3587
	II	0	0	0
$\langle b_{12}^{*12} \rangle / G$	I	0.250	0.246	0.752
$\langle zb_{11}^{*11} \rangle / E$	I	0.3038	0.8524	1.4668
	II	0.2836	0.7854	1.3356
	III	7.1%	8.5%	9.8%
$\langle zb_{22}^{*22} \rangle / E$	I	0.3167	0.8367	1.550
	II	0.916	0.0916	0.0916
$\langle zb_{22}^{*11} \rangle / E$	I	0.095	0.251	0.465
	II	0.0275	0.0275	0.0275
$\langle zb_{12}^{*12} \rangle / G$	I	0.2697	0.2688	0.6083

27.3. Local stress distribution in a cylindrical shell with a corrugated surface

The effective elastic moduli and the local distributions of the functions $b_{\alpha\beta}^{\lambda\mu}$ and $b_{\alpha\beta}^{*\lambda\mu}$ that we have obtained previously for a number of reinforcement types may be utilized for the consideration of various types of boundary value problems associated with shell-like structures. As an example, we consider here the problem of deformation of an axially free cylindrical shell whose outer surface is strengthened by a system of stringers, while the inner surface, a smooth one, is subjected to an internal pressure of q. The radius of the middle surface of the shell is taken to be R.

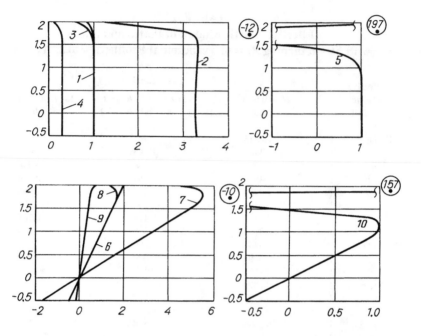

FIGURE 27.9. Local stress distributions over the cross section $\xi_2 = 0$ of the unit cell of Figure 27.8, for $v = 0.3$, $A_2 = 1$, $H = 1.5$, $h_2 = 1$, $a = 0.1$, and $b = 0.2$: curve 1, $b_{11}^{11}/[E/(1 - v^2)]$; curve 2, $b_{22}^{22}/[E/(1 - v^2)]$; curve 3, $b_{11}^{22}/[vE/(1 - v^2)]$; curve 4, $b_{22}^{11}/[vE/(1 - v^2)]$; curve 5, b_{12}^{12}/G; curve 6, $b_{11}^{*11}/[E/(1 - v^2)]$; curve 7, $b_{22}^{*22}/[E/(1 - v^2)]$; curve 8, $b_{11}^{*22}/[vE/(1 - v^2)]$; curve 9, $b_{22}^{*11}/[vE/(1 - v^2)]$; curve 10, b_{12}^{*12}/G.

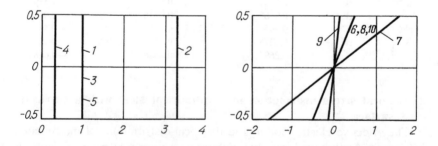

FIGURE 27.10. Local stress distributions over the cross section $\xi_2 = 0.5$ of the unit cell of Figure 27.8, for $v = 0.3$, $A_2 = 1$, $H = 1.5$, $h_2 = 1$, $a = 0.1$, and $b = 0.2$: curve 1, $b_{11}^{11}/[E/(1 - v^2)]$; curve 2, $b_{22}^{22}/[E/(1 - v^2)]$; curve 3, $b_{11}^{22}/[vE/(1 - v^2)]$; curve 4, $b_{22}^{11}/[vE/(1 - v^2)]$; curve 5, b_{12}^{12}/G; curve 6, $b_{11}^{*11}/[E/(1 - v^2)]$; curve 7, $b_{22}^{*22}/[E/(1 - v^2)]$; curve 8, $b_{11}^{*22}/[vE/(1 - v^2)]$; curve 9, $b_{22}^{*11}/[vE/(1 - v^2)]$; curve 10, b_{12}^{*12}/G.

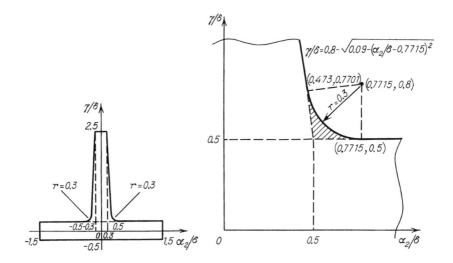

FIGURE 27.11. Unit cell of a shell with trapezoidal ribs and fillets at junctions between carrying surfaces and ribs ($H = 2$, $h_2 = 3$, $a = 0.1$, $b = \frac{1}{6}$).

If the constitutive relations of the homogenized shell, Equations 19.88, are substituted into the equations of motion of a cylindrical shell, Equations 19.95, one finds after some manipulation, that the deflection w is determined by the equation

$$\frac{d^4 w}{d\alpha_1^4} + 2\beta_2 \frac{d^2 w}{d\alpha_1^2} + \beta_0 w = q^* \tag{27.23}$$

where Relations 19.84 have been used and where

$$q^* = \frac{q}{D^*} \qquad \beta_2 = \frac{C^*}{(RD^*)} \qquad \beta_0 = \frac{B^*}{(R^2 D^*)}$$

$$D^* = \delta^3 \left[\langle z b_{11}^{*11} \rangle - \frac{(\langle b_{11}^{*11} \rangle)^2}{\langle b_{11}^{11} \rangle} \right]$$

$$C^* = \delta^2 \left[\frac{\langle b_{22}^{11} \rangle \langle b_{11}^{*11} \rangle}{\langle b_{11}^{11} \rangle - \langle b_{22}^{*11} \rangle} \right]$$

$$B^* = \delta \left[\langle b_{22}^{22} \rangle - \frac{(\langle b_{22}^{11} \rangle)^2}{\langle b_{11}^{11} \rangle} \right]$$

TABLE 27.3
Local Distributions of Functions $b_{\alpha\beta}^{\lambda\mu}$ and $b_{\alpha\beta}^{*\lambda\mu}$ over the Perimeter of the Unit Cell of Figure 27.11

Points on the unit cell perimeter	$b_{11}^{11} / \dfrac{E}{1-\nu^2}$	$b_{22}^{11} / \dfrac{\nu E}{1-\nu^2}$	$b_{11}^{*11} / \dfrac{E}{1-\nu^2}$	$b_{22}^{*11} / \dfrac{\nu E}{1-\nu^2}$	b_{12}^{12} / G	b_{12}^{*12} / G
$\xi_2 = 0$, $z = 2.5$, $\alpha_2 = 0$	0.185	-5.70	1.12	-26.0	187.0	-10.0
$\xi_2 = 0.0625$, $z = 2.5$, $\alpha_2 = 0.1875\delta$	0.147	-4.20	1.22	-22.0	184.0	-0.84
$\xi_2 = 0.1250$, $z = 1.75$, $\alpha_2 = 0.375\delta$	1.02	0.572	1.64	0.131	-3.0	-2.1
$\xi_2 = 0.1875$, $z = 0.5849$, $\alpha_2 = 0.5625\delta$	1.00	0.942	0.580	0.575	0.97	0.521
$\xi_2 = 0.25$, $z = 0.5008$, $\alpha_2 = 0.75\delta$	1.00	0.982	0.501	0.493	1.03	0.52
$\xi_2 = 0.3125$, $z = 0.5$, $\alpha_2 = 0.9375\delta$	1.00	1.02	0.504	0.492	1.08	0.583
$\xi_2 = 0.3750$, $z = 0.5$, $\alpha_2 = 1.125\delta$	0.999	1.06	0.507	0.496	1.10	0.626
$\xi_2 = 0.4375$, $z = 0.5$, $\alpha_2 = 1.3125\delta$	0.998	1.10	0.509	0.500	1.10	0.649
$\xi_2 = 0.5$, $z = 0.5$, $\alpha_2 = 1.5\delta$	0.998	1.11	0.510	0.502	1.10	0.656

For $q^* = $ constant, the general solution of Equation 27.23 is

$$w = e^{\lambda_1 d_1}[C_1 \cos(\lambda_2 \alpha_1) + C_2 \sin(\lambda_2 \alpha_1)]$$

$$+ e^{-\lambda_1 \alpha_1}[C_2 \cos(\lambda_1 \alpha_1) + C_4 \sin(\lambda_2 \alpha_1)] + q^*\beta_0^{-1}$$

where the constants C_1, C_2, and C_3 are to be determined from the end restraint conditions and clearly describe the boundary effects occurring in the shell. If the shell is long, then we have the momentless solution

$$W = q^*\beta_0^{-1} \qquad (27.24)$$

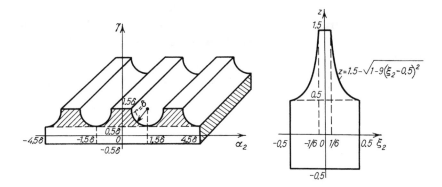

FIGURE 27.12. A shell with parallel grooves of circular cross section; unit cell in the coordinates ξ_2, z.

TABLE 27.4
Effective Elastic Moduli of Plates and Shells
with Parallel Grooves of Circular Cross Section; $v = 0.3$, $A_2 = 1$

Effective elastic moduli	Results based on solution of local problems (I)	Results based on formulas of Korolev (1971) (II)	Percentage changes (%)
$\langle b_{11}^{11} \rangle /E$	1.638	1.575	4%
$\langle b_{22}^{22} \rangle /E$	1.796	1.099	63.3%
$\langle b_{22}^{11} \rangle /E$	0.539	0.330	63.3%
$\langle b_{12}^{12} \rangle /G$	1.266	1.000	26.6%
$\langle b_{11}^{*11} \rangle /E$	0.510	0.437	16.7%
$\langle b_{22}^{*22} \rangle /E$	0.811	0	—
$\langle b_{22}^{*11} \rangle /E$	0.243	0	—
$\langle b_{12}^{*12} \rangle /G$	0.235	0.217	8.3%
$\langle zb_{11}^{*11} \rangle /E$	0.612	0.533	14.8%
$\langle zb_{22}^{*22} \rangle /E$	0.974	0.092	—
$\langle zb_{22}^{*11} \rangle /E$	0.292	0.028	—
$\langle zb_{12}^{*12} \rangle /G$	0.124	—	—

at points remote from the end, and the corresponding strains are given by

$$\varepsilon_{11} = -\frac{\langle b_{22}^{11} \rangle}{\langle b_{11}^{11} \rangle} \frac{q^*}{R\beta_0} \qquad \varepsilon_{22} = \frac{q^*}{R\beta_0} \qquad \tau_{11} = \tau_{22} = 0 \quad (27.25)$$

TABLE 27.5

Local Distributions of Functions $b_{\alpha\beta}^{\lambda\mu}$ and $b_{\alpha\beta}^{*\lambda\mu}$ over the Perimeter of the Unit Cell of Figure 27.12

Points on the unit cell perimeter	$b_{11}^{11}\bigg/\dfrac{E}{1-\nu^2}$	$b_{22}^{11}\bigg/\dfrac{\nu E}{1-\nu^2}$	$b_{11}^{*11}\bigg/\dfrac{E}{1-\nu^2}$	$b_{22}^{*11}\bigg/\dfrac{\nu E}{1-\nu^2}$	b_{12}^{12}/G	b_{12}^{12}/G
$\xi_2 = 0,$ $z = 1.5,$ $\alpha_2 = 0$	1.39	−8.3	2.1	−12.0	−0.91	1.0
$\xi_2 = 0.0625,$ $z = 1.5,$ $\alpha_2 = 0.1875\delta$	1.28	−12.0	1.95	−19.0	−0.8	−0.9
$\xi_2 = 0.1250,$ $z = 1.5,$ $\alpha_2 = 0.375\delta$	1.03	−33.0	1.59	−50.0	−0.73	−0.71
$\xi_2 = 0.1875,$ $z = 1.152,$ $\alpha_2 = 0.5625\delta$	0.961	11.8	1.11	18.0	0.701	0.297
$\xi_2 = 0.25,$ $z = 0.8386,$ $\alpha_2 = 0.75\delta$	0.98	−0.24	0.814	−0.99	0.143	0.024
$\xi_2 = 0.1325,$ $z = 0.6732,$ $\alpha_2 = 0.9375\delta$	0.988	−0.051	0.652	−0.93	0.732	0.547
$\xi_2 = 0.3750,$ $z = 0.573,$ $\alpha_2 = 1.125\delta$	0.991	0.306	0.554	−0.53	1.33	1.06
$\xi_2 = 0.4375,$ $z = 0.5177,$ $\alpha_2 = 1.3135\delta$	0.993	0.495	0.499	−0.33	1.73	1.40
$\xi_2 = 0.5,$ $z = 0.5,$ $\alpha_2 = 1.5\delta$	0.994	0.551	0.481	−0.27	1.87	1.52

The stress component σ_{22}, the most important one in this case, is, from Equation 19.74,

$$\sigma_{22} = b_{22}^{11}\varepsilon_{11} + b_{22}^{22}\varepsilon_{22} \qquad (27.26)$$

or, substituting from Equations 27.25 and making use of Equations 27.9,

$$\sigma_{22} = \frac{qR}{\delta}\tilde{b}_{22}^{22}\left(1 - \frac{\nu\langle\tilde{b}_{22}^{11}\rangle}{\langle\tilde{b}_{11}^{11}\rangle}\right)\left[\langle\tilde{b}_{22}^{11}\rangle - \frac{(\langle\tilde{b}_{22}^{11}\rangle)^2}{\langle\tilde{b}_{11}^{11}\rangle}\right]^{-1} \qquad (27.27)$$

where a tilde denotes the normalization to $E/(1 - \nu^2)$ of the quantity so marked.

For the trapezoidal stringer profile, case II of Table 27.2, Equation 27.27 yields

$$\sigma_{22} = \frac{qR}{\delta} 0.695 \tilde{b}_{22}^{22} \tag{27.28}$$

so that the distribution of the quantity of $\sigma_{22} \delta/(0.695qR)$ coincides with that of $\tilde{b}_{22}^{22}$ as shown in Figures 27.9 for $\xi_2 = 0$ and $\xi_2 = 0.5$, respectively. In particular, it follows from Equation 27.28 that the value assumed by σ_{22} between the stringers is 2.3 times qR/δ, the σ_{22} value in a stringerless shell of thickness δ (see Figure 27.10).

For the cosinusoidal stringer profile (case $h_2 = 3\pi$ in Table 27.1), Equation 27.27 gives

$$\sigma_{22} = \frac{qR}{\delta} \cdot 0.853 \cdot \tilde{b}_{22}^{22} \tag{27.29}$$

The local distributions of the functions $\tilde{b}_{22}^{22}$ are shown for this case in Figures 27.4 and 27.6 for the cross sections of $\xi_2 = 0$ and $\xi_2 = 0.5$, respectively. Analysis of these distributions shows that the use of cosinusoidal stringers results in an even greater concentration of the stress component σ_{22}. For example, at a point at the surface which is equally distant from the two neighboring stringers (and whose coordinates are $\xi_2 = 0.5$, $A_2 = 1$, $z = 0.5$ in Figure 27.3) we obtain $\sigma_{22} = 6.1qR/\delta$ using Equation 27.29 and referring to Figure 27.6.

28. THE FUNDAMENTAL SOLUTION OF THE DOUBLY PERIODIC THREE-DIMENSIONAL PROBLEM OF ELASTICITY

As earlier remarked, the local problems formulated in Section 22 are doubly periodic three-dimensional problems of the theory of elasticity that are set on the unit cell of the structure under study and whose solutions are supposed to satisfy prescribed boundary conditions at the surfaces $z = z^+$, $z = z^-$ and periodicity requirements on the coordinates ξ_1 and ξ_2, with respective periods A_1 and A_2. Obviously enough, problems of this level of complexity are generally amenable only to a numerical solution, for example, by the finite element (FE) method. The application of the FE method to three-dimensional elasticity problems depends on the use of the fundamental solution of the equations of the theory of elasticity for a concentrated force acting in an unbounded elastic medium. For the case of a homogeneous isotropic medium, such a solution was obtained by Lord Kelvin (see, for example, Love 1944) and has been employed in combination with the FE method by a number of recent workers (see Banerjee and

Butterfield 1981, Crouch and Starfield 1983), but as far as the local problems of interest here are concerned, unfortunately, this approach is all but inadequate because of the periodicity of the stress and strain fields in the coordinates ξ_1 and ξ_2. It therefore becomes necessary to construct a periodic fundamental solution of the three-dimensional elasticity problem which would enable a modification of the FE algorithm into a form suitable for the solution of the local problems relevant to study of composites (Kalamkarov et al. 1989).

In the following we confine ourselves to the case of an isotropic medium and set $A_1 = A_2$ and $h_1 = h_2 = h$.

28.1. The derivation of a doubly periodic fundamental solution

In accordance with the way in which the local problems posed by Equations 22.6 and 22.7 and by Equations 22.11 and 22.12 are formulated, by the fundamental doubly periodic solution we mean the solution of the system

$$\frac{1}{h}\frac{\partial \tau_{i\alpha}}{\partial \xi_\alpha} + \frac{\partial \tau_{i3}}{\partial z} - P_i \delta(\xi_1)\delta(\xi_2)\delta(z) \qquad (i = 1, 2, 3)$$

$$\tau_{\alpha\beta} = \frac{c_{44}}{h}\left(\frac{\partial U_\alpha}{\partial \xi_\beta} + \frac{\partial U_\beta}{\partial \xi_\alpha}\right) + c_{12}\delta_{\alpha\beta}\left(\frac{\partial U_\alpha}{h\partial \xi_\gamma} + \frac{\partial U_3}{\partial z}\right)$$

$$\tau_{33} = 2c_{44}\frac{\partial U_3}{\partial z} + c_{12}\left(\frac{1}{h}\frac{\partial U_\gamma}{\partial \xi_\gamma} + \frac{\partial U_3}{\partial z}\right) \qquad (28.1)$$

$$\tau_{3\alpha} = c_{44}\left(\frac{1}{h}\frac{\partial U_3}{\partial \xi_\alpha} + \frac{\partial U_\alpha}{\partial z}\right)$$

in the region $|\xi_\alpha| < A/2$, $|z| < \infty$, with functions U_1, U_2, and U_3 periodic in ξ_1 and ξ_2 with period A. The quantities P_i in the first of Equations 28.1 are the components of the (concentrated) force applied to the origin. As usual, Greek indices range from 1 to 2.

We represent the solution of the system as a sum of three solutions, each corresponding to one of the three components of the force. Setting $P_2 = P_3 = 0$ first and noting that

$$\delta(\xi_\alpha) = \frac{1}{A} + \frac{2}{A}\sum_{n=1}^{\infty}\cos(\beta_n \xi_\alpha) \qquad \left(\beta_n = \frac{2\pi n}{A}\right)$$

we assume the solution of Equations 28.1 to be of the form

$$U_1^{(1)} = \sum_{n=0}^{\infty} \sum_{m=0}^{\infty} U_{1nm}^{(1)}(z)\cos(\beta_n \xi_1)\cos(\beta_m \xi_2)$$

$$U_2^{(1)} = \sum_{n=1}^{\infty} \sum_{m=1}^{\infty} U_{2nm}^{(1)}(z)\sin(\beta_n \xi_1)\sin(\beta_m \xi_2) \qquad (28.2)$$

$$U_3^{(1)} = \sum_{n=1}^{\infty} \sum_{m=0}^{\infty} U_{3nm}^{(1)}(z)\sin(\beta_n \xi_1)\cos(\beta_m \xi_2)$$

and substitute into Equation 28.1 to obtain the following system of ordinary differential equations:

$$c_{44}\frac{d^2 U_{100}^{(1)}}{dz^2} = -\frac{P_1}{A^2}\delta(z)$$

$$c_{44}\left(\frac{d^2 U_{10m}^{(1)}}{dz^2} - \lambda_m^2 U_{10m}^{(1)}\right) = -\frac{2P_1}{A^2}\delta(z)$$

$$c_{44}\frac{d^2 U_{1n0}^{(1)}}{dz^2} + c_{11}\lambda_n^2 U_{1n0}^{(1)}(c_{44} + c_{12})$$

$$\times \lambda_n \frac{dU_{2n0}^{(1)}}{dz} = -\frac{2P_1}{A^2}\delta(z)$$

$$-(c_{44} + c_{12})\lambda_n \frac{dU_{1n0}^{(1)}}{dz} + c_{11}\frac{d^2 U_{3n0}^{(2)}}{dz^2}$$

$$- c_{44}\lambda_n^2 U_{3n0}^{(1)} = 0$$

$$c_{44}\frac{d^2 U_{1nm}^{(1)}}{dz^2} - (c_{11}\lambda_n^2 + c_{44}\lambda_m^2)U_{1nm}^{(1)} + (c_{44} + c_{12})\lambda_n \lambda_m U_{2nm}^{(1)}$$

$$+(c_{44} + c_{12})\lambda_n \frac{dU_{3nm}^{(1)}}{dz} = -\frac{4P_1}{A^2}\delta(z)$$

$$(c_{44} + c_{12})\lambda_n \lambda_m U_{1nm}^{(1)} + c_{44}\frac{d^2 U_{2nm}^{(1)}}{dz^2} - (c_{44}\lambda_n^2 + c_{11}\lambda_m^2)U_{2nm}^{(1)}$$

$$-(c_{44} + c_{12})\lambda_m \frac{dU_{3nm}^{(1)}}{dz} = 0$$

$$-(c_{44} + c_{12})\lambda_n \frac{dU_{1nm}^{(1)}}{dz} + (c_{44} + c_{12})\lambda_m \frac{dU_{2nm}^{(1)}}{dz}$$

$$+ c_{11}\frac{d^2 U_{3nm}}{dz^2} - c_{44}(\lambda_n^2 + \lambda_m^2)U_{3nm}^{(1)} = 0$$

$$\lambda_n = \frac{\beta_n}{h} = \frac{2\pi n}{hA} \qquad\qquad (28.3)$$

Recalling that

$$\delta(z) = \frac{1}{\pi}\int_\pi^\infty \cos(z\eta)\, d\eta$$

we represent the solution to (28.3) in the form

$$\begin{Bmatrix} U_{1n0}^{(1)} \\ U_{10m}^{(1)} \end{Bmatrix} = \frac{2}{\pi}\int_0^\infty \begin{Bmatrix} \tilde{U}_{1n0}^{(1)}(\eta) \\ \tilde{U}_{10m}^{(1)}(\eta) \end{Bmatrix} \cos(z\eta)\, d\eta$$

$$U_{3n0}^{(1)} = \frac{2}{\pi}\int_0^\infty \tilde{U}_{3n0}^{(1)}(\eta)\sin(z\eta)\, d\eta \qquad\qquad (28.4)$$

$$\begin{Bmatrix} U_{1nm}^{(1)} \\ U_{2nm}^{(1)} \end{Bmatrix} = \frac{2}{\pi}\int_0^\infty \begin{Bmatrix} \tilde{U}_{1nm}^{(1)}(\eta) \\ \tilde{U}_{2nm}^{(1)}(\eta) \end{Bmatrix} \cos(z\eta)\, d\eta$$

$$U_{3nm}^{(1)} = \frac{2}{\pi}\int_0^\infty \tilde{U}_{3nm}^{(1)}(\eta)\sin(z\eta)\, d\eta$$

$$(n, m = 1, 2, \ldots)$$

which when substituted into Equations 28.3 yields a system of algebraic equations for $U_{1nm}^{(1)}$, $U_{2nm}^{(1)}$, and $U_{3nm}^{(1)}$. The solutions of Equations 28.3 is obtained by solving the system and subsequently evaluating the integrals involved in Equations 28.4.

Returning to Equations 28.2, the fundamental solution of the problem, in terms of unitary function $\eta(z)$, is

$$U_1^{(1)} = -\frac{P_1}{A^2 c_{44}} z\eta(z) + \frac{P_1}{A^2} \frac{(c_{11} + c_{44})}{2c_{11}c_{44}} \sum_{n=1}^{\infty} \frac{e^{-|z|\lambda_n}}{\lambda_n} \cos(\beta_n \xi_1)$$

$$-\frac{P_1}{A^2} \frac{(c_{11} - c_{44})}{2C_{11}c_{44}} |z| \sum_{n=1}^{\infty} e^{-|z|\lambda_n} \cos(\beta_n \xi_1)$$

$$+\frac{P_1}{A^2 c_{44}} \sum_{m=1}^{\infty} \frac{e^{-|z|\lambda_m}}{\lambda_m} \cos(\beta_m \xi_2)$$

$$+\frac{2P_1}{A^2 c_{44}} \sum_{n=1}^{\infty} \sum_{m=1}^{\infty} \frac{e^{-|z|\sqrt{\lambda_n^2 + \lambda_m 2}}}{\sqrt{\lambda_n^2 + \lambda_m^2}} \cos(\beta_n \xi_1)\cos(\beta_m \xi_2)$$

$$-\frac{P_1(c_{11} - c_{44})}{A^2 c_{11}c_{44}} \sum_{n=1}^{\infty} \sum_{m=1}^{\infty} \left[(\lambda_n^2 + \lambda_m^2)^{-3/2} + |z|(\lambda_n^2 + \lambda_m^2)^{-1} \right]$$

$$\times \lambda_n^2 e^{-\sqrt{\lambda_n^2 + \lambda_m^2}|z|} \cos(\beta_n \xi_1)\cos(\beta_m \xi_2) \tag{28.5}$$

$$U_2^{(1)} = \frac{P_1(c_{11} - c_{44})}{A^2 c_{11}c_{44}} \sum_{n=1}^{\infty} \sum_{m=1}^{\infty} \left[(\lambda_n^2 + \lambda_m^2)^{-3/2} + |z|(\lambda_n^2 + \lambda_m^2)^{-1} \right]$$

$$\times \lambda_n \lambda_m e^{-\sqrt{\lambda_n^2 + \lambda_m^2}|z|} \sin(\beta_n \xi_1)\sin(\beta_m \xi_2) \tag{28.6}$$

$$U_3^{(1)} = \frac{P_1(c_{11} - c_{44})}{A^2 2c_{11}c_{44}} z \sum_{n=1}^{\infty} e^{-|z|\lambda_n} \sin(\beta_n \xi_1)$$

$$+\frac{P_1(c_{11} - c_{44})}{A^2 c_{11}c_{44}} z \sum_{n=1}^{\infty} \sum_{m=1}^{\infty} (\lambda_n^2 + \lambda_m^2)^{-1/2} e^{-\sqrt{\lambda_n^2 + \lambda_m^2}|z|}$$

$$\times \lambda_n \sin(\beta_n \xi_1)\cos(\beta_n \xi_2) \tag{28.7}$$

28.2. Transformation of the doubly periodic fundamental solution

It is essential for our further discussion to transform Equations 28.5 through 28.7 in such a way as to separate singular terms from regular ones. We note first of all that the summation of the single series in these equations can be effected with the aid of the well-known formulas

$$\sum_{n=1}^{\infty} e^{-ny} \cos(n\varphi) = -\frac{1}{2} + \frac{\sinh y}{2(\cosh y - \cos \varphi)}$$

$$\sum_{n=1}^{\infty} e^{-ny} \sin(n\varphi) = \frac{\sin \varphi}{2(\cosh y - \cos \varphi)}$$

$$\sum_{n=1}^{\infty} \frac{1}{n} e^{-ny} \cos(n\varphi) = \frac{y}{2} - \frac{1}{2}\ln(2|\cosh y - \cos \varphi|) \qquad (y > 0, |\varphi| < \pi)$$

given, for example, in Whittaker and Watson (1927).

To transform the double sums in Equations 28.5 through 28.7 we will make use of the two representations of the Jacobi theta function given in the same source,

$$\vartheta_3(\xi | i\tau) = 1 + 2 \sum_{n=1}^{\infty} \exp^{-\pi n^2 \tau} \cos(2n\xi)$$

and

$$\vartheta_2(\xi | i\tau) = \frac{\exp\left(-\dfrac{\xi^2}{\pi \tau}\right)}{\sqrt{\tau}} + \frac{1}{\sqrt{\tau}} \sum_{n=1}^{\infty}$$

$$\times \left[\exp\left(-\frac{(\xi + \pi n)^2}{\pi \tau}\right) + \exp\left(-\frac{(\xi - \pi n)^2}{\pi \tau}\right)\right]$$

Multiplying the equation

$$1 + 2 \sum_{n=1}^{\infty} \exp(-\pi n^2 \tau) \cos(2n\xi)$$

$$= \frac{1}{\sqrt{\tau}} \exp\left(-\frac{\xi^2}{\pi \tau}\right) + \frac{1}{\sqrt{\tau}} \sum_{n=1}^{\infty}$$

$$\times \left[\exp\left(-\frac{(\xi + \pi n)^2}{\pi \tau}\right) + \exp\left(-\frac{(\xi - \pi n)^2}{\pi \tau}\right)\right]$$

by $(1/\tau) \exp(-z^2/\pi\tau)$, take the Laplace transform with respect to τ and noting that (Bateman and Erdélyi 1954)

$$\frac{\exp[-a/(4\tau)]}{\sqrt{\tau}} \doteq \frac{\sqrt{\pi}}{\sqrt{p}} \exp(-\sqrt{ap}) \quad \frac{\exp[-a/(4\tau)]}{\tau} \doteq 2K_0(\sqrt{ap})$$

$$\frac{1}{\sqrt{\tau}} \exp\left(-\frac{a}{4\tau} - \pi n^2\tau\right) \doteq \frac{\sqrt{\pi}}{\sqrt{p+\pi n^2}} \exp\left[-\sqrt{a(p+\pi n^2)}\right]$$

where p is the transform parameter and $K_0(x)$ the MacDonald function, we obtain

$$\frac{\sqrt{\pi}}{\sqrt{p}} e^{-2|z|\sqrt{p/\pi}} + 2\sqrt{\pi} \sum_{n=1}^{\infty} \frac{e^{-2|z|\sqrt{p/\pi+n^2}}}{\sqrt{p+\pi n^2}} \cos(2n\xi)$$

$$= 2K_0\left(2\sqrt{z^1+\xi^2}\sqrt{\frac{p}{\pi}}\right)$$

$$+ 2\sum_{n=1}^{\infty}\left[K_0\left(2\sqrt{z^2+(\xi+\pi n)^2}\sqrt{\frac{p}{\pi}}\right)\right.$$

$$\left. +K_0\left(2\sqrt{z^2+(\xi-\pi n)^2}\sqrt{\frac{p}{\pi}}\right)\right] \qquad (28.8)$$

Setting $p = \pi k^2$ in Equation 28.8 yields

$$\sum_{n=1}^{\infty} \frac{e^{-2|z|\sqrt{k^2+n^2}}}{\sqrt{k^2+n^2}} \cos(2n\xi)$$

$$= \frac{1}{2k} e^{-2|z|k} + K_0\left(2k\sqrt{z^2+\xi^2}\right)$$

$$+ \sum_{n=1}^{\infty}\left[K_0\left(2k\sqrt{z^2+(\xi+\pi n)^2}\right)\right.$$

$$\left. +K_0\left(2k\sqrt{z^2+(\xi-\pi n)^2}\right)\right] \qquad (28.9)$$

Multiplying this by $\cos(2k\eta)$, summing over k, and using the well-known expansion (Prudnikov et al. 1983)

$$\sum_{k=1}^{\infty} K_0(kx)\cos(ka) = \frac{\pi}{2}(x^2 + a^2)^{-1/2} + \frac{1}{2}\left(C + \ln\frac{x}{4\pi}\right)$$

$$+ \frac{\pi}{2}\sum_{k=1}^{\infty}\left[\frac{1}{\sqrt{(2\pi k - a)^2 + x^2}} - \frac{1}{2\pi k}\right]$$

$$+ \frac{\pi}{2}\sum_{k=1}^{\infty}\left[\frac{1}{\sqrt{(2\pi k + a)^2 + x^2}} - \frac{1}{2\pi k}\right] \quad (28.10)$$

where $x > 0$ and C is Euler's constant, the required representation of one of the double sums of interest is

$$\sum_{k=1}^{\infty}\sum_{n=1}^{\infty}\frac{e^{-2|z|\sqrt{k^2 + n^2}}}{\sqrt{k^2 + n^2}}\cos(2n\xi)\cos(2k\eta)$$

$$= \frac{1}{2}\ln\left[\frac{1}{2\pi}\sqrt{2|\cosh(2z) - \cos(2\eta)|(z^2 + \xi^2)}\right.$$

$$+ \frac{\pi}{4}(z^2 + \xi^2 + \eta^2)^{-1/2} + \Phi_0(\xi, \eta, z) \quad \left(|\xi| < \frac{\pi}{2}, |\eta| < \frac{\pi}{2}\right)$$

$$(28.11)$$

where $\Phi_0(\xi, \eta, z)$ is a regular term expressible in the form

$$\Phi_0(\xi, \eta, x) = -\frac{1}{2}(|z| - C)$$

$$+ \frac{\pi}{4}\sum_{k=1}^{\infty}\left[\frac{1}{\sqrt{(\pi k - \eta)^2 + z^2 + \xi^2}} + \frac{1}{\pi k}\right]$$

$$+ \frac{\pi}{4}\sum_{k=1}^{\infty}\left[\frac{1}{\sqrt{(\pi k + \eta)^2 + z^2 + \xi^2}} - \frac{1}{\pi k}\right]$$

$$+ \sum_{k=1}^{\infty}\sum_{n=1}^{\infty}\left[K_0\left(2k\sqrt{z^2 + (\pi n + \eta)^2}\right)\right.$$

$$+ K_0\left(2k\sqrt{z^2 + (\pi n - \eta)^2}\right)\right]\cos(2k\eta) \quad (28.12)$$

The second double-sum representation for Equation 28.5 is found by differentiating Equation 28.9 with respect to k, multiplying by $k\cos(2k\eta)$, and finally summing over k with the aid of the expansion

$$x \sum_{k=1}^{\infty} kK_1(kx)\cos(ka)$$

$$= \frac{\pi}{2}x^2(x^2 + a^2)^{-3/2} - \frac{1}{2}$$

$$+ \frac{\pi}{2} \sum_{k=1}^{\infty} \left[\frac{x^2}{((2\pi k - 2)^2 + x^2)^{3/2}} + \frac{x^2}{((2\pi k + a)^2 + x^2)^{3/2}} \right]$$

which follows from Equation 28.10. The result is

$$\sum_{k=1}^{\infty} \sum_{n=1}^{\infty} \left[(k^2 + n^2)^{-3/2} + 2|z|(k^2 + n^2)^{-1} \right] e^{-2|z|\sqrt{n^2 + k^2}} k^2$$

$$\times \cos(2n\xi)\cos(2k\eta) = \frac{\pi}{4}(z^2 + \eta^2)(z^2 + \xi^2 + \eta^2)^{-3/2}$$

$$+ \frac{1}{2}\ln\left[\sqrt{2|\cosh(2z) - \cos(2\xi)|} \right]$$

$$- \frac{1}{2} \frac{z\sinh(2z)}{\cosh(2z) - \cos(2\xi)} + \Psi_0(\xi, \eta, z)$$

(28.13)

where the regular term $\Psi_0(\xi, \eta, z)$ is of the form

$$\Psi_0(\xi, \eta, z) = -\frac{1}{2} + \frac{\pi}{4} \sum_{k=1}^{\infty} \left[\frac{z^2 + \eta^2}{((\pi k - \xi)^2 + z2 + \eta^2)^{3/2}} \right.$$

$$+ \frac{z^2 + \eta^2}{((\pi k + \xi)^2 + z^2 + \eta^2)^{3/2}} \right]$$

$$+ \sum_{k=1}^{\infty} \sum_{n=1}^{\infty} \left[2k\sqrt{z^2 + (\eta + \pi n)^2} K_1 \right.$$

$$\times \left(2k\sqrt{z^2 + (\eta + \pi n)^2} \right) + 2k\sqrt{z^2 + (\eta - \pi n)^2} K_1$$

$$\left. \times \left(2k\sqrt{z^2 + (\eta - \pi n)^2} \right) \right]$$

(28.14)

To separate singular terms in the double sum in Equation 28.6, we differentiate Equation 28.9 with respect to k and ξ, in that order, and then multiply the result by $\sin(2k\eta)$ and sum over k. Making use of the expansion

$$\sum_{k=1}^{\infty} kK_0(kx)\sin(ka)$$

$$= \frac{\pi}{2}a(x^2 + a^2)^{-3/2}$$

$$+ \frac{\pi}{2}\sum_{k=1}^{\infty}\left[\frac{a - 2\pi k}{\left((a - 2\pi k)^2 + x^2\right)^{3/2}} + \frac{a + 2\pi k}{\left((a + 2\pi k)^2 + x^2\right)^{3/2}}\right]$$

we find

$$\sum_{k=1}^{\infty}\sum_{n=1}^{\infty}\left[(k^2 + n^2)^{-3/2} + 2z(k^2 + n^2)^{-1}\right]$$

$$\times e^{-2z\sqrt{k^2+n^2}}kn\sin(2n\xi)\sin(2k\eta)$$

$$= \frac{\pi}{4}\xi\eta(z^2 + \xi^2 + \eta^2)^{-3/2} + F_0(\xi, \eta, z) \qquad (28.15)$$

where

$$F_0(\xi, \eta, z) = \frac{\pi}{4}\eta\sum_{k=1}^{\infty}\left[\frac{\xi - \pi k}{\left((\xi - \pi k)^2 + z^2 + \eta^2\right)^{3/2}}\right.$$

$$\left. + \frac{\xi + \pi k}{\left((\xi + \pi k)^2 + z^2 + \eta^2\right)}\right]$$

$$+ 2\sum_{k=1}^{\infty}\sum_{n=1}^{\infty}\left[(\eta + \pi n)K_0\left(2k\sqrt{z^2 + (\eta + \pi n)^2}\right)\right.$$

$$\left. + (\eta - \pi n)K_0\left(2k\sqrt{z^2 + (\eta - \pi n)^2}\right)\right]k\sin(2k\xi) \quad (28.16)$$

Using Equations 28.11 through 28.16, we may rearrange Equations 28.5 through 28.7 to read

$$U_1^{(1)} = \frac{P_1 h^2}{16\pi G(1-\nu)}\left[(3-4\nu)(z^2 + h^2\xi_1^2 + h^2\xi_2^2)^{-1/2}\right.$$

$$\left. + h^2\xi_1^2(z^2 + h^2\xi_1^2 + h^2\xi_2^2)^{-3/2}\right]$$

$$+ \frac{P_1}{A^2 G}\left[-z\eta(z) + 2\Phi_3(\xi_1, \xi_2, z) + \frac{1}{2(1-\nu)}\Phi_1(\xi_1, \xi_2, z)\right]$$

$$U_2^{(1)} = \frac{P_1 h^2}{16\pi G(1-\nu)} h^2\xi_1\xi_2(z^2 + h^2\xi_1^2 + h^2\xi_2^2)^{-3/2}$$

$$+ \frac{P_1}{2A^2 G(1-\nu)}\Phi_2(\xi_1, \xi_2, z)$$

$$U_3^{(1)}\frac{P_1 h^2}{16\pi G(1-\nu)} zh\xi_1(z^2 + h^2\xi_1^2 + h^2\xi_2^2)^{-3/2}$$

$$\hspace{6cm}(28.17)$$

$$- \frac{P_1}{2A^2 G(1-\nu)}\frac{z}{h}\frac{\partial}{\partial\xi_1}\Phi_3(\xi_2, \xi_1, z)$$

where

$$\Phi_1(\xi_1, \xi_2, z)$$

$$= \frac{Ah}{4\pi} - \frac{Ah}{8\pi}\sum_{k=1}^{\infty}\left[\frac{A^{-2}h^{-2}(z^2 + h^2\xi_2^2)}{\left((k^2 - \xi_1/A)^2 + A^{-2}h^{-2}(z^2 + h^2\xi_2^2)\right)^{3/2}}\right.$$

$$\left. + \frac{A^{-2}h^{-2}(z^2 + h^2\xi_2^2)}{\left((k + \xi_1/A)^2 + A^{-2}h^{-2}(z^2 + h^2\xi_2^2)\right)^{3/2}}\right]$$

$$- Ah\sum_{k=1}^{\infty}\sum_{m=1}^{\infty}\left[\sqrt{z^2 A^{-2}h^{-2} + \left(\frac{\xi_2}{A+m}\right)^2}\, K_1\right.$$

$$\times\left(2\pi k\sqrt{z^2 A^{-2}h^{-2} + \left(\frac{\xi_2}{A+m}\right)^2}\right)$$

$$+ \sqrt{z^2 A^{-2} h^{-2} + \left(\frac{\xi_2}{A - m} \right)^2} \, K_1$$

$$\times \left(2\pi k \sqrt{z^2 A^{-2} h^{-2} + \left(\frac{\xi_2}{A - m} \right)^2} \right) \right] k \cos(\beta_k \xi_1) \qquad (28.18)$$

$\Phi_2(\xi_1, \xi_2, z)$

$$= \frac{h \xi_2}{8\pi} \sum_{k=1}^{\infty} \frac{\dfrac{\xi_1}{A} - k}{\left(\left(k - \dfrac{\xi_1}{A} \right)^2 + A^{-2} h^{-2} (z^2 + h^2 \xi_2^2) \right)^{3/2}}$$

$$+ \frac{\dfrac{\xi_1}{A} + k}{\left(\left(k + \dfrac{\xi_1}{A} \right)^2 + A^{-2} h^{-2} (z^2 + h^2 \xi_2^2) \right)^{3/2}} \right]$$

$$+ Ah \sum_{k=1}^{\infty} \sum_{m=1}^{\infty} \left[\left(\frac{\xi_2}{A + m} \right) K_0 \left(2\pi k \sqrt{z^2 A^{-2} h^{-2} + \left(\frac{\xi_2}{A + m} \right)^2} \right) \right.$$

$$+ \left(\frac{\xi_2}{A - m} \right) K_0 \left(2\pi k \sqrt{z^2 A^{-2} h^{-2} + \left(\frac{\xi_2}{A - m} \right)^2} \right) \right] k \sin(\beta_k \xi_1)$$

$$(28.19)$$

$\Phi_3(\xi_1, \xi_2, z)$

$$= \frac{Ah}{4\pi} C - \frac{Ah}{4\pi} \left[\ln \sqrt{2 \left| \cosh \left(\frac{2\pi z}{Ah} \right) - \cos \left(\frac{2\pi \xi_1}{A} \right) \right|} \right.$$

$$- \ln \left(\frac{1}{2A} \sqrt{z^2 h^{-2} + \xi_1^2} \right) \right]$$

$$+ \frac{Ah}{8\pi} \sum_{k=1}^{\infty} \left[\frac{1}{\sqrt{\left(k - \dfrac{\xi_2}{A}\right)^2 + A^{-2}h^{-2}(z^2 + h^2\xi_1^2)}} - \frac{1}{k} \right]$$

$$+ \frac{Ah}{8\pi} \sum_{k=1}^{\infty} \left[\frac{1}{\sqrt{\left(k + \dfrac{\xi_2}{A}\right)^2 + A^{-2}h^{-2}(z^2 + h^2\xi_1^2)}} - \frac{1}{k} \right]$$

$$+ \frac{Ah}{2\pi} \sum_{k=1}^{\infty} \sum_{m=1}^{\infty} \left[K_0 \left(2\pi k \sqrt{A^{-2}h^{-2}z^2 + \left(\frac{\xi_1}{A+m} \right)^2} \right) \right.$$

$$\left. + K_0 \left(2\pi k \sqrt{A^{-2}h^{-2}z^2 + \left(\frac{\xi_1}{A-m} \right)^2} \right) \right] \cos(\beta_k \xi_2) \qquad (28.20)$$

It will be understood that the first terms in Equations 28.17 correspond to the well-known singular solution obtained by Lord Kelvin for an unbounded isotropic medium loaded by a force P_1 at point $\xi_1 = \xi_2 = z = 0$; the remaining terms in Equation 28.17 are regular functions at this point. It can be shown also that the expressions

$$\Phi_1(\xi_1, \xi_2, z) - \frac{A^2 h^2}{8\pi} (z^2 + h^2\xi_2^2)(z^2 + h^2\xi_1^2 + h^2\xi_2^2)^{-3/2}$$

$$\Phi_2(\xi_1, \xi_2, z) + \frac{A^2 h^2}{8\pi} h^2\xi_1\xi_2(z^2 + h^2\xi_1^2 + h^2\xi_2^2)^{3/2}$$

and

$$\Phi_3(\xi_1, \xi_2, z) + \frac{A^2 h^2}{8\pi} (z^2 + h^2\xi_1^2 + h^2\xi_2^2)^{-1/2}$$

are periodic in $\xi_{1_{(1)}}$ and $\xi_{2_{(1)}}$ with period A, which secures the periodicity of the displacements U_1, U_2, and U_3 in the same coordinates. We note further, that the equations

$$\Phi_2(\xi_1, \xi_2, z) = \Phi_3(\xi_2, \xi_1, z) \quad \text{and} \quad \Phi_2(\xi_1, \xi_2, z) = \Phi_2(\xi_2, \xi_1, z)$$

hold and that the manner in which the functions $K_0(x)$ and $K_1(x)$ behave at large x secures good convergence for the double series in Equations 28.18 and 28.20.

Thus far we have considered the case when the force P_1 directed along the axis ξ_1 is the only one acting in the body. The cases when either the force P_2 or P_3 is acting, respectively, along the ξ_2 or z axes can be treated in a similar fashion. We write, that is

$$U_1^{(2)} = \sum_{n=1}^{\infty} \sum_{m=1}^{\infty} U_{1nm}^{(2)}(z)\sin(\beta_n \xi_1)\sin(\beta_m \xi_2)$$

$$U_2^{(2)} = \sum_{n=0}^{\infty} \sum_{m=0}^{\infty} U_{2nm}^{(2)}(z)\cos(\beta_n \xi_1)\cos(\beta_m \xi_2)$$

$$U_3^{(2)} = \sum_{n=0}^{\infty} \sum_{m=1}^{\infty} U_{3nm}^{(2)}(z)\cos(|\beta_n \xi_1)\sin(\beta_m \xi_2)$$

and

$$U_1^{(3)} = \sum_{n=1}^{\infty} \sum_{m=0}^{\infty} U_{1nm}^{(3)}(z)\sin(\beta_n \xi_1)\cos(\beta_m \xi_2)$$

$$U_2^{(3)} = \sum_{n=0}^{\infty} \sum_{m=1}^{\infty} U_{2nm}^{(3)}(z)\cos(\beta_n \xi_1)\sin(\beta_m \xi_2)$$

$$U_3^{(3)} = \sum_{n=0}^{\infty} \sum_{m=0}^{\infty} U_{3nm}^{(3)}(z)\cos(\beta_n \xi_1)\cos(\beta_m \xi_2)$$

and obtain by the same kind of argument

$$U_1^{(2)} = U_2^{(1)}\frac{P_2}{P_1}$$

$$U_2^{(2)} = \frac{P_2 h^2}{16\pi G(1 - \nu)}$$
$$\times \left[(3 - 4\nu)(z^2 + h^2\xi_1^2 + h^2\xi_2^2)^{-1/2}\right.$$
$$\left. + h^2\xi_2^2(z^2 + h^2\xi_1^2 + h^2\xi_2^2)^{-3/2}\right]$$
$$+ \frac{P_2}{A^2 G}\left[-z\eta(z) + 2\Phi_3(\xi_2, \xi_1, z)\right.$$
$$\left. + \frac{1}{2(1 - \nu)}\Phi_1(\xi_2, \xi_1, z)\right]$$

$$U_3^{(2)} \frac{P_2 h_2}{16\pi G(1-\nu)} zh\xi_2 (z^2 + h^2\xi_1^2 + h^2\xi_2^2)^{-3/2}$$

$$-\frac{P_2}{2A^2 G(1-\nu)} \frac{z}{h} \frac{\partial}{\partial \xi_2} \Phi_3(\xi 1, \xi 2, z) \tag{28.21}$$

$$U_1^{(3)} = U_3^{(1)} \frac{P_3}{P_1} \qquad U_2^{(3)} = U_3^{(2)} \frac{P_3}{P_1}$$

$$U_3^{(3)} = \frac{P_3 h^2}{16\pi G(1-\nu)} \left[(3-4\nu)(z^2 + h^2\xi_1^2 + h^2\xi_2^2)^{-1/2} \right.$$

$$+ z^2 (z^2 + h^2\xi_1^2 + h^2\xi_2^2)^{-3/2} \Big]$$

$$+ \frac{P_3}{A_2 G} \left[-z\eta(z) + \frac{3-4\nu}{2(1-\nu)} \Phi_3(\xi_2, \xi_1, z) \right.$$

$$\left. - \frac{z}{2(1-\nu)} \frac{\partial}{\partial z} \Phi_3(\xi_2, \xi_1, z) \right] \tag{28.22}$$

where again the first terms correspond to the singular Kelvin solution and the remaining terms are regular functions securing the required periodicity of $U_i^{(2)}$ and $U_i^{(3)}$ with respect to ξ_1 and ξ_2.

Adding together Equations 28.17, 28.21, and 28.22, we are now in a position to express the fundamental solution in the form (Kalamkarov et al., 1989)

$$U_i(x) = \left[F_{ij}(x) + F_{ij}^*(x) \right] P_j \tag{28.23}$$

where

$$x = (x_1, x_2, x_3) \qquad x_1 = h\xi_2 \qquad x_2 = h\xi_2 \qquad x_2 = z$$

$$F_{ij} = \frac{1}{16\pi G(1-\nu)} \left[(3-4\nu) \frac{\delta_{ij}}{r} + \frac{x_i x_j}{r^3} \right] \qquad (r^2 = x_i x_i) \tag{28.24}$$

and the functions $F_{ij}^*(x)$, in terms of the functions $\Phi_i(x_1/h, x_2/h, x)$ $(i = 1, 2, 3)$, are given by

$$
F_{11}^* = \frac{1}{A^2 G}\left[-x_3\eta(x_3) + 2\Phi_3\left(\frac{x_1}{h}, \frac{x_2}{h}, x_3\right) \right.
$$

$$
\left. + \frac{1}{2(1-\nu)}\Phi_1\left(\frac{x_1}{h}, \frac{x_2}{h}, x_3\right) \right]
$$

$$
F_{12}^* = F_{21}^* = \frac{1}{2 A^2 G(1-\nu)}\Phi_2\left(\frac{x_1}{h}, \frac{x_2}{h}, x_3\right)
$$

$$
F_{13}^* = F_{31}^* = -\frac{x_3}{2 A^2 G(1-\nu)}\frac{\partial}{\partial x_2}\Phi_3\left(\frac{x_1}{h}, \frac{x_2}{h}, x_3\right)
$$

$$
F_{22}^* = \frac{1}{A^2 G}\left[-x_3\eta(x_3) + 2\Phi_3\left(\frac{x_1}{h}, \frac{x_2}{h}, x_3\right) \right. \tag{28.25}
$$

$$
\left. + \frac{1}{2(1-\nu)}\Phi_1\left(\frac{x_2}{h}, \frac{x_1}{h}, x_3\right) \right]
$$

$$
F_{23}^* = F_{32}^* = -\frac{x_3}{2 A^2 G(1-\nu)}\frac{\partial}{\partial x_2}\Phi_3\left(\frac{x_1}{h}, \frac{x_2}{h}, x_3\right)
$$

$$
F_{33}^* = \frac{1}{A^2 G}\left[-x3\eta(x_3) + \frac{3-4\nu}{2(1-\nu)}\Phi_3\left(\frac{x_1}{h}, \frac{x_2}{h}, x_3\right) \right.
$$

$$
\left. - \frac{x_3}{2(1-\nu)}\frac{\partial}{\partial x_3}\Phi_3\left(\frac{x_1}{h}, \frac{x_2}{h}, x_3\right) \right]
$$

Combined with Equations 28.18 through 28.20, Equations 28.23 through 28.25 present the fundamental solution of the doubly periodic three-dimensional problem of the theory of elasticity.

28.3. Singly periodic fundamental solution of the plane elasticity problem

Associated with the problem we are going to discuss in this section (i.e., a plane with periodicity in one direction only) are two-dimensional formulations of the problems (b_{11}), (b_{22}), and (b_{11}^*) (b_{22}^*), that is, Equations 22.16 and 22.28, and Equations 22.22 and 22.24, all with $\lambda\mu = 11, 22$.

The fundamental solution periodic in ξ_1 with period A is defined in the plane case as such a solution of the system

$$\frac{1}{h}\frac{\partial \tau_{i2}}{\partial \xi_2} + \frac{\partial \tau_{i3}}{\partial z} = -P_i \delta(\xi_2)\delta(z) \qquad (i = 2, 3)$$

$$\tau_{22} = \frac{c_{11}}{h}\frac{\partial U_2}{\partial \xi_2} + c_{12}\frac{\partial U_3}{\partial z}$$

$$\tau_{33} = \frac{c_{12}}{h}\frac{\partial U_2}{\partial \xi_2} + c_{11}\frac{\partial U_3}{\partial z} \qquad (28.26)$$

$$\tau_{23} = \tau_{33} = c_{44}\left(\frac{1}{h}\frac{\partial U_3}{\partial \xi_2} + \frac{\partial U_2}{\partial z}\right)$$

which is valid in the region $|\xi_2| < A/2$, $|z| < \infty$, with functions U_2 and U_3 periodic in ξ_2 with period A; P_2 and P_3 are the ξ_2- and z- components of the (concentrated) external force acting at the origin.

Without here going into the details of the derivation, the desired solution of Equation 28.26 is

$$U_2(\xi_2, z) = -\frac{P_2}{8G(1 - v)Ah}\left[4(1 - v)z + \frac{3 - 4v}{2\pi}Ah \ln\left(2\left|\cos\left(\frac{2\pi z}{Ah}\right)\right.\right.\right.$$

$$\left.\left.\left. - \cos\left(\frac{2\pi\xi_2}{A}\right)\right|\right) + \frac{z \sinh\left(\dfrac{2\pi z}{Ah}\right)}{\cosh\left(\dfrac{2\pi z}{Ah}\right) - \cos\left(\dfrac{2\pi\xi_2}{A}\right)}\right]$$

$$+ \frac{P_3}{8G(1 - v)Ah}\frac{z \sin\left(\dfrac{2\pi\xi_2}{A}\right)}{\cosh\left(\dfrac{2\pi z}{Ah}\right) - \cos\left(\dfrac{2\pi\xi_2}{A}\right)}$$

$$U_3(\xi_2, z) = \frac{P_2}{8G(1-v)Ah} \frac{z \sin\left(\frac{2\pi\xi_2}{A}\right)}{\left[\cosh\left(\frac{2\pi z}{Ah}\right) - \cos\left(\frac{2\pi\xi_2}{A}\right)\right]}$$

$$+ \frac{P_3}{8G(1-v)Ah}\left[-2(1-2v)z - \frac{3-4v}{2\pi}Ah \ln \right.$$

$$\times \left(2\left|\cosh\left(\frac{2\pi z}{Ah}\right) - \cos\left(\frac{2\pi\xi_2}{A}\right)\right|\right)$$

$$\left. + \frac{z \sinh\left(\frac{2\pi z}{Ah}\right)}{\cosh\left(\frac{2\pi z}{Ah}\right) - \cos\left(\frac{2\pi\xi_2}{A}\right)}\right] \tag{28.27}$$

It should be remarked that the two-dimensional fundamental solution, Equation 28.27, cannot be deduced from the three-dimensional solution given by Equations 28.23 through 28.25 and is therefore of interest by itself. Noting that this solution is only unique up to constant terms, it can be shown (see, e.g., Crouch and Starfield 1983) that in the limit as A tends to infinity, it reduces to the familiar Kelvin solution for the plane deformation of an unbounded isotropic medium loaded by a concentration force P_2, P_3 at the origin.

Appendix

Ufland (1976) presents a useful method for constructing expansions on a composite interval using integral transform inversion formulas. Following Ufland, we consider a boundary value problem for the equation

$$\frac{\partial^2 u}{\partial x^2} = \frac{\partial u}{\partial t} \tag{A.1}$$

on the interval $0 < x < \frac{1}{2}$ under the initial conditions

$$u(x, 0) = f(x) \qquad (0 < x < \tfrac{1}{2})$$

and the boundary conditions

$$\left. \frac{\partial u(x, t)}{\partial x} \right|_{x=0} = 0 \quad \text{and} \quad \left. \frac{\partial u(x, t)}{\partial x} \right|_{x=1/2} = 0 \tag{A.2}$$

We also assume that both the function $u(x, t)$ and its derivative $\partial u(x, t)/\partial x$ undergo discontinuities at the inner point $x = \gamma < \frac{1}{2}$ of the interval $(0, \frac{1}{2})$, that is,

$$u(\gamma - 0, t) = \mu u(\gamma + 0, t) \tag{A.3}$$

$$\left. \frac{\partial u}{\partial x} \right|_{x=\gamma-0} = \nu \left. \frac{\partial u}{\partial x} \right|_{x=\gamma+0} \tag{A.4}$$

where μ and ν are known numbers.

Taking the Laplace transform of Equation A.1 with respect to t and using the initial condition after Equation A.1, we find that

$$\frac{\partial^2 \bar{u}(x, p)}{\partial x^2} + p\bar{u}(x, p) = -f(x) \tag{A.5}$$

where

$$\bar{u}(x, p) = \int_0^\infty u(x, t)e^{-pt}\, dt$$

It is required that the solution to Equation A.5 satisfy Conditions A.2 on $(0, \gamma)$ and $(\gamma, \frac{1}{2})$. We write

$$\bar{u}(x, p)|_{0 \le x < \gamma} = A_1(p)\cosh\left(\sqrt{p}\, x\right) - \frac{1}{\sqrt{p}} \int_x^\gamma f(\xi)$$

$$\times \sinh\left(\sqrt{p}\, \xi\right)\cosh\left(\sqrt{p}\, x\right) d\xi - \frac{1}{\sqrt{p}} \int_0^x f(\xi)$$

$$\times \cosh\left(\sqrt{p}\, \xi\right)\sinh\left(\sqrt{p}\, x\right) d\xi \qquad\qquad (A.6)$$

$$\bar{u}(x, p)|_{\gamma < x < 1/2} = A_2(p)\cosh\left(\sqrt{p}\left(\frac{1}{2} - x\right)\right) - \frac{1}{\sqrt{p}} \int_\gamma^x f(\xi)$$

$$\times \sinh\left(\sqrt{p}\left(\frac{1}{2} - \xi\right)\right)\cosh\left(\sqrt{p}\left(\frac{1}{2} - x\right)\right) d\xi$$

$$- \frac{1}{\sqrt{p}} \int_x^{1/2} f(\xi)\cosh\left(\sqrt{p}\left(\frac{1}{2} - \xi\right)\right)$$

$$\times \sinh\left(\sqrt{p}\left(\frac{1}{2} - x\right)\right) d\xi \qquad\qquad (A.7)$$

and use the contact conditions given by Equations A.3 and A.4 to obtain

$$A_1(p) = \frac{1}{\sqrt{p}\, \omega(p)}\left[\sinh\left(\sqrt{p}\, \gamma\right)\sinh\left(\sqrt{p}\left(\frac{1}{2} - \gamma\right)\right)\right.$$

$$+ \delta \cosh\left(\sqrt{p}\, \gamma\right)\cosh\left(\sqrt{p}\left(\frac{1}{2} - \gamma\right)\right)\right]$$

$$\times \int_0^\gamma f(\xi)\cosh\left(\sqrt{p}\, \xi\right) d\xi$$

$$+ \frac{\mu}{\sqrt{p}\, \omega(p)} \int_\gamma^{1/2} f(\xi)\cosh\left(\sqrt{p}\left(\frac{1}{2} - \xi\right)\right) d\xi \qquad (A.8)$$

$$A_2(p) = \frac{\delta}{\mu\sqrt{p}\,\omega(p)} \int_0^{\gamma} f(\xi)\cosh(\sqrt{p}\,\xi)\,d\xi$$

$$+ \frac{1}{\sqrt{p}\,\omega(p)} \left[\cosh\left(\sqrt{p}\left(\frac{1}{2}-\gamma\right)\right)\cosh(\sqrt{p}\,\gamma)\right.$$

$$\left. + \delta\sinh\left(\sqrt{p}\left(\frac{1}{2}-\gamma\right)\right)\sinh(\sqrt{p}\,\gamma)\right]$$

$$\times \int_{\gamma}^{1/2} f(\xi)\cosh\left(\sqrt{p}\left(\frac{1}{2}-\xi\right)\right)d\xi \qquad \text{(A.9)}$$

where

$$\delta = \frac{\mu}{\nu}$$

$$\omega(p) = \sinh\left(\sqrt{p}\left(\frac{1}{2}-\gamma\right)\right)\cosh(\sqrt{p}\,\gamma)$$

$$+ \delta\cosh\left(\sqrt{p}\left(\frac{1}{2}-\gamma\right)\right)\sinh(\sqrt{p}\,\gamma) \qquad \text{(A.10)}$$

Then the solution to Equation A.5 will have the required discontinuities at $x = \gamma$ if we represent it in the form

$$\bar{u}(x,p)|_{0<x<\gamma} = \mu \int_{\gamma}^{1/2} f(\xi)\frac{1}{\sqrt{p}\,\omega(p)}\cosh\left(\sqrt{p}\left(\frac{1}{2}-\xi\right)\right)$$

$$\times \cosh(\sqrt{p}\,x)\,d\xi + \int_0^x f(\xi)\frac{\cosh(\sqrt{p}\,\xi)}{\sqrt{p}\,\omega(p)}$$

$$\times \left[\sinh\left(\sqrt{p}\left(\frac{1}{2}-\gamma\right)\right)\sinh(\sqrt{p}\,(\gamma-x))\right.$$

$$\left. + \delta\cosh\left(\sqrt{p}\left(\frac{1}{2}-\gamma\right)\right)\cosh(\sqrt{p}\,(\gamma-x))\right]d\xi$$

$$+ \int_x^\gamma f(\xi) \frac{\cosh(\sqrt{p}\,x)}{\sqrt{p}\,\omega(p)} \left[\sinh\left(\sqrt{p} \left(\frac{1}{2} - \gamma \right) \right) \right.$$

$$\times \sinh(\sqrt{p}\,(\gamma - \xi)) + \delta \cosh\left(\sqrt{p} \left(\frac{1}{2} - \gamma \right) \right)$$

$$\left. \times \cosh(\sqrt{p}\,(\gamma - \xi)) \right] d\xi \qquad\qquad \text{(A.11)}$$

$$\bar{u}(x, p)|_{\gamma < x < 1/2} = \frac{\delta}{\mu} \int_0^\gamma f(\xi) \frac{\cosh(\sqrt{p}\,\xi)\cosh(\sqrt{p}\,(\frac{1}{2} - x))}{\sqrt{p}\,\omega(p)} d\xi$$

$$+ \int_\gamma^x f(\xi) \frac{\cosh(\sqrt{p}\,(\frac{1}{2} - x))}{\sqrt{p}\,\omega(p)}$$

$$\times \left[\cosh(\sqrt{p}\,\gamma)\cosh(\sqrt{p}\,(\xi - \gamma)) \right.$$

$$\left. + \delta \sinh(\sqrt{p}\,\gamma)\sinh(\sqrt{p}\,(\xi - \gamma)) \right] d\xi$$

$$+ \int_x^{1/2} f(\xi) \frac{\cosh(\sqrt{p}\,(\frac{1}{2} - \xi))}{\sqrt{p}\,\omega(p)}$$

$$\times \left[\cosh(\sqrt{p}\,\gamma)\cosh(\sqrt{p}\,(x - \gamma)) \right.$$

$$\left. + \delta \sinh(\sqrt{p}\,\gamma)\sinh(\sqrt{p}\,(x - \gamma)) \right] d\xi \qquad \text{(A.12)}$$

From Equations A.11 and A.12, reverting to originals and performing some manipulations, we find

$$u(x, t)|_{0 < x < \gamma} = \frac{1}{(\frac{1}{2} - \gamma + \delta\gamma)} \left(\delta \int_0^\gamma f(\xi)\, d\xi + \mu \int_\gamma^{1/2} f(\xi)\, d\xi \right)$$

$$+ \sum_{n=1}^\infty e^{-p_n^2 t} \cos(p_n x) \left[-2 \frac{C_n^{(1)}}{C_n} \int_0^\gamma f(\xi)\cos(p_n \xi)\, d\xi \right.$$

$$\left. + \frac{2\mu}{C_n} \int_\gamma^{1/2} f(\xi)\cos\left(p_n \left(\frac{1}{2} - \xi \right) \right) d\xi \right]$$

$$u(x,t)|_{\gamma < x < 1/2} = \frac{1}{\mu(\frac{1}{2} - \gamma + \delta\gamma)} \left(\delta \int_0^\gamma f(\xi) \, d\xi + \mu \int_\gamma^{1/2} f(\xi) \, d\xi \right)$$

$$+ \sum_{n=1}^\infty e^{-p_n^2 t} \cos(p_n x) \left(-\frac{C_n^{(2)}}{\mu} \right)$$

$$\times \left[-2 \frac{C_n^{(1)}}{C_n} \int_0^\gamma f(\xi) \cos(p_n \xi) \, d\xi \right.$$

$$\left. + \frac{2\mu}{C_n} \int_\gamma^{1/2} f(\xi) \cos\left(p_n \left(\frac{1}{2} - \xi \right) \right) d\xi \right] \tag{A.14}$$

where p_n are the roots of the equation

$$\sin\left(p_n \left(\frac{1}{2} - \gamma \right) \right) \cos(p_n \gamma) + \delta \cos\left(p_n \left(\frac{1}{2} - \gamma \right) \right) \sin(p_n \gamma) = 0 \tag{A.15}$$

and

$$C_n^{(1)} = \sin\left(p_n \left(\frac{1}{2} - \gamma \right) \right) \sin(p_n \gamma) - \delta \cos\left(p_n \left(\frac{1}{2} - \gamma \right) \right) \cos(p_n \gamma) \tag{A.16}$$

$$C_n^{(2)} = \delta \sin\left(p_n \left(\frac{1}{2} - \gamma \right) \right) \sin(p_n \gamma) - \cos\left(p_n \left(\frac{1}{2} - \gamma \right) \right) \cos(p_n \gamma)$$

$$= -\frac{\cos(p_n \gamma)}{\cos(p_n(\frac{1}{2} - \gamma))} \tag{A.17}$$

$$C_n = \left(\frac{1}{2} - \gamma + \delta\gamma \right) \sin\left(p_n \left(\frac{1}{2} - \gamma \right) \right) \cos(p_n \gamma)$$

$$- \left(\gamma + \frac{\delta}{2} - \delta\gamma \right) \sin\left(p_n \left(\frac{1}{2} - \gamma \right) \right) \sin(p_n \gamma) \tag{A.18}$$

The desired expansion is now found by setting $t = 0$ in Equation A.13 to give

$$f(x) = \frac{1}{2} A_0 \begin{cases} 1 & (0 < x < \gamma) \\ \dfrac{1}{\mu} & \left(\gamma < x < \dfrac{1}{2}\right) \end{cases}$$

$$+ \sum_{n=1}^{\infty} A_n \begin{cases} \cos(p_n x) & (0 < x < \gamma) \\ \dfrac{\cos(p_n \gamma)\cos(p_n(\frac{1}{2} - x))}{\mu \cos(p_n(\frac{1}{2} - \gamma))} & \left(\gamma < x < \dfrac{1}{2}\right) \end{cases} \quad (A.19)$$

with

$$A_n = -\frac{2C_n^{(1)}}{C_n} \int_0^\gamma f(\xi)\cos(p_n \xi)\, d\xi + \frac{2\mu}{C_n} \int_\gamma^{1/2} f(\xi)\cos\left(p_n\left(\frac{1}{2} - \xi\right)\right) d\xi$$

$$(A.20)$$

$$A_0 = \frac{2}{(\frac{1}{2} - \gamma + \delta\gamma)} \left(\delta \int_0^\gamma f(\xi)\, d\xi + \mu \int_\gamma^{1/2} f(\xi)\, d\xi \right) \quad (A.21)$$

It is readily seen that Expansion A.19 defines a function with required discontinuities at $x = \gamma$.

$$f(\gamma - 0) = \mu f(\gamma + 0)$$
$$f'(\gamma - 0) = \nu f'(\gamma + 0) \quad (A.22)$$

If $\mu = \nu = 1$, and $\delta = 1$, it can be proved that Equation A.19 reduces to an ordinary Fourier series on $(0, \frac{1}{2})$,

$$f(x) = \frac{1}{2} A_0 + \sum_{n=1}^{\infty} A_n \cos(2\pi n x) \quad (A.23)$$

where

$$A_n = 4 \int_0^{1/2} f(\xi)\cos(2\pi n \xi)\, d\xi$$

References

Alekhin, V. V., Annin, B. D., and Kolpakov, A. G. (1988) *Synthesis of Laminated Materials and Structures*, Institute of Hydrodynamics, Siberian Branch of the USSR Academy of Sciences, Novosibirsk (in Russian).

Aleksandrov, A. Ya. (1965) On the determination of reduced elastic properties of honeycomb fillers, in *Raschety Elementov Aviats. Konstr.*, Vol. 4, Mashinostroenie, Moscow, pp. 59–70.

Aleksandrov, A. Ya., Bryukker, L. E., Kurshin, L. M., and Prusakov, A. P. (1960) *The Design of Three-Layered Panels*, Oborongiz, Moscow (in Russian).

Alekseev, V. M., Tikhomirov, V. M., and Fomin, S. V. (1979) *Optimal Control*, Nauka, Moscow (in Russian).

Ambartsumyan, S. A. (1974) *General Theory of Anisotropic Shells*, Nauka, Moscow (in Russian).

Ambartsumyan, S. A. (1987) *Theory of Anisotropic Plates*, Nauka, Moscow (in Russian).

Andrianov, I. V., Lesnichaya, V. A., and Manevich, L. I. (1985) *Homogenization Method in the Statics and Dynamics of Ribbed Shells*, Nauka, Moscow (in Russian).

Andrianov, I. V. and Manevich, L. I. (1983) Shell design using the homogenization method, *Uspekhi Mekh.*, **6**, 3–29.

Artola, M. and Duvaut, G. (1977) Homogénéisation d'une plaque reinforcée, *C.R. Acad. Sci., Sér. A*, **284**, 707–710.

Babuška, I. (1976a) Solution of interface problems by homogenization, Part I, *SIAM J. Math. Anal.*, **7**, 603–634.

Babuška, I. (1976b) Solution of interface problems by homogenization, Part II, *SIAM J. Math. Anal.*, **7**, 635–645.

Babuška, I. (1977) Solution of interface problems by homogenization, Part III, *SIAM J. Math. Anal.*, **8**, 923–931.

Bakhvalov, N. S. and Panasenko, G. P. (1984) *Homogenization in Periodic Media. Mathematical Problems of the Mechanics of Composite Materials*, Nauka, Moscow.

Bakhvalov, N. S., Panasenko, G. P., and Shtaras, A. L. (1988) Homogenization of partial differential equations; Theory and applications, *Itogi Nauki i Tekhn. VINITI, Sovr. Probl. Mat. Fund. Napr.*, **34**, 215–241.

Banerjee, P. K. and Butterfield, R. (1981) *Boundary Element Methods in Engineering Science*, McGraw-Hill, New York.

Bateman, H. and Erdélyi, A. (1954) *Tables of Integral Transforms*, Vol. 2, McGraw-Hill, New York.

Bensoussan, A., Lions, J. L., and Papanicolaou, G. (1978) *Asymptotic Analysis for Periodic Structures*, Vol. 5, North-Holland, Amsterdam.

Berdichevskii, V. L. (1983) *Variational Principles in Continuum Mechanics*, Nauka, Moscow (in Russian).

Berlincourt, D. A., Curran, D. R., and Jaffe, H. (1964) Piezoelectric and piezomagnetic materials and their function as transducers, in *Physical Acoustics*, Vol. 1A, Mason, W. P., Ed., Academic, New York.

Birger, I. A. (1961) *Circular Plates and Shells of Rotation*, Oborongiz, Moscow (in Russian).

Bolotin, V. V. and Novichkov, Yu. N. (1980) *Mechanics of Multilayered Structures*, Mashinostroenie, Moscow (in Russian).

Bourgat, J. F. (1979) Numerical experiments of the homogenization method for operators with periodic coefficients, in *Computing Methods in Applied Sciences and Engineering, 1977, 1, Third International Symposium, December 5–9, 1977, Lecture Notes in Math.* Vol. 704, pp. 330–356.

Caillerie, D. (1981a) Equations de la diffusion stationnaire dans un domaine comportant une distribution périodique d'inclusions aplaties de grande conductivité, *C.R. Acad. Sci., Sér. 1*, **292**, 115–118.

Caillerie, D. (1981b) Homogénéisation des équations de la diffusion stationnaire dans les domaines cylindriques aplatis, *RAIRO Anal. Numér*, **15**, 295–319.

Caillerie, D. (1982) Plaques elastique minces à structure périodique de période et d'épaisseur comparables, *C.R. Acad. Sci., Sér. 2*, **294**, 159–162.

Caillerie, D. (1984) Thin elastic and periodic plates, *Math. Meth. Appl. Sci.*, **6**, 159–191.

Caillerie. D. (1987) Nonhomogeneous plate theory and conduction in fibered composites, in *Lecture Notes in Physics*, Vol. 272, Springer, Berlin, pp. 1–62.

Caillerie, D. and Lévy, T. (1983) Application de l'homogénéisation au comportement électromagnétique d'un mélange isolant—conducteur, *C.R. Acad. Sci., Sér. 2*, **296**, 1035–1038.

Cherepanov, G. P. (1983) *Fracture Mechanics of Composite Materials*, Nauka, Moscow (in Russian).

Christensen, R. M. (1979) *Mechanics of Composite Materials*, Wiley, New York.

Ciarlet, Ph. G. and Rabier, P. (1980) Les équations de von Kármán, in *Lecture Notes in Math.*, Vol. 826, Springer, Berlin.

Cioranescu, D. and Paulin, J. (1979) Homogenization in open sets with holes, *J. Math. Anal. Appl.*, **71**, 590–607.

Crouch, S. L. and Starfield, A. M. (1983) *Boundary Element Methods in Solid Mechanics*, Allen & Unwin, London.

Duvaut, G. (1976) Analyse fonctionnelle et méchanique de milieux continus. Application à l'étude des materiaux composites élastiques a structure périodique—homogénéisation, in *Theoretical and Applied Mechanics Preprints Proceedings of the 14th IUTAM Congress (Delft 1976)*, Amsterdam, pp. 11–132.

Duvaut, G. (1977) Comportement macroscopique d'une plaque perforée périodiquement, in *Lecture Notes in Math.*, Vol. 594, Springer, Berlin, 131–145.

Duvaut, G. and Metellus, A. M. (1976) Homogénéisation d'une plaque mince en flexion de structure périodique et symétrique, *C.R. Acad. Sci., Sér. A.*, **283**, 947–950.

Elpatyevskii, A. N. and Gavva, L. M. (1983) *Stress and Strain Fields in Eccentrically Reinforced Panels Including Shear from the Twisting of the Stiffness Ribs*, Moscow Aviation Inst., Moscow (in Russian).

Endogur, A. I., Weinberg, M. V., and Yerusalimskii, K. M. (1986) *Honeycomb Structures: Characterization and Design*, Mashinostroenie, Moscow (in Russian).

Ene, H. I. (1983) On linear thermoelasticity of composite materials, *Int. J. Eng. Sci.*, **21**, 443–448.

Eshelby, J. D. (1957) The determination of the elastic field of an ellipsoidal inclusion, and related problems, *Proc. Roy. Soc. London, Ser. A*, **241**, 376.

Friedlander, I. and Bratukhin, A. (1988) Izvestiya, No. 33, February 1.

Furukawa, T., Ishida, K., and Fukada, E. (1979) Piezoelectric properties in composite systems of polymers and PZT-ceramics, *J. Appl. Phys.*, **50**, 4904–4912.

Goldenblat, I. I. and Kopnov, V. A. (1968) *Strength and Plasticity Criteria in Structural Materials*, Mashinostroenie, Moscow (in Russian).

Gorbachev, V. I. (1979) On the elastic equation of a cylindrical pipe of varying thickness under the action of surface loads and displacements, *Probl. Prochn.*, **5**, 79–83.

Gorbachev, V. I. and Pobedrya, B. E. (1985) On some fracture criteria for composite materials, *Izv. Akad. Nauk Arm. SSR (Mekh.)*, **38**, 30–37.

Grigolyuk, E. I. and Fil'shtinskii, L. A. (1970) *Perforated Plates and Shells*, Nauka, Moscow (in Russian).

Grigolyuk, E. I. and Kabanov, V. V. (1978) *Stability of Shells*, Nauka, Moscow (in Russian).

Guz' A. N. (1986) *Three-Dimensional Stability of Deformable Solids*, Vyshcha Shkola, Kiev (in Russian).

Hashin, Z. (1962) The elastic moduli of heterogeneous materials, *J. Appl. Mech.*, **29**, 143.

Hashin, Z. and Rosen, B. M. (1964) The elastic moduli of fiber–reinforced materials, *J. Appl. Mech.*, **31**, 223.

Hashin, Z. and Shtrikman, S. (1962) A variational approach to the theory of the effective magnetic permeability of multiphase materials, *J. Appl. Phys.*, **33**, 3125.

Iosifyan, G. A., Oleinik, O. A., and Panasenko, G. P. (1982) Asymptotic expansion for an elastic problem with periodic rapidly oscillating coefficients, *Dokl. Akad. Nauk SSSR*, **266**, 18–22.

Ivanenko, O. A. and Fil'shtinskii, L. A. (1986) On the theory of regular piecewise homogeneous structures with a piezoceramic matrix, *Prikl. Mat. i Mekh.*, **50**, 128–135.

Kalamkarov, A. L., Kudryavtsev, B. A., and Parton, V. Z. (1987a) The asymptotic method of homogenization in the mechanics of composites with regular structure, *Itogi Nauki i Tekhn. VINITI. Mekh. Deform. Tv. Tela*, **19**, 78–147.

Kalamkarov, A. L., Kudryavtsev, B. A., and Parton, V. Z. (1987b) A curved composite material layer with wavy surfaces of periodic structure, *Prikl. Mat. i Mekh.*, **51**, 68–75.

Kalamkarov, A. L., Kudryavtsev, B. A., and Parton, V. Z. (1987c) Thermoelasticity of a regularly nonhomogeneous layer with wavy surfaces, *Prikl. Mat. i Mekh.*, **51**, 1000–1008.

Kalamkarov, A. L., Kudryavtsev, B. A., and Parton, V. Z (1987d) Thermoelasticity of strengthened composite plates and shells of regular structure, in *Proceedings of the 14th National Conference on Plate and Shell Theory*, Vol. 2, Tbilisi Univ., Tbilisi, pp. 21–26.

Kalamkarov, A. L., Kudryavtsev, B. A., and Parton, V. Z. (1989) Fundamental solution for a doubly periodic three-dimensional elasticity problem, *Izv. Akad. Nauk SSSR (Mekh. Tv. Tela)*, **3**, 44–50.

Kalamkarov, A. L., Kudryavtsev, B. A., and Parton, V. Z. (1990) Method of a boundary layer in the fracture mechanics of composites of periodic structure, *Prikl. Mat. i Mekh*, **54**, 322–328.

Karimov, A. M. (1986) Free vibrations of an elastic composite layer of periodic structure, *Vestnik MGU (Mat. i. Mekh.)*, **3**, 106–108.

Kerner, E. H. (1956) The electrical conductivity of composite materials, *Proc. Phys. Soc.*, **69**, 802.

Kesavan, S. (1979a) Homogenization of elliptic eigenvalue problems, Part I, *Appl. Math. and Optim.*, **5**, 153–167.

Kesavan, S. (1979b) Homogenization of elliptic eigenvalue problems, Part II, *Appl. Math. and Optim.*, **5**, 197–217.

Knunyants, N. N., Lyapunova, M. A., Manevich, L. I., Oshmyan, V. G., and Shaulov, A. Yu. (1986) Effects of a non-deal adhesive bond on the elastic properties of a dispersion filled composite, *Mekh. Komp. Mater.*, **2**, 231–234.

Kohn, R. V. and Vogelius, M. (1984) A new model for thin plates with rapidly varying thickness, *Int. J. Solids and Struct.*, **20**, 333–350.

Kohn, R. V. and Vogelius, M. (1985) A new model for thin plates with rapidly varying thickness, II: A convergence proof, *Quart. Appl. Math.*, **43**, 1–22.

Kohn, R. V. and Vogelius, M. (1986) A new model for thin plates with rapidly varying thickness, III: Comparison of different scalings, *Quart. Appl. Math.*, **44**, 35–48.

Kolpakov, A. G. and Rakin, S. I. (1986) On the synthesis of one-dimensional composite materials with prescribed characteristics, *Zh. Prikl. Mekh. i Tekhn. Fiz.*, **6**, 143–150.

Korolev, V. I. (1971) *Elastoplastic Deformation in Shells*, Mashinostroenie, Moscow (in Russian).

Kovalenko, A. D. (1970) *Fundamentals of Thermoelasticity*, Naukova Dumka, Kiev (in Russian).

Koshevnikova, M. I. and Kuz'menko, Yu. V. (1984) Effects of temperature on the overall elastic properties of a regularly nonhomogeneous composite, *Fiz. Mikroelektr. Priborov, Moscow*, 106–109.

Kuz'menko, Yu. V., Kozhevnikova, M. I., and Shermergor, T. D. (1982) Determination of the elastic properties of nonhomogeneous materials with regular structure, *Fiz. Mikroelektr. Priborov, Moscow*, 77–81.

Leontyev, N. V. (1984) Reduction problem for the composite material of a superconducting magnetic system, *Prikl. Probl. Prochn. i Plast.*, **27**, 74–79.

Liebowitz, H. (Ed.) (1968) *Fracture*, Vol. 2, *Mathematical Fundamentals*, Academic, New York.

Lions, J. L. (1981) *Some Methods in the Mathematical Analysis of Systems and their Control*, Gordon and Breach, New York.

Lions, J. L. (1985) Remarques sur les problèmes d'homogénéisation dans les milieux à structure périodique et sur quelques problèmes, raides, in *Les Méthodes de l'Homogénéisation: Théorie et Applications en Physique*, Eyrolles, pp. 129–228.

Lions, J. L. (1987) Homogenization and reinforced structures, in *Struct. Contr. Proceedings of the 2nd International Symposium, Waterloo, July, 1985*, Reidel, Dordrecht, pp. 426–445.

Love, A. E. H. (1944) *A Treatise on the Mathematical Theory of Elasticity*, Dover, New York.

Lukovkin, G. M., Volynskii, A. L., and Bakeev, N. F. (1983) Rubber dispersion mechanism for enhancing the impact strength of plastics, *Vysokomolek. Soed. (A)*, **25**, 848–855.

Maksimov, R. D., Plume, E. Z., and Ponomarev, V. M. (1983) Elastic properties of unidirectionally reinforced hybrid composites, *Mekh. Komp. Mater.*, **1**, 13–19.

Malmeister, A. K., Tamužs, V. P., and Teters, G. A. (1980) *Resistance Ability of Polymeric and Composite Materials*, Zinatne, Riga (in Russian).

Manevich, L. I., Oshmyan, V. G., Tovmasyan, Yu. M., and Topolkarev, V. A. (1983) Mathematical modeling of the elastoplastic deformation of dispersion strengthened composite materials, *Dokl. Akad. Nauk SSSR*, **270**, 806–809.

Miloserdova, V. I. (1987) Modeling and design of piezoelectric composites with regular structure, Candidate Degree Thesis, Leningrad.

Mol'kov, V. A. and Pobedrya, B. E. (1985) Effective properties of a unidirectional fiber composite with a periodic structure, *Izv. Akad. Nauk SSSR (Mekh. Tv. Tela)*, **2**, 119–130.

Mol'kov, V. A. and Pobedrya, B. E. (1988) Effective elastic properties of a composite with an elastic contact between the fibers and the binder, *Izv. Akad. Nauk SSSR (Mekh. Tv. Tela)*, **1**, 111–117.

Mushtari, Kh. M. and Galimov, K. Z. (1957) *Nonlinear Theory of Elastic Shells*, Tatknigoizdat, Kazan' (in Russian).

Muskhelishvili, N. I. (1963) *Some Basic Problems of the Mathematical Theory of Elasticity*, Noordhoff, Groningen.

Nayfeh, A. H. (1972) *Perturbation Methods*, Wiley, New York.

Nemat-Nasser, S., Iwakuma, T., and Hejazi, M. (1982) On composites with periodic structure, *Mech. Materials*, **1**, 239–267.

Novozhilov, V. V. (1948) *Fundamentals of Nonlinear Elasticity*, Gostekhizdat, Moscow (in Russian).

Novozhilov, V. V. (1962) *Theory of Thin Shells*, Sudpromgiz, Leningrad (in Russian).

Nowacki, W. (1970) *Teoria Sprezystósci*, Panstwowe Wydawnictwo Naukow, Warszawa.

Numan, K. C. and Keller, J. B. (1984) Effective elasticity tensor of a periodic composite, *J. Mech. and Phys. Solids*, **32**, 259–280.

Obraztsov, I. F., Vasil'ev, V. V., and Bunakov, V. A. (1977) *Optimal Reinforcement of Composite Material Shells of Rotation*, Mashinostroenie, Moscow (in Russian).

Oshmyan, V. G., Knunyants, N. H., Tovmasyan, Yu. M., Topolkarev, V. A., and Manevich, L. I. (1984) Theoretical and experimental study of the static deformations of dispersion filled composites, *Mekh. Komp. Mater.*, **3**, 431–438.

Panasenko, G. P. (1979) Higher order asymptotic analysis of the contact problem for periodic structures, *Mat. Zbornik*, **110**, 505–538.

Panasenko, G. P. and Reztsov, M. V. (1987) Homogenization of the three-dimensional elasticity theory for a nonhomogeneous plate, *Dokl. Akad. Nauk SSSR*, **294**, 1061–1065.

Panin, V. F. (1982) *Honeycomb Filler Structures*, Mashinostroenie, Moscow (in Russian).

Parton, V. Z. (1976) Fracture mechanics for piezoelectric materials, *Acta Astronautica*, Pergamon Press, Vol. 3, 671–683.

Parton, V. Z. (1985) Piezoelectrics as a Base of Electronics Engineering, *Machinostroenie*, **4**, 82–91.

Parton, V. Z. (1992) Fracture Mechanics. *From Theory to Practice*, Gordon and Breach, New York.

Parton, V. Z. and Boriskovsky, V. G. (1989) *Dynamic Fracture Mechanics, Vol. 1: Stationary Cracks*, Hemisphere, New York.

Parton, V. Z. and Boriskovsky, V. G. (1990) *Dynamic Fracture Mechanics, Vol. 2: Propagating Cracks*, Hemisphere, New York.

Parton, V. Z. and Kalamkarov, A. L. (1988a) Thermoelasticity of a regularly nonhomogeneous thin curved layer with rapidly varying thickness, *J. Thermal Stresses*, **11**, 405–420.

Parton, V. Z. and Kalamkarov, A. L. (1988b) On the stability equations for three-dimensional composite material systems, *Dokl. Akad. Nauk SSSR*, **300**, 308–311.

Parton, V. Z. and Kalamkarov, A. L. (1989) Asymptotic studies of process in composite materials, *Mekh. Komp Mater.*, **6**, 993–1000.

Parton, V. Z., Kalamkarov, A. L., and Boriskovsky, V. G. (1990a) On the studies of local field in the approximates of a macrocrack in the composite material of periodic structure, *Fiz.-Khim. Mekh. Mater.* **1**, 3–9.

Parton, V. Z., Kalamkarov, A. L., and Kolpakov, A. G. (1989) On the design of high strength, two-way shell reinforcement systems, *Mekh. Komp. Mater.*, **1**, 129–135.

Parton, V. Z., Kalamkarov, A. L., and Kudryavtsev, B. A. (1986a) Stress and strain fields in a curved, anisotropic nonhomogeneous layer of periodic structure with wavy surfaces, *Dokl. Akad. Nauk SSSR*, **291**, 1330–1335.

Parton V. Z., Kalamkarov, A. L., and Kudryavtsev, B. A. (1986b) Thermoelasticity of nonhomogeneous plates and shells of periodic structure, in strength and fracture. Proceedings of Soviet Symposium with participation countries—members of EEC, Vladimir, pp. 162–163.

Parton, V. Z., Kalamkarov, A. L., and Kudryavtsev, B. A. (1988a) On the study of mono-stresses in the neighbourhood of a crack in a composite material, in *Failure Analysis—Theory and Practice*. Proceedings of the 7th European Conference on Fracture, Budapest, 1988, vol. 1, pp. 427–432.

Parton, V. Z., Kalamkarov, A. L., and Kudryavtsev, B. A. (1989a) Methode d'homogénéisation en nécanique des solides déformables à microstructure régulière, *C. R. Acad. Sci. Paris, Sér 2*, **309**, 641–646.

Parton, V. Z., Kalamkarov, A. L., and Kudryavtsev, B. A. (1989b) Stress and strain fields in reinforced plates and shells of regular structure, *Probl. Prochn.*, **8**, 63–70.

Parton, V. Z., Kalamkarov, A. L., and Kudryavtsev, B. A. (1989c) Method of homogenization in the mechanics of homogeneous materials with regular structure, *Zavod. Labor.*, 55, 62–66.

Parton, V. Z., Kalamkarov, A. L., and Kudryavtsev, B. A. (1989d) On the analysis of the composite periodic shells, in *Future Trends in Applied Mechanics*. Proceedings Intern. Congress in Honour of P. S. Theocaris, 175–183.

Parton, V. Z., Kalamkarov, A. L., and Kudryavtsev, B. A. (1990b) Generalization of integral Fourier transformation in boundary problems of the composite mechanics, *Dokl. Akad. Nauk SSSR*, **315**, 53–56.

Parton, V. Z., Kalamkarov, A. L., and Miloserdova, V. I. (1987) Coupled problem of thermoelasticity for composites with periodic structure, *Mekh. Komp. Mater.*, **7**, 803–807.

Parton, V. Z., Kalamkarov, A. L., and Senik, N. A. (1988b) Use of the homogenization method in the design of ribbed plates and shells with regular structure, *Probl. Prochn.*, **8**, 101–106.

Parton, V. Z. and Kudryavtsev, B. A. (1986) On the fracture of laminated composites, *Fiz.-Khim. Mekh. Mater.*, **1**, 76–84.

Parton, V. Z. and Kudryavtsev, B. A. (1988) *Electromagnetoelasticity Piezoelectrics and Electrically Conductive Solids*, Gordon and Breach, New York.

Parton, V. Z. and Morozov, E. M. (1978) *Elastic-plastic Fracture Mechanics*, Mir, Moscow.

Parton, V. Z. and Morozov, E. M. (1989) *Mechanics of Elastic-Plastic Fracture*, Second Edition, Hemisphere, New York.

Paşa, G. I. (1983) A convergence theorem for periodic media with thermoelastical properties, *Int. J. Eng. Sci.*, **21**, 1313–1319.

Pobedrya, B. E. (1983a) On elastic composites, *Mekh. Komp. Mater.*, **2**, 216–222.

Pobedrya, B. E. (1983b) On the numerical analysis of the mechanical behavior of a deformable nonhomogeneous solid, *Vestnik MGU (Mat. i Mekh.)*, **4**, 78–85.

Pobedrya, B. E. (1984) *Mechanics of Composite Materials*, Moscow State Univ., Moscow (in Russian).

Pobedrya, B. E. (1985) Zeroth approximation theory in the mechanics of a nonhomogeneous deformable solid, in *Mathematical Methods and Physical and Mechanical Fields* (Kieve) Vol. 22, pp. 34–40.

Pobedrya, B. E. and Gorbachev, V. I. (1977) On static problems in the elasticity of composite materials, *Vestnik MGU (Mat. i Mekh.)*, **5**, 101–110.

Pobedrya, B. E. and Sheshenin, S. V. (1979) On the influence matrix, *Vestnik MGU (Mat. i. Mekh.)*, **6**, 76–81.

Pobedrya, B. E. and Sheshenin, S. V. (1981) Some problems of the equilibrium of an elastic parallelepiped, *Izv. Akad. Nauk SSSR (Mekh. Tv. Tela)*, **1**, 74–86.

Podlipenets, A. N. (1984) Harmonic wave propagation in orthotropic materials with periodic structure, *Prikl. Mekh.*, **20**, 20–24.

Podstrigach, Ya. S., Lomakin, V. A., and Kolyano, Yu. M. (1984) *Thermoelasticity of Nonhomogeneous Structures*, Nauka, Moscow (in Russian).

Podstrigach, Ya. S. and Shvets, R. N. (1978) *Thermoelasticity of Thin Shells*, Naukova, Dumka, Kiev (in Russian).

Pontryagin, L. S., Boltyanskii, V. G., Gamkrelidze, R. V., and Mishchenko, E. F. (1976) *Mathematical Theory of Optimal Processes*, Nauka, Moscow (in Russian).

Prudnikov, A. P., Brychkov, Yu. A., and Marichev, O. I. (1983) *Integrals and Series. Special Functions*, Nauka, Moscow (in Russian).

Pshenichnov, G. I. (1982) *Theory of Thin Elastic Network Plates and Shells*, Nauka, Moscow (in Russian).

Russel, W. B. and Acrivos, A. (1972) On the effective moduli of composite materials: slender rigid inclusions at dilute concentrations, *Z. Angw. Math. und Phys.*, **23**, 434.

Sanchez-Palencia, E. (1980) Non-homogeneous media and vibration theory, in *Lecture Notes in Physics*, Vol. 127, Springer, Berlin.

Sanchez-Palencia, E. (1987) Boundary layers and edge effects in composites, in *Lecture Notes in Physics*, Vol. 272, Springer, Berlin.

Sedov, L. I. (1972) *A Course in Continuum Mechanics*, Wolters-Noordhoff, Groningen.

Sendeckyj, G. P. (1974) Elastic behavior of composites, in *Composite Materials*, Vol. 2, *Mechanics of Composite Materials*, Academic, New York.

Sgubini, S., Graziani, F., and Agneni, A. (1983) Elastic waves propagation in periodic composite materials, *Fibre Sci. and Technol.*, **19**, 1–13.

Shakhtakhtinskii, M. G., Guseinov, B. A., Kurbanov, M. A., and Gazaryan, Yu. N. (1985) Effects of polarization conditions on the piezoelectric properties of polymeric composites, *Dokl. Akad. Nauk Az. SSR*, **11**, 20–24.

Shakhtakhtinskii, M. G., Kurbanov, M. A. et al. (1987) Polymeric composites with high values of piezoelectric coupling and sensitivity, *Vysokomol. Soed.*, *T(B)*, **29**, 3–5.

Sheshenin, S. V. (1980) Overall moduli of one composite, *Vestnik MGU* (*Mat. i Mekh.*), **6**, 79–83.

Shul'ga, N. A. (1984) Shear wave propagation in stratified media with periodic properties, *Prikl. Mekh.*, **20**, 116–119.

Tamužs, V. P. and Protasov, V. D. (Eds.) (1986) *Fracture of Composite Material Constructions*, Zinatne, Riga (in Russian).

Tartar, L. (1978) Nonlinear constitutive relations and homogenization, in *Contemporary Developments in Continuum Mechanics and Partial Differential Equations, Proceedings of the International Symposium, Rio de Janeiro, 1977*, North-Holland, Amsterdam, pp. 472–484.

Ufland, Ya. S. (1976) On some new integral transforms and their application to problems in mathematical physics, in *Problems of Mathematical Physics*, Nauka, Moscow, pp. 93–105.

Van Fo Fy, G. A. (1971) *Theory of Reinforced Materials with Coatings*, Naukova Dumka, Kiev (in Russian).

Vanin, G. A. (1977) Integral parameters of a piezoelectric composite medium under longitudinal shear, *Dokl. Akad. Nauk Ukr. SSR* (*A*), **10**, 894–897.

Vanin, G. A. (1985) *Micromechanics of Composite Materials*, Naukova Dumka, Kiev (in Russian).

Vanin, G. A. and Semenyuk, N. P. (1987) *Stability of Composite Material Shells Containing Defects*, Naukova Dumka, Kiev (in Russian).

Vasil'ev, V. V. (1988) *Mechanics of Composite Material Constructions*, Mashinostroenie, Moscow (in Russian).

Whittaker, E. T. and Watson, G. N. (1927) *A Course in Modern Analysis, Part II*, Cambridge University Press.

Zheludev, I. S. (1968) *Physics of Crystalline Dielectrics*, Nauka, Moscow (in Russian).

Zobnin, A. I., Kudryavtsev, B. A., and Parton, V. Z. (1988) Linearized equations of motion for a viscous fluid in a porous medium of periodic structure, *Izv. Akad. Nauk SSSR* (*Mekh. Zhidk. i Gaza*), **2**, 123–130.

INDEX

A

Adiabatic processes, elastic potential in, 17–18

Anisotropy, 22–24
in laminated shells, 262–263

Asymptotic analysis
of curved, thin, corrugated, regularly non-homogeneous shells, 250–255
of wavy-surfaced regularly nonhomogeneous shells, 221–233

Asymptotic homogenization
with Dirichlet conditions, 57–68
local problems and effective coefficients
for laminated materials, 78–84
in special situations, 84–86
for square with circular hole, 86–90
Weierstrass' functions in, 86–87
nonperiodic effects and, 69–70
in perforated media, 62–66
of periodic (regular) structures, nonhomogenity dimension in, 56
with piecewise-smooth coefficients, 62
in regions with wavy boundaries
in infinite cylinder, 76–78
macroscopic heat conduction in, 73–74
microscopic heat conduction in, 70–73, 75–76
Sanchez-Palencia proof in, 71–73
spectral analysis with, 66–67
by two-scale expansion method, 56–62

B

Boundary material displacement, 103–104
Boundary stress, 97–103
Buckling, 137
Bulk moduli, 34, 38–43, 53–54

C

Ceramic, polarized, 168–175, 182
Chebyshev polynomials, 330–335
Circular hole problems, 86–90
Compliance tensor, 113
Composite cylinders model, 35–36
Composite spheres model (Hashan), 34
Continuum mechanics
and composite modeling
by Hashin/Shtrikman variational principle, 38–46
materials characteristics and, 32–38
of periodic structure, 46–54
elastic potential: stress/strain relations, 16–20
of electroelasticity/piezoelectric media, 27–31
as expressed in text, 1–2
of heat conduction/thermoelasticity, 21–27
moving particle kinematics: deformation theory, 2–9
point of stress: equilibrium equations, 9–15
Corrugated surfaces, 337–347, see also Shells
Cracks, macro-, 126–136

D

Deformation
basic relations in, 2–9
local stresses in macro-, 163
in nonlinear, nonhomogeneous periodic solid, 96–97
nonstationary thermal and mechanical, 24–25
stability prediction and, 139
Dilute suspensions, 33
Dirichlet conditions, n periodic composites, 57–68
Discontinuity, 150
Duhamel-Neumann law of anisotropy, 23, 148–149